ISBN 978-0-260-29471-5
PIBN 11003116

FB &c Ltd, Dalton House, 60 Windsor Avenue, London, SW19 2RR.
Company number 08720141. Registered in England and Wales.

For support please visit www.forgottenbooks.com

PRACTICAL AND MENTAL

ARITHMETIC,

ON A NEW PLAN;

IN WHICH

MENTAL ARITHMETIC IS COMBINED WITH THE USE OF THE SLATE:

CONTAINING

A COMPLETE SYSTEM

FOR ALL PRACTICAL PURPOSES;

BEING IN

DOLLARS AND CENTS.

STEREOTYPE EDITION,

REVISED AND ENLARGED,

WITH

EXERCISES FOR THE SLATE.

TO WHICH IS ADDED,

A PRACTICAL SYSTEM OF BOOK-KEEPING

BY ROSWELL C. SMITH.

BOSTON:
RICHARDSON, LORD AND HOLBROOK,
NO. 133, WASHINGTON STREET.

1830.

DISTRICT OF MASSACHUSETTS, *to wit:*

District Clerk's Office.

Be it remembered, That on the twenty-first day of January, A. D. 1829 in the fifty-third year of the Independence of the United States of America, *Richardson & Lord*, and *S. G. Goodrich*, of the said district, have deposited in this office the title of a book, the right whereof they claim as proprietors, in the words following, to wit:

"Practical and Mental Arithmetic, on a New Plan, in which Mental Arithmetic is combined with the Use of the Slate: containing a Complete System for all practical Purposes; being in Dollars and Cents. Stereotype Edition, revised and enlarged, with Exercises for the Slate. To which is added, a Practical System of Book-Keeping. By Roswell C. Smith."

In conformity to the act of the Congress of the United States, entitled, "An Act for the encouragement of learning, by securing the copies of maps, charts, and books, to the authors and proprietors of such copies, during the times therein mentioned;" and also to an act, entitled, "An Act supplementary to an act, entitled, An Act for the encouragement of learning, by securing the copies of maps, charts, and books, to the authors and proprietors of such copies during the times therein mentioned; and extending the benefits thereof to the arts of designing, engraving, and etching historical and other prints."

JNO. W. DAVIS,

Clerk of the District of Massachusetts.

RECOMMENDATIONS.

From the Jan. No. for 1828 *of the* JOURNAL OF EDUCATION

"A careful examination of this valuable work will show that its author has compiled it, as all books for school use ought to be compiled, from the results of actual experiment and observation in the school-room. It is entirely a practical work, combining the merits of Colburn's system with copious practice on the slate.

"Two circumstances enhance very much the value of this book. It is very comprehensive, containing twice the usual quantity of matter in works of this class; while, by judicious attention to arrangement and printing, it is rendered, perhaps, the cheapest book in this department of education. The brief system of Book-Keeping, attached to the Arithmetic, will be a valuable aid to more complete instruction in common schools, to which the work is, in other respects, so peculiarly adapted.

"There are several very valuable peculiarities in this work, for which we cannot, in a notice, find sufficient space. We would recommend a careful examination of the book to all teachers who are desirous of combining good theory with copious and rigid practice."

From the Report of the SCHOOL-COMMITTEE OF PROVIDENCE.

"The books at present used in the schools are, in the opinion of your Committee, altogether above the range of thought of the pupils. Works of a narrative character would be better understood, would be more interesting, and would, of course, teach the pupil to read with more taste and judgment. The boy who pores, in utter disgust, over the book which he reads in schools, will hasten home to read with avidity his story-book. The true wisdom would then be, to introduce the story-book into school, and thus render his place of education the place of his amusement.

"Nevertheless, as this subject is one in which time and judgment are necessary for a selection, and as a change of this sort, through all the schools, would be productive of considerable additional expense, your Committee would recommend that no change, at present, be made in books, excepting only the Arithmetic. If a school, by way of experiment, be established on the monitorial plan, various school-books can be tried there, and, after a fair opportunity of testing the merits of several, those can be selected which seem best adapted to accomplish the purposes of education. Your Committee are, however, of opinion, that it would be expedient to introduce the system of Arithmetic published by Mr. Smith [subsequently adopted] into all the Public Grammar Schools; and, also, that all the scholars in arithmetic be taught by classes, and not individually, as is now the prevalent mode."

The above Report was signed by the following named gentlemen:—

Rev. F. WAYLAND, Jr., D.D. *Pres. Brown Univ.,* (Chairman.)
Rev. THOMAS T. WATERMAN.
WILLIAM T. GRINNELL, Esq.

Dated *April* 24, 1828.

This work is recommended by the State-Commissioners of Vermont to be adopted throughout that State. It is likewise introduced into the public and private schools of Hartford, Conn. by the concurrence both of committees and teachers, and in like manner in various other places.

PREFACE

TO THE THIRD EDITION.

WHEN a new work is offered to the public, especially on a subject abounding with treatises, like this, the inquiry is very naturally made, "Does this work contain any thing new?" "Are there not a hundred others as good as this?" To the first inquiry it is replied, that there are many things which are believed to be new; and, as to the second, a candid public, after a careful examination of its contents, and not till then, it is hoped, must decide. Another inquiry may still be made: "Is this edition different from the preceding?" The answer is, Yes, in many respects. The *present* edition professes to be strictly on the Pestalozzian, or inductive plan of teaching. This, however, is not claimed as a novelty. In this respect, it resembles many other systems. The novelty of this work will be found to consist in adhering more closely to the true spirit of the Pestalozzian plan; consequently, in differing from other systems, it differs less from the Pestalozzian. This similarity will now be shown.

1. *The Pestalozzian professes to unite a complete system of mental with written arithmetic. So does this.*

2. *That rejects no rules, but simply illustrates them by mental questions. So does this.*

3. *That commences with examples for children as simple as this, is as extensive, and ends with questions adapted to minds as mature.*

Here it may be asked, "In what respect, then, is this different from that?" To this question it is answered, In the execution of our common plan.

The following are a few of the prominent characteristics of this work, in which it is thought to differ from all others.

1. *The interrogative system is generally adopted throughout this work.*

2. *The common rules of arithmetic are exhibited so as to correspond with the occurrences in actual business. Under this head is reckoned the application of Ratio to practical purposes, Fellowship, &c.*

3. *There is a constant recapitulation of the subject attended to, styled "Questions on the foregoing."*

4. *The mode of giving the individual results without points, then the aggregate of these results, with points, for an answer, by which the relative value of the whole is determined, thus furnishing a complete test of the knowledge of the pupil. This is a characteristic difference between this and the former editions.*

5. *A new rule for calculating interest for days with months*

1*

6. *The mode of introducing and conducting the subject of Proportion.*

7. *The adoption of the federal coin, to the exclusion of sterling money, except by itself.*

8. *The arithmetical tables are practically illustrated, previously and subsequently to their insertion.*

9. *As this mode of teaching recognises no authority but that of reason, it was found necessary to illustrate the rule for the extraction of the cube root, by means of blocks, which accompany this work.*

These are some of the predominant traits of this work. Others might be mentioned, but, by the examination of these, the reader will be qualified to decide on their comparative value.

As, in this work, the common rules of arithmetic are retained, perhaps the reader is ready to propose a question frequently asked, "What is the use of so many rules?" "Why not proscribe them?" The reader must here be reminded, that these rules are taught differently, in this system, from the common method. The pupil is first to satisfy himself of the truth of several distinct mathematical principles. These deductions, or truths, are then generalized; that is, briefly summed in the form of a rule, which, for convenience' sake, is *named*. Is there any impropriety in this? On the contrary, is there not a great convenience in it? Should the pupil be left to form his own rules, it is more than probable he might mistake the most concise and practical one. Besides, different minds view things differently, and draw different conclusions. Is there no benefit, then, in *helping* the pupil to the most concise and practical method of solving the various problems incident to a business life?

Some have even gone so far as to condemn the Rule of Three, or Proportion, and almost all the successive rules growing out of it. With more reason, they might condemn Long Division, and even Short Division; and, in fact, all the common and fundamental rules of arithmetic, except Addition; for these may all be traced to that. The only question then is, "To what extent shall we go?" To this it is replied, As far as convenience requires. As the Rule of Three is generally taught, it must be confessed, that almost any thing else, provided the mind of the pupil be exercised, would be a good substitute. But when taught as it should be, and the scholar is led on in the same train of thought that originated the rule, and thus effectually made to see, that it is simply a convenient method of arriving at the result of both Multiplication and Division combined, its necessity may be advocated with as much reason as any fundamental rule. As taught in this work, it actually saves more figures than Short, compared with Long Division. Here, then, on the ground of convenience, it would be reasonable to infer, that its retention was more necessary than either. But, waiving its utility in this respect, there is another view to be taken of this subject, and that not the least in importance, viz. the ideas of beauty arising from viewing the harmonious relations of numbers. Here is a delightful field for an inquisitive mind. It here imbibes truths as lasting as life. When the utility and convenience of this rule are once conceded, all the other rules growing out of this will demand a place, and for the same reason.

It may, perhaps, be asked by many, "Why not take the principle without the name?" To this it is again replied, Convenience forbids. The name, the pupil will see, is only an aggregate term, given to a process imbodying several distinct principles. And is there no convenience in this? Shall the pupil, when in actual business, be obliged to call off his mind from all other pursuits, to trace a train of deductions arising from abstract reasoning, when his attention is most needed on other subjects. With as much propriety the name of *captain* may be dispensed with; for, although the general, by merely summoning his captain, may summon 100 men, still he might call on each separately, *although* not quite so conveniently With these remarks, the subject will be dismissed, merely adding, by way of request, that the reader will defer his decision till he has examined the doctrine of Proportion, Fellowship, &c., as taught in this work.

The APPENDIX contains many useful rules, although a knowledge of these is not absolutely essential to the more common purposes of life. Under this head are reckoned Alligation, Roots, Progression, Permutation, Annuities, &c. The propriety of scholars becoming acquainted, some time or other, with these rules, has long since been settled; the only question is, with regard to the expediency of introducing them into our arithmetics, and not reserving them for our algebras. In reply to this, the writer would ask, whether it can be supposed, the developement of these truths, by figures, will invigorate, strengthen, and expand the mind less than by letters? Is not a more extensive knowledge of the power of figures desirable, aside from the improvement of the mind, and the practical utility which these rules afford? Besides, there always will, in some nook or other, spring up some poor boy of mathematical genius, who will be desirous of extending his researches to more abstruse subjects. Must he, as well as all others, be taxed with an additional expense to procure a system, containing the same principles, only for the sake of discovering them by letters?

Position, perhaps, may be said to be entirely useless. The same may be said of the doctrine of Equations by algebra. If the former be taught rationally, what great superiority can be claimed for the one over the other? Is it not obvious, then, that it is as beneficial to the pupil to discipline his mind by the acquisition of useful and practical knowledge, which may be in the possession of almost every learner, as to reserve this interesting portion of mathematics for a favoured few, and, in the mean time, to divert the attention of the pupil to less useful subjects?

The blocks, illustrative of the rule for the Cube Root, will satisfactorily account for many results in other rules; as, for instance, in Decimals, Mensuration, &c., which the pupil, by any other means, might fail to perceive. By observing these, he will see the reason why his product, in decimals, should be less than either factor; as, for instance, why the solid contents of a half an inch cube should be less than half as much as an inch cube. In this case, the factors are each half an inch, but the solid contents are much less than half a solid inch.

In this work, the author has endeavoured to make every part conform to this maxim, viz. THAT NAMES SHOULD SUCCEED IDEAS. This method of communicating knowledge is diametrically opposed to that which obtains, in many places, at the present day. The former, by first giving ideas, allures the pupil into a luminous comprehension of the subject; while the latter astounds him, at first, with a pompous name, to which he seldom affixes any definite ideas, and it is exceedingly problematical whether he ever will. In addition to this is the fact, that, by the last mentioned method, when the name is given and the process shown, not a single reason of any operation is adduced; but the pupil is dogmatically told he must proceed thus and so, and he will come out so and so. This mode of teaching is very much as if a merchant of this city should direct his clerk, without intrusting him with any business, first to go to South Boston, then to the state-house, afterwards to the market, and then to return, leaving him to surmise, if he can, the cause of all this peregrination. Many are fools enough to take this jaunt pleasantly; others are restiff, and some fractious. This sentiment is fully sustained by an article in Miss Edgeworth's works, from which the following extract is made: "A child's seeming stupidity, in learning arithmetic, may, perhaps, be a proof of intelligence and good sense. It is easy to make a boy, who does not reason, repeat, by rote, any technical rules, which a common writing master, with magisterial solemnity, may lay down for him; but a child who reasons will not be thus easily managed; he stops, frowns, hesitates, questions his master, is wretched and refractory, until he can discover *why* he is to proceed in such and such a manner; he is not content with seeing his preceptor make figures and lines on the slate, and perform wondrous operations with the self-complacent dexterity of a conjurer; he is not content to be led to the treasures of science blindfold; he would tear the bandage from his eyes, that he might know the way to them again."

In confirmation of the preceding remarks, and as fully expressive of the author's views on this subject, the following quotation is taken from the preface to Pestalozzi's system.

"The PESTALOZZIAN plan of teaching ARITHMETIC, as one of the great branches of the mathematics, when communicated to children upon the principles detailed in the following pages, needs not fear a comparison with her more favoured sister, GEOMETRY, either in precision of ideas, in clearness and certainty of demonstration, in practical utility, or in the sublime deductions of the most interesting truths.

"In the regular order of instruction, arithmetic ought to take precedence of geometry, as it has a more immediate connexion with it than some are willing to admit. It is the science which the mind makes use of in measuring all things that are capable of augmentation or diminution; and, when rationally taught, affords to the youthful mind the most advantageous exercise of its reasoning powers, and that for which the human intellect becomes early ripe, while the more advanced parts of it may try the energies of the most vigorous and matured understanding."

January, 1829 THE AUTHOR.

TABLE OF CONTENTS.

SIMPLE NUMBERS.

Mental Exercises.

Mental Exercises combined with Exercises for the Slate.

COMPOUND NUMBERS.

FRACTIONS.

APPENDIX.

Miscellaneous Examples.

BOOK-KEEPING.

SUGGESTIONS TO TEACHERS

ON THE METHOD OF USING THIS WORK.

For a course of mental arithmetic, adapted to the capacities of very young pupils, they may take the mental exercises in each rule, as far as the first example for the slate. This course is not meant to include any of the exercises styled "Questions on the foregoing."

This course embraces the whole of the first 20 pages, together with the arithmetical tables, extending to the Appendix. The necessity of impressing these tables on the minds of pupils at an early age is sufficiently obvious. When the pupil is perfect master of this course, as will, most probably, be the case after one or two reviews, the teacher will find no difficulty in making him understand the operations by slate. He may then take the whole in course.

In every school, it would be well to institute classes; and as there are seldom any answers given to the mental questions, the pupils may be allowed to read in their turns the questions from the book; thus giving the teacher no further trouble than occasional corrections. By this, the reader will perceive, that the work may be used to advantage in monitorial schools, as the former editions have been. In large schools these corrections may be made by an advanced scholar, instead of the teacher. Whenever an advanced scholar takes up the book with a view of profiting from it, he should omit nothing as he progresses, but make it his practice to qualify himself to answer any question, in the mental exercises, rules, or respecting the reason of the operations.

Teachers will find it to be a useful occupation for their scholars, to assign them a morning lesson, to be recited as soon as they come into school. With little exertion on the part of teachers, pupils in this way may be made assiduous and ambitious, very much to their advantage, and to the credit of their teachers.

The mental questions, under the head of "Questions on the foregoing," will, intelligently answered, furnish to committees an admirable test of the pupil's knowledge of this subject.

The Appendix is designed for those who have time and opportunity to devote to the study of the more abstruse parts of mathematics.

Note. Lest some may mistake the object of the figures in the parentheses, it may here be remarked, that these figures are separate answers, left without assigning any value to them, reserving this particular for the discretion of the pupil, which he must necessarily exercise, in order to obtain the answer which follows, that being the aggregate of the whole.

The above directions are those which seem the best to the author; but as every intelligent teacher has a way of his own, which, though not intrinsically the best, is, perhaps, the best for him, the subject is respectfully submitted to his own choice.

ARITHMETIC.

MENTAL EXERCISES.

ADDITION.

¶ I.* 1. How many little fingers have you on your right hand? How many on your left? How many on both?

2. How many eyes have you?

3. If you have two apples in one hand, and one in the other, how many have you in both? How many are two and one, then, put together?

4. How many do your ears and eyes make, counted together?

5. If you have two nuts in one hand, and two in the other, how many have you in both? How many do two and two make, put together?

6. If you have three pins in one hand, and James puts another in, how many will you have in your hand? How many are three and one then?

7. If you have three pins in one hand, and James puts two more in, how many will you have in your hand? How many are three and two then?

8. If you have four apples in one pocket, and two in the other, how many will you have in both? How many are four and two then?

9. Thomas has four cents, and William has three; how many have they both together? How many are four and three then?

10. You have five pins in one hand, and three in the other; how many have you in both? How many are five and three then?

11. You have four nuts in one hand, and four in the other; how many have you in both? How many are four and four then?

---

* The questions in ¶ I and ¶ II are intended for very young children. Older pupils may omit these. But the two remaining sections, and the four tables, *will claim* an attentive perusal.

12. If you count the fingers and thumb on one hand, and only the fingers on the other, how many will they make? How many are five and four then?

13. How many fingers and thumbs have you on both hands?

14. James has five marbles, and Thomas five; how many have they both? How many are five and five then?

15. How many cents would it take to buy two whistles, if one cost six cents, and the other four? How many are six and four then?

16. If you have eight pins on one sleeve, and two on the other, how many will you have on both? How many are eight and two then?

17. How many legs have two cats and a bird?

18. If I should give you six cents, and you should find five, how many would you have then? How many are six and five then?

19. If you count all your fingers, thumbs, and nose, how many will they make?

20. If you buy a picture-book for ten cents, and a pear for two cents, how many cents will pay for both? How many are ten and two then?

21. How much money would you have, if your father should give you seven cents, and your brother six? How many are seven and six then?

22. If you have seven pins in one hand, and seven in the other, how many will you have in both? How many are seven and seven then?

23. A man bought a chair for three dollars, and a looking-glass for twelve; how much did he give for both? How many are three and twelve then?

24. You give thirteen cents for a spelling-book, and three for an inkstand; how much do they come to? How many are thirteen and three?

25. Count one hundred.

One	1	Fourteen	14
Two	2	Fifteen	15
Three	3	Sixteen	16
Four	4	Seventeen	17
Five	5	Eighteen	18
Six	6	Nineteen	19
Seven	7	Twenty	20
Eight	8	Twenty-one	21
Nine	9	Twenty-two	22
Ten	10	Twenty-three	23
Eleven	11	Twenty-four	24
Twelve	12	Twenty-five	25
Thirteen	13	Twenty-six	26

Twenty-seven............27	Fifty 50
Twenty-eight 28	Sixty.....................[illegible]
Twenty-nine.............29	Seventy ....[illegible]
Thirty 30	Eighty 80
Thirty-one, &c. 31	Ninety 90
Forty 40	One hundred.............100

Note. The pupil is to recite the above, with the written numbers covered over. The answers to the following questions are to be given by writing them down on the slate at recitation, to test the pupil's knowledge of numbers from one to one hundred.

26. Write down in proper figures, Four; Seven; Eight; Twelve; Eighteen; Twenty-two; Thirty-two; Forty-five; Forty-nine; Fifty-six; Fifty-nine; Sixty-three; Seventy-five; Eighty-seven; Ninety-two; Ninety-seven; Ninety-nine.

27. James has seventy-eight cents, and Rufus eighty-seven cents; which has the most?

28. Thomas has fifty-nine dollars, and William sixty-nine; which has the most? Which is the most, eighty-nine, or ninety-nine? Forty-seven, or seventy-four?

29. Repeat the

ADDITION TABLE.

2 and 1 are 3	3 and 1 are 4	4 and 1 are 5
2 and 2 are 4	3 and 2 are 5	4 and 2 are 6
2 and 3 are 5	3 and 3 are 6	4 and 3 are 7
2 and 4 are 6	3 and 4 are 7	4 and 4 are 8
2 and 5 are 7	3 and 5 are 8	4 and 5 are 9
2 and 6 are 8	3 and 6 are 9	4 and 6 are 10
2 and 7 are 9	3 and 7 are 10	4 and 7 are 11
2 and 8 are 10	3 and 8 are 11	4 and 8 are 12
2 and 9 are 11	3 and 9 are 12	4 and 9 are 13
2 and 10 are 12	3 and 10 are 13	4 and 10 are 14
2 and 11 are 13	3 and 11 are 14	4 and 11 are 15
2 and 12 are 14	3 and 12 are 15	4 and 12 are 16
5 and 1 are 6	6 and 1 are 7	7 and 1 are 8
5 and 2 are 7	6 and 2 are 8	7 and 2 are 9
5 and 3 are 8	6 and 3 are 9	7 and 3 are 10
5 and 4 are 9	6 and 4 are 10	7 and 4 are 11
5 and 5 are 10	6 and 5 are 11	7 and 5 are 12
5 and 6 are 11	6 and 6 are 12	7 and 6 are 13
5 and 7 are 12	6 and 7 are 13	7 and 7 are 14
5 and 8 are 13	6 and 8 are 14	7 and 8 are 15
5 and 9 are 14	6 and 9 are 15	7 and 9 are 16
5 and 10 are 15	6 and 10 are 16	7 and 10 are 17
5 and 11 are 16	6 and 11 are 17	7 and 11 are 18
5 and 12 are 17	6 and 12 are 18	7 and 12 are 19

8 and 1 are 9	9 and 1 are 10	10 and 1 are 11
8 and 2 are 10	9 and 2 are 11	10 and 2 are 12
8 and 3 are 11	9 and 3 are 12	10 and 3 are 13
8 and 4 are 12	9 and 4 are 13	10 and 4 are 14
8 and 5 are 13	9 and 5 are 14	10 and 5 are 15
8 and 6 are 14	9 and 6 are 15	10 and 6 are 16
8 and 7 are 15	9 and 7 are 16	10 and 7 are 17
8 and 8 are 16	9 and 8 are 17	10 and 8 are 18
8 and 9 are 17	9 and 9 are 18	10 and 9 are 19
8 and 10 are 18	9 and 10 are 19	10 and 10 are 20
8 and 11 are 19	9 and 11 are 20	10 and 11 are 21
8 and 12 are 20	9 and 12 are 21	10 and 12 are 22

11 and 1 are 12	11 and 9 are 20	12 and 5 are 17
11 and 2 are 13	11 and 10 are 21	12 and 6 are 18
11 and 3 are 14	11 and 11 are 22	12 and 7 are 19
11 and 4 are 15	11 and 12 are 23	12 and 8 are 20
11 and 5 are 16	12 and 1 are 13	12 and 9 are 21
11 and 6 are 17	12 and 2 are 14	12 and 10 are 22
11 and 7 are 18	12 and 3 are 15	12 and 11 are 23
11 and 8 are 19	12 and 4 are 16	12 and 12 are 24

Questions on the Table.

30. How many are 2 and 5? 2 and 7? 2 and 10? 2 and 12? 3 and 3? 3 and 9? 3 and 12? 4 and 2? 4 and 6? 4 and 8? 4 and 10? 4 and 12? 5 and 3? 5 and 5? 5 and 9? 5 and 11? 6 and 4? 6 and 7? 6 and 10? 6 and 12? 7 and 2? 7 and 4? 7 and 7? 7 and 9? 7 and 12? 8 and 2? 8 and 5? 8 and 7? 8 and 9? 8 and 10? 8 and 12? 9 and 6? 9 and 9? 9 and 12? 10 and 3? 10 and 4? 10 and 6? 10 and 8? 10 and 11? 10 and 12? 11 and 3? 11 and 5? 11 and 6? 11 and 9? 11 and 12? 12 and 3? 12 and 6? 12 and 9? 12 and 12?

Note. The design of the foregoing and following questions is to prevent the scholar from resting satisfied with saying his table merely by rote, which frequently happens. For, if he can count, he will say it, without making a single addition in his mind.

31. You borrow 12 dollars at one time, and 2 at another; how much have you borrowed in all? How many are 12 and 2?

32. William has 11 cents, and James 11; how many do they both have? How many are 11 and 11?

33. A man bought a cart for 13 dollars, and a plough for 7 dollars; how much did he pay for both? How many are 13 and 7?

34. A man bought 10 bushels of rye for 15 dollars, 6 bushels of apples for 6 dollars; how much did he pay for both? How many are 15 and 6?

35. William has 4 marbles in one pocket, 6 in the other, and

3 in his right hand; how many has he in all? How many are 4, 6 and 3?

36. Peter gave to his companions apples as follows; to James 7, to Henry 9, to William 10; how many did he give away? How many are 7, 9 and 10?

37. Rufus has 12 cents, James 12, and Thomas 2; if Rufus and James should give Thomas all their cents, how many would Thomas have? How many are 12, 12 and 2?

38. You give 16 cents for a knife, 4 cents for an inkstand, and 5 for a lead pencil; how much will all of them come to? How many are 16, 4 and 5?

39. Your brother William gave you 19 cents, your brother John 10, and your cousin 2; how many did you have given you in all? How many are 19, 10 and 2?

40. How many are 6 and 4? 16 and 4? 26 and 4? 36 and 4? 46 and 4? 56 and 4? 66 and 4? 76 and 4? 86 and 4? 96 and 4? 10 and 5? 20 and 5? 40 and 5? 70 and 5? 80 and 5? 6 and 10? 6 and 40? 6 and 70? 7 and 3? 17 and 3? 37 and 3? 57 and 3? 77 and 3? 97 and 3? 5 and 5? 5 and 10? 5 and 15? 5 and 20? 25 and 5? 30 and 5? 45 and 5? 60 and 5? 75 and 5? 95 and 5? 8 and 4? 18 and 4? 28 and 4? 38 and 4? 48 and 4? 58 and 4? 68 and 4? 78 and 4? 88 and 4? 98 and 4? 9 and 3? 19 and 3? 29 and 3? 49 and 3? 79 and 3? 89 and 3? 6 and 5? 6 and 15? 6 and 25? 6 and 35? 6 and 45? 6 and 65? 6 and 85? 6 and 95?

SUBTRACTION.

¶ II. 1. If you should lose one finger from one hand, how many would you have left on that hand? How many are 4 less 1? Why? *Ans.* Because 1 and 3 are 4.

2. If you have 5 cents, and give away 2, how many will you have left? How many are 5 less 2 then? Why?

3. If you shut both your little fingers, and leave the other fingers open, how many will be open? How many are 8 less 2? Why?

4 If you have 8 cents, and lose 3; how many will you have left? How many are 8 less 3? Why?

5 If you have 9 cents in a box, and take out 4, how many will be left in the box? How many are 9 less 4, or 4 from 9? Why?

6. You borrow 8 pins, and pay 4; how many do you still owe? How many are 4 from 8 then? Why?

7. If you have 12 dollars, and lose 2, how many will you have left? How many are 2 from 12 then? Why?

8. A man, owing 20 dollars, paid 16; how many remain to be paid? How many are 16 from 20 then? Why?

9. You gave 18 cents for an inkstand, and sold it for 16 cents, did you make or lose, and how much? How many are 16 from 18 then? Why?

10. Your papa gave you 9 dollars, and you gave your brother 5; how many had you left? How many are 5 from 9 then? Why?

11. William bought a knife for 20 cents, and sold it for 22; how much did he make in trading? How many are 20 from 22 then? Why?

12. A man bought a barrel of molasses for 15 dollars, and sold it for 19; how much more than he gave for it did he sell it for? How many are 15 from 19 then? Why?

13. William has apples in both pockets; in one pocket he has 11, in the other 18; how many has he in one pocket more than in the other? How many are 11 from 18 then? Why?

14. A boy gave 17 cents for some picture-books, which were worth no more than 10 cents; how much more than their worth did he give for them? How many are 10 from 17? Why?

15. A man bought a cow for 13 dollars, and a calf for 3; how much more did the cow cost than the calf? How many are 3 from 13? Why?

16. A man bought a barrel of flour for 17 dollars, and, not proving so good as he expected, he could sell it for no more than 13 dollars; how much did he lose on it? How many are 13 from 17? Why?

17. A man bought a barrel of beef for 20 dollars, and, being damaged, he is obliged to lose 12 dollars on the sale of it; how much did he sell it for? How many are 12 from 20 then? Why?

18. How many legs will 4 chairs have to stand on, if 1 have 3 broken legs? How many are 3 from 16? Why?

19. 20 birds light on a tree; if 6 fly off, how many are left on the tree? How many are 6 from 20? Why?

20. Suppose you and William lose a finger apiece, how many fingers will you both have then? How many are 2 from 16? Why?

21. If you have 25 cents, and give 20 for a knife, and the rest for some marbles, how many cents will the marbles cost? How much more will the knife cost than the marbles? How many are 20 from 25? 5 from 20? Why?

22. A poor man had 16 bushels of rye given him; his eldest son gave him 10 bushels, and the youngest the rest; how many bushels did the youngest give him? How many did the elder give him more than the younger? How many are 10 from 16? *6 from 10?* Why?

23. 28 boys were sliding on the ice, which breaking, all but 4 fell in and perished; how many lost their lives? How many are 4 from 28? Why?

24. Repeat the

SUBTRACTION TABLE.

1 from 1 leaves 0	2 from 2 leaves 0	3 from 3 leaves 0
1 from 2 leaves 1	2 from 3 leaves 1	3 from 4 leaves 1
1 from 3 leaves 2	2 from 4 leaves 2	3 from 5 leaves 2
1 from 4 leaves 3	2 from 5 leaves 3	3 from 6 leaves 3
1 from 5 leaves 4	2 from 6 leaves 4	3 from 7 leaves 4
1 from 6 leaves 5	2 from 7 leaves 5	3 from 8 leaves 5
1 from 7 leaves 6	2 from 8 leaves 6	3 from 9 leaves 6
1 from 8 leaves 7	2 from 9 leaves 7	3 from 10 leaves 7
1 from 9 leaves 8	2 from 10 leaves 8	3 from 11 leaves 8
1 from 10 leaves 9	2 from 11 leaves 9	3 from 12 leaves 9
1 from 11 leaves 10	2 from 12 leaves 10	3 from 13 leaves 10
1 from 12 leaves 11	2 from 13 leaves 11	3 from 14 leaves 11
1 from 13 leaves 12	2 from 14 leaves 12	3 from 15 leaves 12

4 from 4 leaves 0	5 from 5 leaves 0	6 from 6 leaves 0
4 from 5 leaves 1	5 from 6 leaves 1	6 from 7 leaves 1
4 from 6 leaves 2	5 from 7 leaves 2	6 from 8 leaves 2
4 from 7 leaves 3	5 from 8 leaves 3	6 from 9 leaves 3
4 from 8 leaves 4	5 from 9 leaves 4	6 from 10 leaves 4
4 from 9 leaves 5	5 from 10 leaves 5	6 from 11 leaves 5
4 from 10 leaves 6	5 from 11 leaves 6	6 from 12 leaves 6
4 from 11 leaves 7	5 from 12 leaves 7	6 from 13 leaves 7
4 from 12 leaves 8	5 from 13 leaves 8	6 from 14 leaves 8
4 from 13 leaves 9	5 from 14 leaves 9	6 from 15 leaves 9
4 from 14 leaves 10	5 from 15 leaves 10	6 from 16 leaves 10
4 from 15 leaves 11	5 from 16 leaves 11	6 from 17 leaves 11
4 from 16 leaves 12	5 from 17 leaves 12	6 from 18 leaves 12

7 from 7 leaves 0	8 from 8 leaves 0	9 from 9 leaves 0
7 from 8 leaves 1	8 from 9 leaves 1	9 from 10 leaves 1
7 from 9 leaves 2	8 from 10 leaves 2	9 from 11 leaves 2
7 from 10 leaves 3	8 from 11 leaves 3	9 from 12 leaves 3
7 from 11 leaves 4	8 from 12 leaves 4	9 from 13 leaves 4
7 from 12 leaves 5	8 from 13 leaves 5	9 from 14 leaves 5
7 from 13 leaves 6	8 from 14 leaves 6	9 from 15 leaves 6
7 from 14 leaves 7	8 from 15 leaves 7	9 from 16 leaves 7
7 from 15 leaves 8	8 from 16 leaves 8	9 from 17 leaves 8
7 from 16 leaves 9	8 from 17 leaves 9	9 from 18 leaves 9
7 from 17 leaves 10	8 from 18 leaves 10	9 from 19 leaves 10
7 from 18 leaves 11	8 from 19 leaves 11	9 from 20 leaves 11
7 from 19 leaves 12	8 from 20 leaves 12	9 from 21 leaves 12

10 from 10 leaves 0	11 from 11 leaves 0	12 from 12 leaves 0
10 from 11 leaves 1	11 from 12 leaves 1	12 from 13 leaves 1
10 from 12 leaves 2	11 from 13 leaves 2	12 from 14 leaves 2
10 from 13 leaves 3	11 from 14 leaves 3	12 from 15 leaves 3
10 from 14 leaves 4	11 from 15 leaves 4	12 from 16 leaves 4
10 from 15 leaves 5	11 from 16 leaves 5	12 from 17 leaves 5
10 from 16 leaves 6	11 from 17 leaves 6	12 from 18 leaves 6
10 from 17 leaves 7	11 from 18 leaves 7	12 from 19 leaves 7
10 from 18 leaves 8	11 from 19 leaves 8	12 from 20 leaves 8
10 from 19 leaves 9	11 from 20 leaves 9	12 from 21 leaves 9
10 from 20 leaves 10	11 from 21 leaves 10	12 from 22 leaves 10
10 from 21 leaves 11	11 from 22 leaves 11	12 from 23 leaves 11
10 from 22 leaves 12	11 from 23 leaves 12	12 from 24 leaves 12

Questions on the Table.

25. How many does 2 from 8 leave? 2 from 10? 2 from 12? 2 from 15? 2 from 20? 2 from 24? 3 from 7? 3 from 10? 3 from 12? 3 from 18? 3 from 19? 4 from 8? 4 from 9? 4 from 13? 4 from 15? 4 from 18? 4 from 20? 5 from 10? 5 from 14? 5 from 17? 5 from 20? 5 from 25? 6 from 12? 6 from 18? 6 from 20? 6 from 26? 7 from 14? 7 from 21? 7 from 23? 8 from 10? 8 from 12? 8 from 15? 8 from 16? 8 from 19? 8 from 20? 9 from 12? 9 from 15? 9 from 18? 9 from 20? 9 from 22? 10 from 15? 10 from 17? 10 from 19? 10 from 20? 10 from 22? 10 from 25? 11 from 15? 11 from 18? 11 from 19? 11 from 22? 12 from 14? 12 from 16? 12 from 19? 12 from 24?

Practical Questions on the Table.

26. If you buy 15 cents' worth of tape, and give the shopkeeper a pistareen, or twenty cent bit, how many cents must you have in change? How many are 15 from 20? Why?

27. If you had 17 fingers, how many would you have more than you have now? How many are 8 from 17? Why?

28. A man had to travel 24 miles, but has travelled all but 4; how many miles has he journeyed? How many are 4 from 24? Why?

29. 20 children are in a class, and the 8 best are put into a higher class; how many are left in the lower class? How many are 8 from 20? Why?

30. If you have 25 cents, and should give 10 cents for a ruler, and 10 for a top, how many cents will you have left? How many do 10 and 10 from 25 leave? Why?

31. You have 16 apples, and give 5 to your sister, 5 to your brother; how many will you have left? How many do 5 and 5 from 16 leave? Why?

32. A man bought a mirror for 12 dollars, for which he gave 6 bushels of corn, worth 5 dollars, 3 bushels of potatoes, worth 1 dollar, and the rest in money; how much did he pay? How many do 5 and 1 from 12 leave? Why?

33. The distance from Boston to Walpole is 20 miles; after you have arrived at Dedham, which is 11 miles from Boston, how many more miles will you have to travel to reach Walpole? How many are 11 from 20? Why?

MULTIPLICATION.

¶ III. 1. If I give you 2 pins at one time, and 2 at another, how many pins shall I give you? How many are 2 times 2 then?

2. How many legs have 2 chairs? How many are 2 times 4?

3. How many eyes have 6 birds? How many 7? How many 8? How many are 2 times 6? 2 times 7? 2 times 8?

4. I hold my hand out, and you put 3 pins in it, William 3, and James 3; how many pins will I have? How many are 3 times 3?

5. If I put in your pocket 4 apples, at 3 different times, how many apples will you have in your pocket? How many are 3 times 4?

6. If I should give you 4 apples at 4 different times, how many apples will you have? How many are 4 times 4?

7. If I give 2 cents for 1 orange, how many cents must I give for 8? How many are 2 times 8?

8. How many cents will buy 10 marbles, if 1 cost 3 cents? How many are 3 times 10?

9. If you give 4 cents for a yard of tape, how many cents will buy 3 yards? How many 4? 5? 6? 7? How many are 4 times 3? 4 times 4? 4 times 5? 4 times 6? 4 times 7?

10. What will 5 picture-books come to, at 2 cents apiece? What will 6? 7? 8? 9? 10? 11? 12? How many are 5 times 2? 6 times 2? 7 times 2? 8 times 2? 9 times 2? 10 times 2? 11 times 2? 12 times 2?

11. What will 2 marbles cost, at 3 cents apiece? Will 3 marbles? Will 4? Will 5? Will 6? Will 7? Will 8? Will 9? Will 10? Will 11? How many are 3 times 2? 3 times 3? 3 times 4? 3 times 5? 3 times 6? 3 times 7? 3 times 8? 3 times 9? 3 times 10? 3 times 11?

12. Repeat the Multiplication Table.

MULTIPLICATION TABLE.

2 times 1 are 2	3 times 1 are 3	4 times 1 are 4
2 times 2 are 4	3 times 2 are 6	4 times 2 are 8
2 times 3 are 6	3 times 3 are 9	4 times 3 are 12
2 times 4 are 8	3 times 4 are 12	4 times 4 are 16
2 times 5 are 10	3 times 5 are 15	4 times 5 are 20
2 times 6 are 12	3 times 6 are 18	4 times 6 are 24
2 times 7 are 14	3 times 7 are 21	4 times 7 are 28
2 times 8 are 16	3 times 8 are 24	4 times 8 are 32
2 times 9 are 18	3 times 9 are 27	4 times 9 are 36
2 times 10 are 20	3 times 10 are 30	4 times 10 are 40
2 times 11 are 22	3 times 11 are 33	4 times 11 are 44
2 times 12 are 24	3 times 12 are 36	4 times 12 are 48
5 times 1 are 5	6 times 1 are 6	7 times 1 are 7
5 times 2 are 10	6 times 2 are 12	7 times 2 are 14
5 times 3 are 15	6 times 3 are 18	7 times 3 are 21
5 times 4 are 20	6 times 4 are 24	7 times 4 are 28
5 times 5 are 25	6 times 5 are 30	7 times 5 are 35
5 times 6 are 30	6 times 6 are 36	7 times 6 are 42
5 times 7 are 35	6 times 7 are 42	7 times 7 are 49
5 times 8 are 40	6 times 8 are 48	7 times 8 are 56
5 times 9 are 45	6 times 9 are 54	7 times 9 are 63
5 times 10 are 50	6 times 10 are 60	7 times 10 are 70
5 times 11 are 55	6 times 11 are 66	7 times 11 are 77
5 times 12 are 60	6 times 12 are 72	7 times 12 are 84
8 times 1 are 8	9 times 1 are 9	10 times 1 are 10
8 times 2 are 16	9 times 2 are 18	10 times 2 are 20
8 times 3 are 24	9 times 3 are 27	10 times 3 are 30
8 times 4 are 32	9 times 4 are 36	10 times 4 are 40
8 times 5 are 40	9 times 5 are 45	10 times 5 are 50
8 times 6 are 48	9 times 6 are 54	10 times 6 are 60
8 times 7 are 56	9 times 7 are 63	10 times 7 are 70
8 times 8 are 64	9 times 8 are 72	10 times 8 are 80
8 times 9 are 72	9 times 9 are 81	10 times 9 are 90
8 times 10 are 80	9 times 10 are 90	10 times 10 are 100
8 times 11 are 88	9 times 11 are 99	10 times 11 are 110
8 times 12 are 96	9 times 12 are 108	10 times 12 are 120
11 times 1 are 11	11 times 9 are 99	12 times 5 are 60
11 times 2 are 22	11 times 10 are 110	12 times 6 are 72
11 times 3 are 33	11 times 11 are 121	12 times 7 are 84
11 times 4 are 44	11 times 12 are 132	12 times 8 are 96
11 times 5 are 55	12 times 1 are 12	12 times 9 are 108
11 times 6 are 66	12 times 2 are 24	12 times 10 are 120
11 times 7 are 77	12 times 3 are 36	12 times 11 are 132
11 times 8 are 88	12 times 4 are 48	12 times 12 are [illegible]

Practical Questions on the Table.

13. How many cents will buy 8 books, if 1 cost 2 cents? How many will buy 3 books? 5 books? 8 books? 10 books? 12 books?

14. How many cents will 10 yards of ribbon come to at 2 cents for 1 yard? At 3 cents? 5 cents? 9 cents? 12 cents?

15. What are 2 barrels of flour worth, if 1 be worth 11 dollars? What are 3 barrels worth? What are 5? What are 7? What are 9? What are 11? What are 12?

16. What will 7 pair of shoes come to, at 5 dollars a pair? What will 8 pair? What will 10 pair? What will 12 pair?

17. What will 9 yards of broadcloth come to, at 6 dollars a yard? At 7 dollars? At 20 dollars?

18. There are 8 furlongs in one mile; how many are there in 6 miles? In 7? In 9? In 11? In 12?

19. There are 12 inches in one foot; how many are there in 2 feet? In 5 feet? In 6 feet? In 12 feet?

20. If a man earn 7 dollars in one week, how many dollars will he earn in 2 weeks? In 4? In 6? In 8? In 10? In 11? In 12?

21. If 1 bushel of clover-seed cost 12 dollars, what will 2 bushels cost? What will 3 bushels? 5 bushels? 7 bushels? 9 bushels? 11 bushels? 12 bushels?

22. If you travel 5 miles in 1 hour, how far can you travel in 2 hours? In 4? In 8? In 10? In 12?

23. William and James performed a piece of work together in 6 days; how many days will it take William to do the same work alone?

24. If you pay eight dollars for 1 quarter's tuition, what will 2 quarters come to? What will 3? 5? 7? 9? 11? 12?

25. If the interest of 1 dollar for 1 year is 6 cents, what is the interest of 2 dollars for the same time? Of 3? Of 6? Of 8? Of 10? Of 12?

26. If you pay 3 dollars for 1 week's board, what will 2 weeks' come to? What will 3? 5? 8? 10? 12?

27. If you give 5 apples for 1 orange, how many apples will buy 2 oranges? How many 3? How many 5? How many 6? How many 9? How many 10? How many 12?

DIVISION.

¶ IV. 1. Divide 6 apples between 2 boys, and tell me how many each will have? How many times 2 in 6? Why?

Ans. Because 2 times 3 are 6.

2. Divide 10 pins between 5 boys, and tell me how many each will have? How many times 5 in 10? Why?

3. If you wish to divide 8 oranges between your 2 little sisters, how many would each have? How many times 2 in 8? Why?

4. A man divides 14 peaches between 7 of his children; how many will they have apiece? How many times 7 in 14? Why?

5. 14 cents were given to 2 poor boys? how many cents is that for each boy? How many times 2 in 14? Why?

6. If 1 orange cost 6 cents, how many oranges will 18 cents buy? How many times 6 in 18? Why?

7. If it cost 6 cents to go in and see the wax figures, how many times can you go in for 30 cents? How many times 6 in 30? Why?

8. 8 boys found 48 cents, which they agreed to divide equally between them; how many will each have? How many times 8 in 48? Why?

9. I sold 8 lead pencils for 80 cents? how much is that apiece? How many times 8 in 80? Why?

10. 10 men found a pocket-book containing 100 dollars; how many dollars will each have, if the money be equally divided between them? How many times 10 in 100? Why?

11. There are 4 weeks in a month; how much will a man have a week, that has 48 dollars a month? How many times 4 in 48? Why?

12. 12 men by contract are to have 96 dollars for performing a piece of work; how many dollars is each man's part? How many times 12 in 96? Why?

13. There are 4 quarts in a gallon; what is a quart of molasses worth, when a gallon is worth 32 cents? How many times 4 in 32? Why?

14. An older brother distributed 60 picture-books between his 6 younger brothers; how many did each have? How many times 6 in 60? Why?

15. 108 cents are to be equally divided between 9 children? how many will that be apiece? How many times 9 in 108? Why?

16. 132 bushels of corn are to be divided equally between 12 poor men; how many will each man have? How many times 12 in 132? Why?

17. 12 men engage to do a piece of work for 144 dollars; what will be each man's part of the money? How many times 12 in 144? Why?

18. Repeat the Division Table.

DIVISION TABLE.

2 in 2 — 1 time	4 in 4 — 1 time	6 in 6 — 1	8 in 8 — 1
2 in 4 — 2 times	4 in 8 — 2 times	6 in 12 — 2	8 in 16 — 2
2 in 6 — 3 times	4 in 12 — 3 times	6 in 18 — 3	8 in 24 — 3
2 in 8 — 4 times	4 in 16 — 4 times	6 in 24 — 4	8 in 32 — 4
2 in 10 — 5 times	4 in 20 — 5 times	6 in 30 — 5	8 in 40 — 5
2 in 12 — 6 times	4 in 24 — 6 times	6 in 36 — 6	8 in 48 — 6
2 in 14 — 7 times	4 in 28 — 7 times	6 in 42 — 7	8 in 56 — 7
2 in 16 — 8 times	4 in 32 — 8 times	6 in 48 — 8	8 in 64 — 8
2 in 18 — 9 times	4 in 36 — 9 times	6 in 54 — 9	8 in 72 — 9
2 in 20 — 10 times	4 in 40 — 10 times	6 in 60 — 10	8 in 80 — 10
2 in 22 — 11 times	4 in 44 — 11 times	6 in 66 — 11	8 in 88 — 11
2 in 24 — 12 times	4 in 48 — 12 times	6 in 72 — 12	8 in 96 — 12

3 in 3 — 1 time	5 in 5 — 1 time	7 in 7 — 1	9 in 9 — 1
3 in 6 — 2 times	5 in 10 — 2 times	7 in 14 — 2	9 in 18 — 2
3 in 9 — 3 times	5 in 15 — 3 times	7 in 21 — 3	9 in 27 — 3
3 in 12 — 4 times	5 in 20 — 4 times	7 in 28 — 4	9 in 36 — 4
3 in 15 — 5 times	5 in 25 — 5 times	7 in 35 — 5	9 in 45 — 5
3 in 18 — 6 times	5 in 30 — 6 times	7 in 42 — 6	9 in 54 — 6
3 in 21 — 7 times	5 in 35 — 7 times	7 in 49 — 7	9 in 63 — 7
3 in 24 — 8 times	5 in 40 — 8 times	7 in 56 — 8	9 in 72 — 8
3 in 27 — 9 times	5 in 45 — 9 times	7 in 63 — 9	9 in 81 — 9
3 in 30 — 10 times	5 in 50 — 10 times	7 in 70 — 10	9 in 90 — 10
3 in 33 — 11 times	5 in 55 — 11 times	7 in 77 — 11	9 in 99 — 11
3 in 36 — 12 times	5 in 60 — 12 times	7 in 84 — 12	9 in 108 — 12

10 in 10 — 1 time	11 in 11 — 1 time	12 in 12 — 1 time
10 in 20 — 2 times	11 in 22 — 2 times	12 in 24 — 2 times
10 in 30 — 3 times	11 in 33 — 3 times	12 in 36 — 3 times
10 in 40 — 4 times	11 in 44 — 4 times	12 in 48 — 4 times
10 in 50 — 5 times	11 in 55 — 5 times	12 in 60 — 5 times
10 in 60 — 6 times	11 in 66 — 6 times	12 in 72 — 6 times
10 in 70 — 7 times	11 in 77 — 7 times	12 in 84 — 7 times
10 in 80 — 8 times	11 in 88 — 8 times	12 in 96 — 8 times
10 in 90 — 9 times	11 in 99 — 9 times	12 in 108 — 9 times
10 in 100 — 10 times	11 in 110 — 10 times	12 in 120 — 10 times
10 in 110 — 11 times	11 in 121 — 11 times	12 in 132 — 11 times
10 in 120 — 12 times	11 in 132 — 12 times	12 in 144 — 12 times

Practical Questions on the Table.

19. If 12 yards of tape cost 24 cents, what will 1 yard cost?

20. If you give 2 cents for an apple, how many can you buy for 4 cents? How many for 6 cents? For 10 cents? For 14 cents? For 18 cents? For 20 cents?

21. If 1 lead pencil cost 3 cents, how many can you buy for

6 cents? For 9 cents? For 18 cents? For 21 cents? For 24 cents? For 30 cents? For 36 cents?

22. If 4 cents will buy 1 orange, how many oranges will 8 cents buy? How many 16 cents? How many 24 cents? How many 32 cents? How many 40 cents? How many 48 cents?

23. If the stage fare be 5 cents a mile, how far may you be carried for 10 cents? For 15? For 20? For 25? For 30? For 35? For 40? For 50? For 60?

24. If 6 cents will buy 1 pine-apple, how many will 12 cents buy? Will 24? Will 36? Will 42? Will 48? Will 60?

25. If a small slate cost 7 cents, how many slates will 14 cents buy? Will 28? Will 35? Will 56? Will 63?

26. If a writing-book cost 8 cents, how many writing-books will 16 cents buy? 24 cents? 40 cents? 56 cents? 80 cents? 96 cents?

27. How many spelling-books will 18 cents buy, if 1 cost 9 cents? Will 27? Will 36? Will 45? Will 54? Will 72?

28. How many fish can you buy for 20 cents, if 1 cost 10 cents? How many for 40 cents? For 60 cents? For 100 cents? For 110 cents? For 120 cents?

29. If you pay 11 cents for an inkstand, how many can you buy for 22 cents? For 33 cents? For 55 cents? For 88 cents? For 110 cents? For 132 cents?

30. How many pounds of butter can you buy for 24 cents, when the price is 12 cents for 1 pound? How many pounds for 36 cents? For 60 cents? For 108 cents? For 132 cents? For 144 cents?

Practical Questions on the foregoing.

1. A boy, having 18 apples, gave them to his companions, as follows; to William 4, to Rufus 6, and to Thomas 5; how many did he give away in all, and how many had he left?

2. Thomas gave to one of his companions 6 peaches, to another 3, to another 2, and sold 3; how many had he at first?

3. A man bought a wagon for 17 dollars, and gave 5 dollars to have it repaired, then sold it for 26 dollars; how much did he make by the bargain?

4. A man bought a horse for 25 dollars, and, to pay for it, gave 6 bushels of rye, worth 6 dollars, and the rest in money; how much money did he pay?

5. Rufus, having 20 cents, bought a book for 12 cents, and a knife for 6 cents? how much more did the book cost than the knife? and how many cents had he left?

6. What is the cost of 5 yards of cloth, at 4 dollars a yard? At 3 dollars? At 7 dollars? At 2 dollars? At 8 dollars? At 9 dollars? At 12 dollars?

7. If 1 lemon be worth 3 apples, how many lemons are 6 apples worth? Are 12 apples worth? Are 18 apples worth? Are 24 apples worth? Are 36 apples worth?

8. How many barrels of flour, at 8 dollars a barrel, can you buy for 16 dollars? For 48 dollars? For 96 dollars? For 80 dollars?

9. How many are 2, 3, and 5? Are 4, 2, and 6? Are 8, 3, and 2? Are 9, 3, and 4? Are 10, 8, and 2? Are 5, 4, 3, and 2? Are 4, 3, 2, and 1? Are 7, 6, 3, and 2? Are 8, 9, and 10? Are 12, 11, 10, and 9?

10. How many are 6 times 3? 6 times 4? 6 times 7? 7 times 8? 9 times 7? 12 times 7? 9 times 5? 8 times 7? 7 times 6? 7 times 9? 12 times 11? 8 times 5? 3 times 7? 12 times 12?

11. How many times 2 in 12? 2 in 18? 2 in 24? 3 in 6? 3 in 12? 3 in 36? 4 in 20? 4 in 32? 4 in 48? 5 in 25? 5 in 35? 5 in 60? 6 in 36? 6 in 48? 6 in 72? 7 in 14? 7 in 56? 7 in 84? 8 in 40? 8 in 96? 9 in 36? 9 in 108? 11 in 22? 11 in 55? 11 in 132? 12 in 144?

Note. Younger pupils should be required to review, and dwell on the preceding questions for illustration, and the tables, till their solutions be made perfectly familiar.

NUMERATION.

¶ V. *Q.* When I say to you, Give me that book, do I mean one book or more than one?

Q. When we speak of a single thing, then, what is it called? *A.* A unit, or one.

Q. What are one unit and one more, or one and one, called?

Q. What are two units and one more, or two and one, called?

Q. What are three units and one more, or three and one, called?

Q. What are four units and one more, or four and one, called?

Q. What are five units and one more, or five and one, called?

Q. What are six units and one more, or six and one, called?

Q. What are seven units and one more, or seven and one, called?

Q. What are eight units and one more, or eight and one, called?

Q. What are nine units and one more, or nine and one, called?

Q. Now, to be obliged always to write these numbers out in words, would be very troublesome; to prevent this, how do we sometimes express the numbers one, two, &c. up to thousands, *millions, &c.* *A.* By letters.

Q. What does the letter I stand for? *A.* One.
Q. What does the letter V stand for? *A.* Five.
Q. What does the letter X stand for? *A.* Ten.
Q. What does the letter L stand for? *A.* Fifty.
Q. What does the letter C stand for? *A.* One hundred.
Q. What does the letter D stand for? *A.* Five hundred.
Q. What does the letter M stand for? *A.* One thousand.

Q. You said that V stands for five; suppose you place the letter I before the V, thus, IV, what will both these letters stand for then? *A.* Only four.

Q. What, then, may be considered as a rule for determining the value of these letters? *A.* A letter standing for a smaller number, and before a larger, takes out its value from the larger.

Q. One X stands for ten; what do two XX's stand for? *A.* Twenty.

Q. What, then, is the value of a letter repeated? *A.* It repeats the value as often as it is used.

Q. How many letters do we use for expressing numbers? *A.* Seven. Name them. *A.* I, V, X, L, C, D, M.

Q. What is this method of expressing numbers by letters called? *A* The Roman method.

Q. Why called Roman? *A.* Because the Romans invented and used it.

Repeat the

ROMAN TABLE.

One,	I.	Thirty,	XXX.
Two,	II.	Forty,	XL
Three,	III.	Fifty,	L
Four,	IV.	Sixty,	LX.
Five,	V.	Seventy,	LXX.
Six,	VI.	Eighty,	LXXX.
Seven,	VII.	Ninety,	XC.
Eight,	VIII.	One hundred,	C.
Nine,	IX.	Two hundred,	CC.
Ten,	X.	Three hundred,	CCC.
Eleven,	XI.	Four hundred,	CCCC.
Twelve,	XII.	Five hundred,	D.
Thirteen,	XIII.	Six hundred,	DC.
Fourteen,	XIV.	Seven hundred,	DCC.
Fifteen,	XV.	Eight hundred,	DCCC.
Sixteen,	XVI.	Nine hundred,	DCCCC.
Seventeen,	XVII.	One thousand,	M.
Eighteen,	XVIII.	Fifteen hundred,	MD.
Nineteen,	XIX.	Sixteen hundred,	MDC.
Twenty,	XX.	Two thousand,	MM.

Eighteen hundred and twenty-eight, ... MDCCCXXVIII.

¶ VI. We have a shorter method still, which is in very general use, as will appear by observing what follows:—

A unit, or one, is written ....1.
Two, ...2.
Three, ...3.
Four, ...4.
Five, ...5.
Six, ...6.
Seven, ...7.
Eight, ...8.
Nine, ...9.

Q. What are these characters called? A. Figures.

Q. By what other name are they sometimes called? A. The 9 digits.

Q. What is this method of expressing numbers called? A. The Arabic method.

Q. Why so called? A. Because the Arabs are supposed to have invented it.*

Let me see you write down on the slate, in figures, the numbers one, two, three, four, five, six, seven, eight, nine.

Q. To express ten, as we have no one character that will do it, what two characters do we make use of to represent this number? A. The first character, 1, and 0, or cipher; thus, 10.

Q. What place does the 0, or cipher, in this case take? A. The units' place.

Q. What place does the figure 1 take? A. A new place.

Q. What is this new place called? A. The tens' place.

Q. Write down in figures, on the slate, the number ten; now take away the 1, and what will be left? A. Nothing but 0, or cipher.

Q. What is the value of this 0, or cipher, thus standing alone? A. No value.

Q. Now place the 0 at the right of the figure 1, and what will it become? A. Ten, (10.)

Q. How many times is the figure 1 increased by the 0, or cipher? A. Ten times.

Q. What effect, then, has a cipher in all cases when placed at the right of figures? A. It increases the value ten times.

Q. In what proportion is this increase said to be? A. Tenfold proportion.

---

* Q. How was it obtained from the Arabs? A. The Moors communicated it to the Spaniards, and John of Basingstoke, Archdeacon of Leicester, introduced it into England; hence its introduction into our own country.

Q. About what time was it introduced into England? A. About the middle of the eleventh century.

Q. How extensively is it now used? A. All over the civilized world.

As you have probably learned by this time how to write down ten in figures, by the help of a cipher, and learned also the value of this cipher, we will now proceed to higher numbers; and to begin: *let me see you write down in figures, on the slate, the following numbers, viz.*

One ten and one unit, or eleven,................11
One ten and two units, or twelve, ...12
One ten and three units, or thirteen, ...13
One ten and four units, or fourteen, ...14
One ten and five units, or fifteen, ...15
One ten and six units, or sixteen, ...16
One ten and seven units, or seventeen, ...17
One ten and eight units, or eighteen, ...18
One ten and nine units, or nineteen, ...19
Two tens,.............or twenty,..................20
Three tens,...........or thirty,..................30
Four tens, ...or forty, ...40
Five tens,.............or fifty,...................50
Six tens, ...or sixty, ...60
Seven tens,...........or seventy,...............70
Eight tens,............or eighty,.................80
Nine tens, ...or ninety, ...90
Ten tens, ...or one hundred,..........100

Q. Here we see the value of the cipher again; for, by placing a cipher at the right of ten, it becomes one hundred, (100,) that is, ten tens: should we place another cipher still at the right of the 100, (thus, 1000,) what would it become? *A.* One thousand, (1000).

Q. From what you have now seen of the value of figures, what may 2 and 5 be made to stand for? *A.* 25 or 52.

Q. What is this different value called, which arises from the figures being placed or located differently? *A.* Their local value.

Q. What would be the value of the five written alone? *A.* Simply 5.

Q. What is the value, then, of a figure standing alone? *A.* The simple value.

Q. How many values do figures appear to have? *A.* Two.

Q. What are they? *A.* Simple and local.

Q. Now, as it takes 10 units to make one ten, or one in the next left hand place, and 10 tens to make 100, how do figures appear to increase by being removed one place farther to the left? *A.* In a tenfold proportion from right to left.

You must have acquired, by this time, some considerable knowledge of figures: let me examine you a little; and, in the *first place, let me see* you write down on the slate the figure 8.

Q. What do you call it? *A.* 8 units.

Write at the left of the 8 the figure 2, (thus, 28.)

Q. What do you call them both, and how are they read?

A. 8 units and 2 tens read twenty-eight.

Write at the left of the 28 the figure 8, (thus, 828.)

Q. What do you call the three figures now, and how are they read? *A.* 8 units, 2 tens, and 8 hundreds, read eight hundred and twenty-eight.

Write at the left of 828 the figure 1, (thus, 1828.)

Q. What do you call the 4 figures now, and how read?

A. 8 units, 2 tens, 8 hundreds, and 1 thousands, read one thousand eight hundred and twenty-eight.

Q. We have now been combining, or placing figures together till we have obtained the number 1828, representing the number of years it is since Christ appeared on earth, to the present time. We might continue to put figures together in this way that would express higher numbers still, up to billions, &c That you may be able to form some idea of the power of figures, let me tell you that there is not a billion of seconds in thirty thousand years; notwithstanding there are 60 seconds in every minute, 60 minutes in every hour, 24 hours in every day, and in a solar year, 365 days, 5 hours, 48 minutes, and about 48 seconds. Should we continue to go on as we began, in combining more figures still, it would be very inconvenient; to avoid this we have a rule by which we can read almost any number of figures, ever so large. What is this rule called? *A.* Numeration.

Q. What is the reading, or expressing a number by figures as now shown, called? *A.* Notation or Numeration.

RULE.

I. *From the above illustrations, how does it appear that you must begin to numerate?* *A.* Begin at the right hand.

II. *At which hand would you begin to read?* *A.* The left.

III. *What is the first figure at the right hand, or first place, called?* *A.* Units.

What is the second figure, or second place, called? *A.* Tens

What is the third place called? *A.* Hundreds.

What is the fourth place called? *A.* Thousands.

IV. *In reading, what value do you give those figures which were called units in numerating?* *A.* Units.

V. *What value do you give tens?* *A.* Tens.

VI. *What value do you give hundreds, thousands, &c* *A.* Hundreds, thousands, &c.

1. Repeat the Numeration Table.

NUMERATION TABLE.

Billions.	Hundreds of Thous. of Mills.	Tens of Thousands of Mills.	Thousands of Millions.	Hundreds of Millions.	Tens of Millions.	Millions.	Hundreds of Thousands.	Tens of Thousands.	Thousands.	Hundreds.	Tens.	Units.	
1	0	0	0	0	0	0	0	0	0	0	0	0,	*read*—One billion.
	2	0	0	0	0	0	0	0	0	0	0	0,	*read*—Two hundred thous. mills.
		3	0	0	0	0	0	0	0	0	0	0,	*read*—Thirty thousand millions
			4	0	0	0	0	0	0	0	0	0,	*read*—Four thousand millions.
				5	0	0	0	0	0	0	0	0,	*read*—Five hundred millions.
					6	0	0	0	0	0	0	0,	*read*—Sixty millions.
						7	0	0	0	0	0	0,	*read*—Seven millions.
							8	0	0	0	0	0,	*read*—Eight hundred thousand.
								9	0	0	0	0,	*read*—Ninety thousand.
									1	0	0	0,	*read*—One thousand.
										2	0	0,	*read*—Two hundred.
											3	0,	*read*—Thirty.
												4,	*read*—Four.

Questions on the Table.

Here let the teacher cover over the written numbers only on the right of the table above, and ask the pupil the following questions, viz. What is the value of 4? Of 3 and one cipher? Of 2 and 2 ciphers? and so on, up to the top of the table.

Q. What is the meaning of *annex?* *A.* To place after.

Q. What is the meaning of *prefix?* *A.* To place before.

Note.—Let the scholar write down in figures, the answers to the following questions on his slate at recitation.

Q. How much does 1, with 1 cipher annexed, stand for? *A.* Ten.

Q. Why? *A. Because the* 1 *is tens when I numerate.* Numerate the 10 and see.

Q. What does 1 with 3 ciphers stand for? *A.* One thousand.

Q. Why? *A. Because when I numerate by saying units, tens, hundreds, thousands, the* 1 *comes thousands.*

Q. What does 5 with five ciphers stand for? *A.* Five hundred thousand. Why? *A. Because when I numerate, the 5 comes hundreds of thousands.*

Numerate and see.

Q. What does 8 with 6 ciphers stand for? *A.* 8 millions.

Q. Why? *A. In numerating, the 8 comes millions.*

Numerate and see.

Q. How do you read the figures 624? *A.* Six hundred and twenty-four.

Q. Why do you say 6 hundred?

Q. What does 6278 stand for? *A.* Six thousand two hundred and seventy-eight.

Q. How do you know that the 6 is 6 thousand?

Q. How do you read the figures 56768? How do you read 27365? How do you read 654212?

Express in words the following numbers.

Note.—The pupil may learn the value of each succeeding number by a former one.

8 = Eight.
9
30 = Thirty.
70
239 = Two hundred and thirty-nine.
629
5005 = Five thousand and five.
7007
30002 = Thirty thousand and two.
50009
623029 = Six hundred twenty-three thousand and twenty-nine.
928028
6000066 = Six millions and sixty-six.
8000099
75000100 = Seventy-five millions and one hundred.
83000800

Express in figures the following numbers.

Sixty.—One hundred and twenty-five.

Three thousand three hundred and thirty-three.

Three millions, three hundred thirty-three thousand, three hundred and thirty-three.

Thirty millions.

Three hundred millions and twenty-five.

Repeat the following Numeration Table, and the corresponding value of the figures prefixed.

53 Quatrillions.
53 C. of Thousands of Trillions.
53 X. of Thousands of Trillions.
53 Thousands of Trillions.
53 C. of Trillions.
53 X. of Trillions.
53 Trillions.
53 C. of Thousands of Billions.
53 X. of Thousands of Billions.
53 Thousands of Billions.
53 C. of Billions.
53 X. of Billions.
53 Billions.
53 C. of Thousands of Millions.
53 X. of Thousands of Millions.
53 Thousands of Millions.
53 C. of Millions.
53 X. of Millions.
53 Millions.
53 C. of Thousands.
53 X. of Thousands.
53 Thousands.
53 Hundreds.
53 Tens.
53 Units.

SIMPLE ADDITION.

¶ **VII.** 1. You bought an orange for 9 cents, and a melon for 15 cents; what did you pay for both?

2. James bought a top for 6 cents, a knife for 12 cents, and an inkstand for 8 cents; how much did they all come to?

3. Harry and James lost some money; James lost 20 cents, and Harry 12; how much did both lose?

4. A boy laid out 10 cents in marbles, 8 cents in quills, and 6 cents for a slate-pencil; how much did he lay out in all?

5. You give 40 cents for a Practical Arithmetic, 8 cents for a ruler, 9 cents for an inkstand, and lose 6 cents; how much money has gone from you?

6. A man gave his children money in the following manner; to his oldest 3 dollars, to James 5 dollars, to Thomas 9 dollars, and to his two daughters, 4 dollars apiece; how much did he give away?

7. A boy bought 20 marbles for 20 cents, 6 peaches for 8 cents, and 3 apples for 2 cents; how much money did he lay out?

8. A man bought a cart for 6 dollars, a plough for 2 dollars, a pair of steers for 9 dollars, and 2 acres of land for 8 dollars; how much did he lay out in all?

9. How old would you be, were your age double what it now is?

10. If you had three times as many fingers and thumbs as you have now, how many would you have in all?

11. How many quarters to an apple, or any thing?

12. How many thirds to an apple, or any thing?

13. If an apple, a number, or any thing, is divided into 4 equal parts, what would one of those parts be called? *A.* One quarter, or $\frac{1}{4}$.

14. In the above, if divided into 3 equal parts, what would one part be called?

15. If an apple, or any thing, is divided into 5 equal parts, what would one part be called? *A.* One fifth, or $\frac{1}{5}$.

16. What would 2 parts be called? *A.* Two fifths, or $\frac{2}{5}$.

17. What would 4 parts be called?

18. How many parts does it take to make 5 fifths? *A.* 5.

19. How many parts does it take to make the whole? *A.* 5.

20. Why is $\frac{5}{5}$ the whole? *A.* Because the whole of the apple was divided into 5 equal parts.

21. If $\frac{1}{4}$ of an apple cost 2 cents, what will a whole apple cost?

22. If $\frac{1}{5}$ of an apple cost 1 cent, what will the whole cost?

Note A.—Of the two following tables, the first is to be added from left to right, thus, 1 and 2 are 3; then the next line, thus, 1 and 2 are 3, and 3 are 6; then the next line, thus, 1 and 2 are 3, and 3 are 6, and 4 are 10; and thus with all the lines.

The second is to be added from left to right, in the same manner.

The learner, in reciting either, is not to look on the book; the order of the figures being such as to render it unnecessary.

23. *What is the sum of the following numbers?*

1	2											*A.* 3
1	2	3										*A.* 6
1	2	3	4									*A.* 10
1	2	3	4	5								*A.* 15
1	2	3	4	5	6							*A.* 21
1	2	3	4	5	6	7						*A.* 28
1	2	3	4	5	6	7	8					*A.* 36
1	2	3	4	5	6	7	8	9				*A.* 45
1	2	3	4	5	6	7	8	9	10			*A.* 55
1	2	3	4	5	6	7	8	9	10	11		*A.* 66
1	2	3	4	5	6	7	8	9	10	11	12	*A.* 78

2	2	2	2	2	2	2	2	2	2	2	2	*A.* 24
3	3	3	3	3	3	3	3	3	3	3	3	*A.* 36
4	4	4	4	4	4	4	4	4	4	4	4	*A.* 48
5	5	5	5	5	5	5	5	5	5	5	5	*A.* 60
6	6	6	6	6	6	6	6	6	6	6	6	*A.* 72
7	7	7	7	7	7	7	7	7	7	7	7	*A.* 84
8	8	8	8	8	8	8	8	8	8	8	8	*A.* 96
9	9	9	9	9	9	9	9	9	9	9	9	*A.* 108
10	10	10	10	10	10	10	10	10	10	10	10	*A.* 120
11	11	11	11	11	11	11	11	11	11	11	11	*A.* 132
12	12	12	12	12	12	12	12	12	12	12	12	*A.* 144

24. If $\frac{1}{4}$ of an apple be worth 1 cent, how much is a whole apple worth?

25. If $\frac{1}{4}$ of a vessel be worth 1000 dollars, how much is the whole vessel worth? How much is $\frac{1}{2}$ worth?

26. If you give 800 dollars for $\frac{1}{3}$ of a house, how much is the whole house worth? How much is $\frac{2}{3}$ worth?

27. If $\frac{1}{5}$ of an apple cost 2 cents, what is the whole apple worth?

28. If $\frac{1}{6}$ of a factory be worth 2000 dollars, what is the whole worth?

29. 16 boys, throwing stones at an apple-tree, beat off a number of apples: says one boy, My part is $\frac{1}{16}$, and I am entitled to one apple; how many apples is $\frac{2}{16}$ then? How many $\frac{4}{16}$? How many $\frac{6}{16}$? How many $\frac{10}{16}$? How many $\frac{12}{16}$? How many $\frac{16}{16}$?

30. 16 men caught so many fish, that they could not count them; a bystander told one man that his part was 100, just $\frac{1}{16}$ of the whole; how many fish would $\frac{2}{16}$ be? How many $\frac{4}{16}$? How many $\frac{6}{16}$? How many $\frac{10}{16}$? How many $\frac{12}{16}$? How many fish did they catch in all?

31. John was born twenty years after James; how old will James be when John is 21?

32. When Joseph was 21, he married a wife that was 10 years old when he was born; how old was the wife when Joseph married?

33. How many are 7 and 9? 47 and 9? 87 and 9? 37 and 9? 7 and 5? 27 and 5? 57 and 5? 8 and 7? 48 and 7? 68 and 7? 58 and 7? 78 and 7? 8 and 8? 28 and 8? 48 and 8? 58 and 8? 78 and 8? 98 and 8? 9 and 9? 39 and 9? 59 and 9? 79 and 9? 69 and 9? 6 and 6? 36 and 6? 56 and 6? 76 and 6? 96 and 6? 106 and 6?

Note A.—After the manner of the last examples, the pupil should be taught to perform the following. Should he hesitate in any instance, as, for example, in adding 8 to 88, just say to him, 8 and 8 are 16, and he will soon see that 88 and 8 are 96, there being a 6 in both cases. By this means, if he can add any two numbers together, both under 10, he may be taught to add any number smaller than 10 to any number larger than 10.

Note B.—The following may be added by calling each finger a figure. Care should be exercised, lest the learner give the total amount from the book, without making the individual additions for himself.

34. Add 5 twos, and 5 threes together: thus—2 and 2 are 4 and 2 are 6, and 2 are 8, and 2 are 10, and 3 are 13, and 3 are 16, and 3 are 19, and 3 are 22, and 3 are 25.

Add	5 fours,	5 threes,	and 5 twos together.		*A.* 45
"	5 fives,	5 fours,	and 5 threes	"	*A.* 60.
"	5 sixes,	5 fives,	and 5 fours	"	*A.* 75.
"	5 sevens,	5 sixes,	and 5 fives	"	*A.* 90
"	5 eights,	5 sevens,	and 5 sixes	"	*A.* 105.
"	5 nines,	5 eights,	and 5 sevens	"	*A.* 120.
"	10 ones		and 10 twos	"	*A.* 30.
"	10 threes		and 10 twos	"	*A.* 50.
"	10 fours		and 10 threes	"	*A.* 70.
"	10 fives		and 10 fours	"	*A.* 90

Add 10 sixes and 10 fives together.		*A.* 110.
" 10 sevens and 10 sixes	"	*A.* 130.
" 10 eights and 10 sevens	"	*A.* 150.
" 10 nines and 10 eights	"	*A.* 170.

Q. What is this which you have now been doing called? *A.* Addition.

Q. What, then, may the putting together of two or more numbers, and finding how much they make, be called? *A.* Addition.

Q. What is the putting together, or collecting of several numbers, of the same name, or denomination, called? *A.* Simple Addition.

Q. What do you mean by the same name, or denomination? *A.* All pounds, all dollars, all cents, or all drams, &c.

How many are 20 *and* 30? *What do you call the* 50? *A.* Amount.

Note C.—It is thought advisable, whenever the operation of the first sum in any rule is given, to direct the pupil, after a careful attention to the questions in the book, to copy the sum on his slate, and from this to answer the questions respecting the work, without looking on the book at recitation.

Operation by Slate illustrated.

1. A man bought a cart for 25 dollars, a yoke of oxen for 69 dollars, and a plough for 7 dollars. What did he give for the whole?

OPERATION.

Cart,	25 dollars.
Oxen,	69 dollars.
Plough,	7 dollars.
	——
Amount,	101 dollars.

In writing this example down, why do you place the 7 (*units*) *under the* 9 (*units*)? *why not place it under the* 6 (*tens*)? *A.* Because, if I should, the 7 units would become 7 tens, or 70; that is, the 7 dollars would become 70 dollars.

How do you obtain the 1 *in the Ans.?* *A.* I say 7 (units) and 9 (units) are 16 (units), and 5 more (units) are 21 (units), or 2 tens and 1 unit, writing the 1 in the units' place.

What do you do with the 2 *tens?* *A.* I carry, or add them to the 6 (tens), in the next higher place, where all the tens belong.

What is this adding of the 2 *tens called?* *A.* Carrying one for every ten.

How do you proceed to get the 10 *in the Ans.?* *A.* The 2 (tens) to carry, and 6 (tens) make 8 (tens), and the 2 (tens) over the 6 (tens) are 10 (tens).

From these illustrations we derive the following

RULE.

I. *How do you write the numbers down?* *A.* Units under units, tens under tens, &c. with a line underneath.

II. *At which hand do you begin to add?* *A.* The right.

III. *If the amount of any column be 9, or less, how do you proceed?* *A.* Set it down.

IV. *If it be more than 9, what do you do?* *A.* I set down the right hand figure, and carry the left hand figure, or figures, to the next column.

V. *Which figure would you write down, and which carry in* 18, 10, 13, 36, 81, 94, 108, 58, 67, 125?

VI. *What do you do with the amount of the last column?* *A.* I set the whole of it down.

PROOF. *How do you prove the operation?* *A.* Begin at the top, and add the figures downward in the same manner as they were added upwards.

What must this amount be like? *A.* The first amount.

More Exercises for the Slate.

Note D.—To save the trouble of having the pupil running up continually to his teacher to know if his sums be right, and to prevent the learner from copying the answers on his slate from the book, indirect answers will be given through the book, when that can conveniently be done: in other cases direct answers will be given.

2. A man bought a suit of clothes for 57 dollars, a pair of boots for 8 dollars, and a secretary for 28 dollars. What did he give for the whole? *A.* 93.

3. In an orchard, 20 trees bear pears, 54 bear peaches, and 6 bear plums. How many are there in the orchard? *A.* 80.

4. A man bought a barrel of flour for 10 dollars, a barrel of molasses for 29 dollars, and a barrel of rum for 36 dollars. How much did he pay for all the articles? *A.* 75.

5. James bought at one time 89 marbles, at another time 54, at another 60, and at another 75. How many did he buy in all? *A.* 278.

6. A man gives 89 dollars for $\frac{1}{3}$ of a chaise; how much must he give for the whole at that rate? *A.* 267.

7. You expend for a gold watch 165 dollars, for a chaise 225 dollars, for a new suit of clothes 80 dollars, and give your father 400 dollars. How much money have you parted with in all? *A.* 870.

8. *If* $\frac{1}{3}$ *of a vessel be worth 2265 dollars, what is the whole vessel worth?* *A.* 6795 dollars.

9. What is the whole stock which a man has in trade worth, if $\frac{1}{4}$ be worth 3500 pounds? *A.* 14000 pounds.

10. If one man own $\frac{1}{8}$ of a bank, and his part cost 26000 dollars, what would the whole be worth at that rate? *A.* 208000 dollars.

11. Suppose my neighbour should borrow of me at one time 656 dollars, at another 50 dollars, at another 3656 dollars, and at another 5000 dollars; how much should I lend him in all? *A.* 9362 dollars.

12. A merchant owes 617 dollars to Messrs. B. & T. C. Hoppin, 516 dollars to Messrs. B. & C. Dyer, 600 dollars to the Exchange Bank, 1000 dollars to the Union Bank; I demand how much he owes in the whole? *A.* 2733 dollars.

13. A merchant bought at one time 600 barrels of beef, at another 500 barrels, at another 416 barrels; how many barrels did he buy in the whole? *A.* 1516 barrels.

14. James was born A. D. 1800; what year of our Lord will it be, when James is 37 years of age? *A.* 1837.

15. Gen. George Washington was born A. D. 1732, and lived 67 years; in what year did he die? *A.* 1799.

16. From the creation of the world to the flood was 1656 years; from thence to the building of Solomon's temple, 1336 years; thence to the birth of our Saviour, 1008 years; in what year of the world was the birth of Christ? *A.* Anno Mundi 4000.

(17.)	(18.)	(19.)	(20.)	(21.)
Dollars.	Drams.	Dimes.	Mills.	Shillings.
35	313	1645	132132	4 5 6 7 3 2 1 2 3 2 1 2
64	280	0321	245123	1 2 1 2 1 2 1 2 1 2 1 2
21	741	4610	521085	1 2 3 4 1 2 3 4 1 2 3 4
18	240	5386	603898	2 3 4 2 3 4 2 3 4 2 3 4
12	391	5210	789783	8 9 8 7 6 8 9 8 7 6 5 1
———	———	———	———	———
*———	———	———	———	———

(22.)	(23.)
Eagles.	Dollars.
2 7 1 1 3 5 2 6 0 3 5 7	1 2 3 4 5 6 7 8 9 1 2 3 4 5 6 7 8 9
3 1 1 0 7 0 5 8 2 1 2 1	9 8 7 6 5 4 3 2 1 9 8 7 6 5 4 3 2 0
6 2 5 5 1 5 3 7 4 2 3 2	5 5 4 4 3 3 2 2 1 1 2 2 3 3 4 4 2 2
8 1 3 6 0 2 1 1 5 6 0 1	6 6 5 5 4 4 3 3 2 2 1 1 0 0 1 1 3 3
5 3 5 1 2 6 1 0 0 8 1 2	7 7 6 6 5 5 4 4 3 3 2 2 1 1 0 0 1 1
———	———
———	———

* The teacher will observe that the amounts of the several sums are divided by 3, and the quotients given for the answers will be found in sum No 27, each quotient being set against the No. of the sum, that he may more readily tell, if the sum be right.

(24)
Pennyweights.

```
987654321
 98765432
  9876543
   987654
    98765
     1234
      123
       12
        1
---------

---------
```

(25)
Degrees.

```
6542103459652
        13000
    852670000
       895965
 823245261785
     12312612
          721
           21
            3
-------------

-------------
```

(26.)
Furlongs.

```
34567230137954321056513
82130549865203156821342
13221423001360421210050
22343152243132002303213
53004311322411213132120
```

(27.)

```
A. to No. 17.                                   50
A. to No. 18.                                  655
A. to No. 19.                                 5724
A. to No. 20.                               764007
A. to No. 21.                         611453269181
A. to No. 22.                         852149811041
A. to No. 23.                  1035914702588852225
A. to No. 24.                            365794695
A. to No. 25.                        2455404871253
A. to No 26.               68422222190020371507746
                     ------------------------------
                                                    ×3
                     ------------------------------
Total amount.            205269774325927004629731
```

28. Add 8541, 1256, 3560, and 2456 together. *A.* 15813.

29. Add 15000 dolls. 2500 dolls. 36594 dolls. 29321 dolls. together. *A.* 83415 dolls.

30. Add 11000 mills, 1100 mills, 110 mills, and 11 mills together. *A.* 12221 mills.

31. Add 555555 ounces, 3333 ounces, 66 ounces, 4444444 ounces, and 22222 ounces together. *A.* 5025620.

32. What is the sum of the following numbers? viz.

Twenty-five, Three hundred sixty-five, Two thousand one *hundred and forty*-five, Eighty-nine thousand, Four hundred eighty-five, Nine millions and six, Ninety millions and nine thousand. *A.* 99101026.

SIMPLE SUBTRACTION.

¶ **VIII.** 1. George had 10 apples, and gave 6 of them to William; how many did he have left? Why? *A. Because* 4 *and* 6 *are* 10.

2. Rufus, having 20 dollars, gave 12 to James; how many had he left? Why?

3. A man, owing 30 dollars, paid 20; how many did he then owe?

4. A man, having 100 dollars, lost 50 of them; how many had he left?

5. A merchant bought a piece of cloth for 120 dollars, and sold it for 140 dollars; how much did he make by the bargain?

6. From 100 take 20; take 10; take 40; take 60; take 70; take 80; take 90; take 95; take 85; take 75; take 5; take 15.

7. John, having 75 apples, gave 20 to his oldest brother, 20 to his youngest, and 20 to his sister; how many had he left?

8. Harry had 25 marbles in both pockets; he lost 9 out of one pocket, and 7 out of the other; how many had he left?

9. William has two pockets, both of which will hold 75 peaches; he has in one 15, and in the other 45; how many more will both hold?

10. A boy, returning with a basket full of oranges, containing 100, and meeting his cousin by the way, gave him 20; how many did he carry home?

11. Two boys were playing at marbles; each had 20 when they began; John lost 5; how many did each have then? When the unfortunate boy had lost all but 2, how many had James won from John?

12. You bought 100 new marbles for 50 cents, and sold Peter 10 for 15 cents; Harry 6 for 10 cents, and Thomas 34 for 20 cents; how many marbles had you remaining? and how much more did you pay for them than what you sold came to?

13. How many quarters to an apple, or any thing? How many thirds? How many fifths? How many sixths? Sevenths?

14. If you had 4 pencils, and should give away $\frac{1}{4}$, how many would you have left?

15. If you had 3 cents, and should give away $\frac{1}{3}$, how many would you have left?

16. If you had 8 pencils, and should give away $\frac{1}{8}$, how many would you have left?

4 *

17. How many would you have left each time, if you should give away $\frac{2}{8}$, $\frac{3}{8}$, $\frac{4}{8}$, $\frac{6}{8}$, $\frac{7}{8}$?

18. If you had 16 marbles, and should give away $\frac{1}{16}$, $\frac{2}{16}$, $\frac{3}{16}$, $\frac{4}{16}$, $\frac{7}{16}$, $\frac{8}{16}$, $\frac{10}{16}$, $\frac{12}{16}$, $\frac{16}{16}$, how many would you have left each time?

Q. What is this which you have now been doing called? *A.* Subtraction.

Q. What, then, is the taking of one number from another of the same name, or denomination, called? *A.* Simple Subtraction?

Q. What do you mean by the same name, or denomination? *A.* When the numbers are either all dollars, or all days, or all shillings, or all seconds, &c.

Q. In Addition, you recollect that you were required to put together two or more numbers, to find their amount; now it seems that we are to take one number from another to find their difference: *how, then, does Subtraction appear to differ from Addition?* *A.* It is exactly the opposite of Addition.

Q. What is the largest number called? *A.* The Minuend

Q. What is the smaller number called? *A.* The Subtrahend.

Q. What is that which is left after subtracting called? *A.* The Difference, or Remainder.

Q. From the above, how many parts do there appear to be in Subtraction, and what are they? *A.* Three—Minuend, Subtrahend, and Difference.

Operation by Slate illustrated.

1. A man, having 387 dollars, lost 134 dollars; how many had he left?

OPERATION.

He had	387	dollars, *the Minuend.*
He lost	134	dollars, *the Subtrahend.*
Had left	253	dollars, *the Remainder.*

In this example how do you obtain the 3, 5, and 2, in the Remainder? *A.* I say 4 (units) from 7 (units) leaves the 3 (units); 3 tens from 8 (tens) leaves the 5 (tens), and 1 (hundreds) from 3 (hundreds) leaves the 2 hundreds.

2. A man bought a wagon for 62 dollars, and a harness for 39 dollars; what did the wagon cost him more than the harness?

OPERATION.

Wagon,	62	dollars.
Harness,	39	dollars.
Difference,	23	dollars.
Proof,	62	dollars.

In this example we have a little difficulty in attempting to subtract as before, by saying 9 (units) *from* 2 (units); *but suppose we take one ten from the* 6 *tens, the next upper figure, which would leave* 5 (tens), *and join or add this* 1 (ten), *that is,* 10 *units, to the* 2 *units, making* 12 *units; how would you, then, proceed to get the* 3? *A.* I would say, 9 (units) from 12 (units) leaves 3 (units). *Now, as we took* 1 *ten from the* 6 *tens, it is evident that we must call the* 6 *tens* 5 *tens, and say,* 3 *tens from* 5 *tens leave* 2 *tens; but suppose that, instead of making the upper figure* 1 *less, calling it* 5, *we should make the lower figure one more, calling it* 4, *what would be the result, and how would you proceed? A.* I would say, 1 to carry to 3 makes 4, and 4 from 6 leaves 2, the same as before.

What is this taking 1 *from* 6, *and adding it to* 2, *the upper figure, called? A.* Borrowing ten.

Proof.—*If* 8 *from* 14 *leaves* 6, *because* 6 *and* 8 *are* 14, *how would you proceed to prove the operation? A.* I add 23 (the Difference) to 39, (the Subtrahend,) making 62, an amount like the Minuend—therefore right.

From these illustrations we derive the following

RULE.

I. *How do you write the numbers down? A.* The less under the greater.

II. *How do you place units, tens, &c.? A.* Units under units, tens under tens, &c.

III. *At which hand do you begin to subtract? A.* The right.

IV. *How do you subtract each figure in the lower line? A* From the figure above it.

V. *What do you set down? A.* The Difference.

VI. *If the lower figure be greater than that above it, what do you do? A.* Add ten to the upper figure.

VII. *What do you do then? A.* From this amount take the lower figure.

VIII. *What do you set down? A.* The Difference.

IX. *How many do you carry in all cases, when the lower figure is greater than that above it? A.* One.

Proof.—*Which numbers do you add together to prove the operation? A.* The Difference and Subtrahend.

What must the amount be like? A. The Minuend.

More Exercises for the Slate.

3. A man, having 98 dollars, paid away 49; how many had he left? *A.* 49 dollars.

4. James bought 78 marbles, and lost 29 of them; how many had he left? *A.* 49 marbles.

5. A man paid 175 dollars for a gold watch, and 55 dollars for a horse; how much more did he pay for the watch than for the horse? *A.* 120 dollars.

6. A man bought a chaise for 215 dollars, and to pay for it gave a wagon, worth 37 dollars, and the rest in money; how much money did he pay? *A.* 178 dollars.

7. A merchant bought a piece of cloth, containing 489 yards, and sold 365 yards; how many yards had he left? *A.* 124 yards.

8. If you have 20 dollars in your pocket, and owe 15 dollars, how many dollars will you have left in your pocket, when your debts are paid?

9. If you have 2560 dollars' worth of stock, and owe 1500 dollars, how much worth of stock will you have, after your debts are paid? *A.* 1060 dollars.

10. America was discovered by Christopher Columbus in 1492; how many years had elapsed at the time when hostilities commenced, in the revolutionary war, 1775? *A.* 283 years.

11. Gen. Washington was born in 1732, and died in 1799; how old was he? *A.* 67 years.

12. William has 15 cents; he owes Rufus 6 cents, and James 4; now how many will he have left, after paying Rufus and James?

13. A merchant owes to the Exchange Bank 2365 dollars, to the Bank of North America 15000 dollars, and his whole stock is worth no more than 42000 dollars; how much will he have left, after paying both banks? *A.* 24635.

14. If you buy 20 apples for 40 cents, and sell 15 for 35 cents, how many apples will you have left, and how much will they cost you?

15. A grocer buys 560 bushels of rye for 530 dollars, and sells 200 bushels for 400 dollars; how many bushels will he have left, and what will they cost him? *A.* 360 bushels, and they cost him 130 dollars.

16. A wine merchant bought 3600 gallons of wine, and sold at one time 2400 gallons, at another 1000 gallons; how many gallons has he on hand? *A.* 200.

17. From 200 take 150; take 190.

OPERATION.

200	200	50
150	190	10
50	10	60 *Ans.*

The pupil should be required in all rules to prove his results It is of practical importance; besides, it occasions less trouble to the teacher in detecting mistakes.

18. From 99 take 22; take 55. *A.* 121.
19. From 176 take 58; take 42. *A.* 252.
20. From 176 take 90; take 100. *A.* 162.
21. From 1000 take 700; take 550. *A.* 750.
22. From 1000 take 600; take 400. *A.* 1000.
23. From 1500 take 1000; take 1200. *A.* 800.
24. From 1500 take 900; take 350. *A.* 1750.
25. From 2538 take 1624; take 299. *A.* 3153.
26. From 2538 take 999; take 2000. *A.* 2077.
27. From 7836542 take 7000; take 70. *A.* 15666014.
28. From 80000 take 79999; take 78888. *A.* 1113.
29. From 80000 take 5000; take 12345. *A.* 142655.
30. From 900000 take 1; take 10. *A.* 1799989.
31. From 900000 take 100; take 1000. *A.* 1798900.
32. From 900000 take 10000; take 100. *A.* 1789900.
33. From 1000000 take 1; take 10. *A.* 1999989.
34. From nine millions take 3. *A.* 8999997.
35. From nineteen millions take nineteen. *A.* 18999981.
36. From forty millions take one million. *A.* 39000000.

SIMPLE MULTIPLICATION.

IX. 1. What will 3 books come to, at 20 cents apiece? Why? *A. Because* 20 *and* 20 *are* 40, *and* 20 *are* 60, *that is,* 3 *times* 20 *are* 60.

2. What will 3 bushels of apples come to at 30 cents a bushel? Why? *A. Because* 30 *and* 30 *are* 60, *and* 30 *more are* 90, *that is,* 3 *times* 30 *are* 90.

3. What will 2 cows come to, at 10 dollars a head? At 12 dollars? At 14 dollars? At 18 dollars? At 20 dollars? At 25 dollars? How many are 2 times 10, then? 2 times 12? 2 times 14? 2 times 18? 2 times 20? 2 times 25?

4. What will 30 yards of cloth come to, at 2 cents per yard? What will 14 yards? 16 yards? 12 yards? 25 yards? 30 yards? 60 yards? 80 yards? How many are 2 times 30? 14? 16? 12? 25? 30? 60? 80?

5. What will 3 yards of cloth come to, at 10 cents per yard? What will 4 yards? 6 yards? 12 yards? 20 yards? 30 yards? 60 yards? 80 yards? How many are 10 times 3? 4? 6? 12? 20? 30? 60? 80?

6. What will 4 oranges cost, at 10 cents apiece? Why? *A. Because* 10 *and* 10 *are* 20, *and* 10 *are* 30, *and* 10 *more are* 40, *that is,* 4 *times* 10 *are* 40.

Q. What, then, is Multiplication a quick way of performing? *A.* Many additions.

Q. What is the number called, which is to be added to itself, or repeated several times? *A.* The Multiplicand.

Q. What is the number, which shows how many times the multiplicand is to be repeated, called? *A.* The Multiplier.

Q. What are both multiplier and multiplicand sometimes called? *A.* Factors, or Terms.

Q. What is the result, or number found by multiplying, called? *A.* The Product.

When the Multiplier is 12, or less.

Operation by Slate illustrated.

¶ X. 1. How much will 4 barrels of pork come to at 17 dollars a barrel?

OPERATION.

Multiplicand,	17	dollars.
Multiplier,	4	barrels.
	—	
Product,	68	dollars.

Since 4 *times* 7 *are the same as* 7 *times* 4, *we see that it makes no difference which number is considered the multiplier: why is the* 4, *then, made the multiplier in this example?* *A.* For the sake of convenience, it being the smaller number. *How do you get the* 8 *units in the product?* *A.* I say, 4 times 7 (units) are 28 (units), or 2 tens and 8 units, writing down the 8 units, and carrying the 2 (tens) as in Addition.

How do you obtain the 6 (*tens*)? *A.* I say, 4 times 1 (ten) are 4 (tens), and 2 (tens), to carry, make 6 (tens).

PROOF.

17	
3	1 less than 4, the former multiplier.
—	
51	
17	the multiplicand.
—	
68	like the result of the other operation,—therefore right.

PROOF.—*As* 3 *times* 17 *and* 1 *time* 17 *evidently make* 4 *times* 17, *how can you prove the above operation?*

A. I can multiply 17 by 3, one less than 4, making 51; then add 17 (the multiplicand) to 51, making 68; which, being like the result in the first operation, proves the work to be right.

n these remarks and illustrations we derive the following

RULE.

How are the terms to be placed? *A.* The less under the ter, with units under units, tens under tens, &c.

At which hand do you begin to multiply? *A.* At the hand.

. *How are the figures of the multiplicand to be multiplied e multiplier?* A. Separately.

. *How do you carry and write down?* *A.* As in Simple tion.

More Exercises for the Slate.

What will 125 pounds of cheese cost, at 6 cents a pound? 0 cents.

What will 420 pounds of pork come to, at 9 cents a pound? 80 cents.

What will 167 barrels of flour come to, at 9 dollars a bar- *A.* 1503 dollars.

What will be the price of 8 hogsheads of wine, at 129 doler hogshead? *A.* 1032 dollars.

A merchant sold 895 oranges at 11 cents apiece; what ey come to? *A.* 9845 cents.

What will 236 lemons come to, at 8 cents apiece? At ts?

236	236	1888
6	8	1416
1416	1888	3304 *Ans.*

Multiply 120 by 2; by 3.	*A.* 600.
Multiply 1211 by 5; by 6.	*A.* 13321.
Multiply 1211 by 7; by 8.	*A.* 18165.
Multiply 65321 by 9; by 6.	*A.* 979815.
Multiply 65321 by 8; by 10.	*A.* 1175778.
Multiply 123456 by 11; by 4.	*A.* 1851840.
Multiply 123456 by 3; by 5.	*A.* 987648.
Multiply 345612 by 3; by 8.	*A.* 3801732
Multiply 345612 by 12; by 7.	*A.* 6566628.
Multiply 123450006789 by 3; by 4.	*A.* 86415047523.
Multiply 123450006789 by 5; by 6.	*A.* 135795074679.
Multiply 236120013 by 2; by 3.	*A.* 1180600065

XI. When the Multiplier is more than 12.

ere are 365 days in one year; how many are there in ?

OPERATION.

```
  365  days.
   36  years.
 ----
 2190
1095
-----
13140  days, Ans.
```

How do you get the 2190? *A.* By multiplying 365 by 6, as in ¶ X.

How do you obtain the 1095? *A.* By multiplying 365 by 3, as before.

Now, as you have seen that figures increase their value ten times, by being removed one place farther towards the left, can you tell me why 1095 *is so removed, thereby making the* 5 *to fall under the multiplier* 3? *A.* Multiplying by the 3 tens, that is, 30 units, gives a product 10 times greater than if the 3 tens were 3 units; hence the 5 must be placed in the tens' place.

To explain why you add the two products together.

What does multiplying 365 *by* 6 *give? A.* The number of days in 6 years.

What does multiplying 365 *by* 3 *tens give? A.* As the 3 (tens) are 30 units, multiplying by 30 will, of course, give the number of days in 30 years.

Why, then, do you add these two products together? A. To get the whole number of days in 30 and 6 years, that is, in 36 years.

From this example we derive the following

RULE.

I. *How do you write the numbers down, and multiply? A.* As in ¶ X.

II. *Where do you write the first figure in each product? A.* Directly under its multiplier.

III. *How do you proceed next? A.* Add all the products together as they stand.

More Exercises for the Slate.

2. What will 315 bushels of rye cost, at 42 cents a bushel? *A.* 13230 cts.

3. There are 63 gallons in a hogshead; how many gallons are there in 25 hogsheads? *A.* 1575 gallons.

4. What will 426 gallons of molasses cost, at 46 cents a gallon? *A.* 19596 cts.

5. If 363 men can do a piece of work in 137 days, how long will it take one man to do the same? *A.* 49731 days.

6. What will 234 barrels of beef come to, at 15 dollars a barrel? At 13 dollars a barrel? *A.* 6552 dollars.

7. *If a man* receive 256 cents for one day's work, how many

cents will he receive at that rate for 17 days? For 29 days? *A.* 11776 cents. For 116 days? For 105 days? *A.* 56576.

8. If the stage run 9 miles an hour, how many miles will it run at that rate in 12 hours? In 19 hours? *A.* 279 miles. In 25 hours? In 36 hours? *A.* 549 miles.

9. If a man save 161 dollars in a year, how much will it amount to in 19 years? In 35 years? *A.* 8694 dollars.

10. Multiply 62123 by 13. *A.* 807599.

11. Multiply 35432 by 14; by 15. *A.* 1027528.

12. Multiply 65217 by 16; by 17. *A.* 2152161.

13. Multiply 207812 by 19; by 21. *A.* 8312480.

14. Multiply 207812 by 25; by 35. *A.* 12468720.

15. Multiply 32100421 by 65; by 85. *A.* 4815063150.

16. Multiply 32100421 by 27; by 33. *A.* 1926025260.

17. Write down one thousand, multiply it by 25, add five thousand to the product, subtract twenty-nine thousand nine hundred and ninety-nine from the amount, and see if the remainder be 1.

¶ XXII. When the Multiplier is 10, 100, 1000, &c.

How many are 10 times 5? Now, if we annex a cipher to the 5, thus, 50, it will produce the same effect: why is this? *A.* Because, by annexing a cipher to 5, the 5 is removed to the tens' place; hence the value is increased 10 times.

What effect would two ciphers have, or three ciphers, &c.? *A.* Two ciphers would remove any figure two places towards the left, and of course increase it 100 times, and so on for 1000, &c.

RULE. *What, then, appears to be the rule?* *A.* Annex to the multiplicand all the ciphers in the multiplier.

Exercises for the Slate.

1. What will 36 bushels of rye cost, at 100 cents a bushel? *A.* 3600 cents.

2. What will 100 bushels of salt cost, at 87 cents a bushel? What will 1000 bushels? What will 10000 bushels? What will 10 bushels? *A.* 966570 cents.

3. Multiply 8978 by 10; by 100; by 1000; by 10000; by 100000; by 1000000. *A.* 9975545580.

¶ XXIII. When there are ciphers at the right hand of either or both the Factors.

RULE. *From the illustrations given, ¶XII., how does it appear that we can multiply?* *A.* Multiply without the ciphers first, and annex them to the product.

Exercises for the Slate.

(1.)	
Multiplicand, 63500 50 ——— 3175000	*How do you get the 3 ciphers in the product?* *A.* There are 2 at the right of the multiplicand, and 1 at the right of the multiplier, making 3.

2. Multiply 62123000 by 130000. *A.* 8075990000000.
3. Multiply 35432000 by 256000. *A.* 9070592000000.
4. Multiply 6789354270000 by 685300. *A.* 4652744481231000000.
5. Multiply 78954398765 by 7235000. *A.* 571235075064775000.
6. Multiply 123456789 by 123450000. *A.* 15240740602050000.
7. Multiply 1234567890 by 1234560000. *A.* 1524148134278400000.

¶ XIV. When the Multiplier is a Composite Number.

How many are 5 *times* 8? 7 *times* 9? 12 *times* 12?

What are these products, 40, 63, 144, *called?* *A.* Composite Numbers.

What are the multiplying numbers, 5 *and* 8, 7 *and* 9, 12 *and* 12, *called?* *A.* The Component Parts.

What are the component parts of 36? *Of* 72? *Of* 100? *Of* 27? *Of* 15? *Of* 35? *Of* 132? *Of* 144?

What, then, is the product of any two numbers called? *A.* A Composite Number.

---

1. What will 14 barrels of flour cost, at 8 dollars a barrel?

OPERATION.	
8 dollars. 7 barrels. —— 56 dollars. 2 times 7 are 14. —— 112 dollars, *Ans.*	*What does multiplying* 8 *dollars by* 7 *barrels give?* *A.* The price of 7 barrels at 8 dollars a barrel, making 56 dollars. *How much more will* 14 *barrels cost than* 7 *barrels?* *A.* 2 times as much as 7, that is, 2 times 56, making 112 dollars.

RULE I. *How, then, would you begin to multiply?* *A.* By one of the component parts first.

II. *What would you multiply this product by?* *A.* By the other component part.

More Exercises for the Slate.

2. What will 36 hundred weight of sugar cost, at 29 dollars a hundred? *A.* 1044 dollars.

3. Multiply 3065428 by 35. *A.* 107289980.

4. Multiply 4078945 by 96. *A.* 391578720.

5. Multiply 18934 by 108. *A.* 2044872.

6. Multiply 45678 by 144. *A.* 6577632.

SIMPLE DIVISION.

¶ XV. 1. If you divide 12 apples equally between two boys, how many will each have? How many times 2 in 12, then? *Why? A. Because 2 times 6 are 12.*

2. How many oranges, at 8 cents apiece, can you buy for 48 cents? For 96 cents? How many times 8 in 48? 8 in 96? *Why?*

3. A man bought 8 lemons for 80 cents; how much did he give apiece? How many times 8 in 80? *Why, or proof?*

4. How many gallons of brandy, at 3 dollars a gallon, can be bought for 36 dollars? For 60 dollars? For 90 dollars? For 300 dollars? *Why?*

5. Four boys found a bag containing 48 silver dollars; how many will they have apiece, if it be divided equally?

6. When oranges are 2 cents apiece, how many will 8 cents buy? Will 16 cents buy? Will 32 cents? Will 36 cents? Will 48 cents? Will 100 cents?

7. If you pay 9 cents for one pound of sugar, how many pounds can you buy for 45 cents? For 54 cents? For 108 cents?

8. How much is one half ($\frac{1}{2}$) of 4? Of 8? Of 16? Of 20? Of 24? Of 30? Of 100? Of 200?

9. Harry had 16 apples, and gave one half ($\frac{1}{2}$) of them to Thomas; how many did Thomas receive?

10. How much is one third ($\frac{1}{3}$) of 6? Of 24? Of 30? Of 36?

11. How much is one half ($\frac{1}{2}$) of 8? One third ($\frac{1}{3}$) of 24? One fourth ($\frac{1}{4}$) of 16? One fifth ($\frac{1}{5}$) of 35? One sixth ($\frac{1}{6}$) of 24? One seventh ($\frac{1}{7}$) of 35? One eighth ($\frac{1}{8}$) of 56? One ninth ($\frac{1}{9}$) of 108? One twelfth ($\frac{1}{12}$) of 144?

12. How many times 4 in 40? 3 in 60? 5 in 100? 6 in 1200? 8 in 480?

Q. What is this method of finding how many times one number is contained in another, or of dividing a number into equal parts, called? *A.* Division.

Q. What is the method of finding how many times one number is contained in another of only one name, or denomination, called. *A.* Simple Division.

Q. What is the number given to divide by called? *A.* The Divisor.

Q. What is the number to be divided called? *A.* The Dividend.

Q. What is the number of times that the divisor is contained in the dividend called? *A.* The Quotient.

Q. What is that which is sometimes left after dividing, or, after the operation is performed, called? *A.* The Remainder, which must always be less than the Divisor.

Q. Of what name, or denomination, is the remainder? *A.* The same as the dividend.

Q. If your dividend, for instance, be ounces, what will your remainder be? *A.* Ounces.

Q. How many times 4 in 40? *and why?*

Q. From this example what does division appear to be the opposite of? *A.* Multiplication.

	12 *oranges.*
One to each makes,	4
1st time he had	8 *left.*
One to each makes	4
2nd time he had	4 *left.*
One to each makes	4
3d time he had	0 *left.*

Q. James, having 12 oranges, was desirous of dividing them equally among his 4 little sisters, and, in order to do this, he handed them at first one apiece; how many had he left?

Q. When he handed them another apiece, how many had he left?

Q. When he handed them one more apiece, how many had he left?

Q. From these illustrations how does it appear that a number may be divided into equal parts? *A.* By Subtraction.

Q. How many times did James give to each of his sisters an orange apiece?

Q. How many times did you subtract? *A.* Three times.

Q. How many times 4 in 12?

Q. By this we see that the quotient represents the number of subtractions: now, if the quotient were 4000, how many times would it be necessary to take the divisor from the dividend to perform division by subtraction? *A.* 4000 times.

Q. What, then, is Division a quick way of performing? ***A.*** **Many subtractions.**

SHORT DIVISION.

¶ XVI. *Q.* What is Short Division? *A.* When the divisor is 12, or less.

1. How many oranges, at 3 cents apiece, may be bought for 657 cents?

OPERATION.

Dividend.
Divisor, 3)657 cents.

Quotient, 219 oranges, *Ans.*

How do you obtain the 2 (hundreds) *in the quotient?* *A.* I begin on the left of the dividend, and say, 3, the divisor, is contained in 6 (hundreds) 2 (hundreds) times, that is, 200 times, writing the 2 (hundreds) down under the 6 (hundreds).

How do you get the 1 (*ten*)? *A.* 3 in 5 (tens) 1 time, and 2 (tens) left.

What do you do with the 2 *which is left?* *A.* I join, or carry it as 2 tens, that is, 20 units, to the 7 units, making 27.

How do you proceed to get the 9, *then?* *A.* 3 in 27, 9 times.

PROOF.

Quotient, 219
Divisor, 3

Dividend, 657

How many times 6 *in* 30? *and why?*

How, then, would you proceed to prove the foregoing example? *A.* I would multiply 219 (the quotient) by 3 (the divisor), making 657 (the dividend)—therefore right.

From the illustrations now given we derive the following

RULE.

I. *At which hand of the dividend do you place the divisor?* *A.* At the left.

II. *How many figures do you take first?* *A.* Enough to contain the divisor once, or more.

III. *What do you set down underneath?* *A.* The quotient.

IV. *If there should be a remainder, how would you proceed?* *A.* I join, or carry it to the next figure of the dividend, as so many tens. *For example, suppose* 3 *remain, and the next figure be* 8, *how would you say?* *A.* I would say, 3 (to carry) to 8, makes 38.

V. *How do you proceed if the divisor be not contained in the next figure of the dividend?* *A.* Write a cipher in the quotient, and join this figure to the figure next to it, as so many tens.

PROOF.—*Which terms do you multiply together to prove the operation?* *A.* The divisor and quotient.

What is to be done with the remainder, if there be any? *A* Add it to this product.

What must the amount be like? *A.* The dividend.

More Exercises for the Slate.

2. Rufus divided 42 oranges equally between his two little brothers; how many had they apiece? *A.* 21.

3. If 3 bushels of apples cost 360 cents, how much is that a bushel? *A.* 120 cents.

4. How many months are there in 452 weeks, there being 4 weeks in each month? *A.* 113 months.

5. A man, having 416 dollars, laid them all out in cider, at 4 dollars a barrel; how much cider did he buy? *A.* 104 barrels.

6. A man bought 6 oxen for 318 dollars; how much did he pay a head? *A.* 53 dollars.

7. How much flour, at 7 dollars a barrel, can be bought for 1512 dollars? *A.* 216 barrels.

8. At 8 cents apiece, how many oranges will 8896 cents buy? *A.* 1112 oranges.

9. At 10 dollars a barrel, how many barrels of flour may be bought for 1720 dollars? *A.* 172 barrels.

10. 12 men by contract are to receive 1500 dollars for a job of work; how many dollars will be each man's part, if they be divided equally among them? *A.* 125 dollars.

11. 2 men, trading in company, gained 2450 dollars; how much was each man's part? *A.* 1225 dollars.

12. At 3 dollars a barrel, how many barrels of pork can be bought for 5463 dollars? *A.* 1821 bbls.

Note.—The total remainder is found by adding together what remains after each operation.

13. Divide 256587 by 2. *A.* 128293—1 *rem.*
14. Divide 378567 by 2; by 3. *A.* 315472—1 *rem.*
15. Divide 278934 by 2; by 3. *A.* 232445.
16. Divide 256788 by 3; by 4. *A.* 149793.
17. Divide 256788 by 5; by 6. *A.* 94155—3 *rem.*
18. Divide 65342167 by 4; by 5. *A.* 29403974—5 *rem.*
19. Divide 65342167 by 6; by 7. *A.* 20224956—3 *rem.*
20. Divide 523467898 by 4; by 6. *A.* 218111623—6 *rem.*
21. Divide 523467898 by 7; by 8. *A.* 140214615—4 *rem.*
22. Divide 2653286 by 7; by 8. *A.* 710700—12 *rem.*
23. Divide 2653286 by 9; by 10. *A.* 560137—11 *rem.*
24. Divide 52345 by 9; by 10. *A.* 11050—6 *rem.*
25. Divide 52345 by 11; by 12. *A.* 9120—8 *rem.*

The operation, thus far, has been carried on partly in the mind, and partly by writing the numbers down; but oftentimes the divisor will be too large to be thus performed. When, therefore, we write the operation out at length, what is the process called? *A.* Long Division.

LONG DIVISION.

¶ XVII. 1. A man, dying, left 957 dollars to be divided equally among his 4 sons; what was each son's part?

Long Division.
OPERATION.

```
            Dividend. Quotient.
Divisor, 4 ) 957 ( 239¼
             8
             --
             15
             12
             --
              37
              36
              --
               1
```

Short Division.
OPERATION.

```
4 ) 957
    ----
    239¼ Quotient.
```

As Long and Short Division are exactly alike, except in Short Division the whole operation is not written down, to begin, then, in Short Division, we should say, 4 in 9, 2 times, and 1 over. This we discover by saying in the mind, 4 times 2 are 8, and 8 from 9 leaves 1. Now, to express in figures this operation, we may write the numbers where we please: where, then, for the sake of convenience, may the 2 (times, the quotient figure) be placed? *A.* At the right hand of the dividend.

We are next to say, 4 times 2 are 8: this 8, you know, must be subtracted from 9: where would it be convenient to place the 8? *A.* Under the 9.

By taking 8 from 9, we have 1 remaining, which we should, in Short Division, carry or join to 5, the next figure of the dividend; how can we do this now? *A.* By joining or bringing down the 5 to the right hand of the 1, making 15.

How do you get the 3 in the quotient? *A.* I say, 4 in 15, 3 times.

How do you proceed next? *A.* I say, 3 times 4 are 12; and 12 from 15 leaves 3.

What do you do with the 3? *A.* I bring down 7 of the dividend to the right hand of the 3, making 37.

How do you get the 9 in the quotient? *A.* I say, 4 times 9 are 36, and subtracting 36 from 37 leaves 1, remainder.

It now appears that each son has 239 dollars, and there is 1 dollar still remaining undivided: to explain the division of this, tell me how many quarters there are in a dollar? *A.* Four.

Now, as there are 4 sons to share equally this dollar, how much ought each son to have? *A.* $\frac{1}{4}$, or one quarter of a dollar apiece.

In this expression, $\frac{1}{4}$, we use the remainder, 1, and the divisor, 4: *how, then, may division be carried out more exactly?* *A.* By writing the divisor under the remainder with a line between.

From these remarks and illustrations we derive the following

RULE.

I. *How do you begin to divide?* *A.* As in Short Division.

II. *How many steps are there?* *A.* Four.

III. *What are they?* *A.* 1st. Find how many times;
2d. Multiply;
3d. Subtract;
4th. Bring down.

IV. *Where do you write the quotient?* *A.* At the right hand of the dividend.

V. *In performing the operation, whenever you have subtracted, what must the remainder be, less than?* *A.* Than the divisor.

VI. *When you have brought down a figure, and the divisor is not contained in the new dividend thus formed, what is to be done?* *A.* Place a cipher in the quotient, and bring down another figure; after which divide as before.

PROOF.—*How do you prove the operations?* *A.* As in Short Division.

More Exercises for the Slate.

2. A man wishes to divide 626 dollars equally among 5 men; how many will that be apiece? *A.* 125$\frac{1}{5}$ dollars, or 125 dollars and 20 cents.

3. There are 7 days in one week; how many weeks are there in 877 days? *A.* 125$\frac{2}{7}$ weeks.

4. A man, having 5520 bushels of corn, wishes to put it into bins, each holding 16 bushels; how many bins will it take? *A.* 345 bins.

5. Four boys had gathered 113 bushels of walnuts; in dividing them equally, how many will each have? *A.* 28$\frac{1}{4}$ bushels

6. If a man is to travel 1201 miles in 12 months, how many is that a month? *A.* $100\frac{1}{12}$ miles.

7. If 1600 bushels of corn are to be divided equally among 40 men, how many is that apiece? *A.* 40 bushels.

8. 27000 dollars are to be divided equally among 30 soldiers; how many will each have? *A.* 900 dollars.

9 The salary of the president of the United States is 25000 dollars a year; how much is that a day, reckoning 365 days to the year? *A.* $68\frac{180}{365}$ dollars.

10. A regiment of soldiers, consisting of 500 men, are allowed 1000 pounds of pork per day; how much is each man's part? *A.* 2 pounds.

11. James says that he has a half bushel that holds 27000 beans; how many will that be apiece for 9 boys, if they be divided equally? How many apiece for 27 boys? *A.* 4000 beans. 12. For 36 boys? For 54 boys? *A.* 1250 beans.

13. Divide 29876543 by 13. *A.* $2298195\frac{8}{13}$
14. Divide 6283459 by 29. *A.* 216671.
15. Divide 37895429 by 112. *A.* $338352\frac{5}{112}$.
16. Divide 29070 by 15; by 18. *A.* 3553.
17. Divide 29070 by 19; by 17. *A.* 3240.
18. Divide 10368 by 27; by 36. *A.* 672.
19. Divide 10368 by 54; by 18. *A.* 768.
20. Divide 2688 by 112; by 224. *A.* 36.
21. Divide 101442075 by 4025. *A.* 25203.

¶ XVIII. When the divisor is a composite number.

1. *Bought 20 yards of cloth for 80 dollars; how much was that a yard? Now, as 2 times 10 are 20 (a composite number), it is plain that, if there had been but 10 yards, the cost of 1 yard would be 8 dollars, for 10 in 80, 8 times; but, as there are 2 times 10 yards, it is evident that the cost of 1 yard will be but one half (½) as much: how much, then, will it be?*

RULE. *What, then, appears to be the rule for dividing by a composite number?* *A.* Divide by one of its component parts first, and this quotient by the other.

Exercises for the Slate.

2. Divide 1152 dollars among 24 men.

OPERATION.

4) 1152
6) 288
Ans. 48 dollars.

4 times 6 are 24.

3. Divide 2520 by 63. *A.* 40.

4. Divide 5040 by 28. *A.* 180. By 15. *A.* 336. By 24. *A.* 210. By 84. *A.* 60. By 35. *A* 144. By 72. *A.* 70

¶ XIX. To divide by 10, 100, 1000, &c.

In ¶ XII. *it was observed, that annexing* 1 *cipher to any number multiplied it by* 10, 2 *ciphers by* 100, *&c. Now, Division being the reverse of Multiplication, what will be the effect, if we cut off a cipher at the right of any number?* *A.* It must decrease, or divide it by 10.

What will be the effect, if we cut off two ciphers? *A.* It will be the same as dividing by 100.

Why does it have this effect? *A.* By cutting off one cipher or figure at the right, the tens take the units' place, and hundreds the tens' place, and so on.

RULE I. *What, then, is the rule for dividing by* 10, 100, *&c.?* *A.* Cut off as many places or figures at the right hand of the dividend, as there are ciphers in the divisor.

II. *What are the figures cut off?* *A.* The remainder

III. *What are the other figures?* *A.* The quotient.

Exercises for the Slate.

1. A prize, valued at 25526 dollars, is to be equally divided among 100 men; what will be each man's part?

OPERATION.

255|26

$255\frac{26}{100}$ dollars.

2. Divide 1786582 by 10000. *A.* $178\frac{6582}{10000}$.

3. Divide 87653428 by 10; by 100; by 1000; by 10000; by 100000; by 1000000. *A.* Remainder to each, $\frac{8}{10}$, $\frac{28}{100}$, $\frac{428}{1000}$, $\frac{3428}{10000}$, $\frac{53428}{100000}$, $\frac{653428}{1000000}$. Quotients, total, 9739257.

¶ XX. When there are ciphers at the right hand of the Divisor.

1. Divide 4960 dollars among 80 men.

OPERATION.

8 times 10 are 8|0)496|0

62 dollars.

In this example we have a divisor, 80, *which is a composite number; (thus,* 8 *times* 10 *are* 80;) *how, then, may we proceed to divide by* 10, *one of the component parts?* *A.* By cutting off one place at the right hand of the dividend, as in ¶ XIX.

How do you obtain the 62? *A.* By dividing the 496 by 8, as usual.

RULE I. *As any number, which has a cipher or ciphers at the right, can be produced by two other numbers, one of which may be either* 10, 100, 1000, *&c., how, then, would you proceed to di* *vide when there are ciphers at the right of the divisor?* *A.* Cut them off, and the same number of figures from the right of the dividend.

II. *How do you divide the remaining figures of the dividend?* *A.* As usual.

III. *What is to be done with the figures of the dividend which are cut off?* *A.* Bring them down to the right hand of the remainder.

Exercises for the Slate.

2. How many oxen, at 30 dollars a head, may be bought for 38040 dollars? *A.* 1268.

3. Divide 783567 by 2100. *A.* $\frac{267}{2100}$ rem.

4. Divide 2082784895876 by 1200000. *A.* $\frac{95876}{1200000}$ rem.

5. Divide 7942851265321 by 12500000. *A.* $\frac{1265321}{12500000}$ rem.

6. Divide 18515952875 by 112000. *A.* $\frac{875}{112000}$ rem.

Miscellaneous Questions on the foregoing.

Q. *What is the subject which you have now been attending to called?* *A.* Arithmetic.

Q. *From what you have seen of it, how would you define it?* *A.* It teaches the various methods of computing by numbers.

Q. *What rules have you now been through?* *A.* Notation or Numeration, Addition, Subtraction, Multiplication, and Division.

Q. *How many rules do these make?*

Q. *What are these rules sometimes called?* *A.* The fundamental rules of arithmetic.

Q. *Why?* *A.* Because they are the foundation of all the other rules.

Q. *To denote the operation of these different rules, we have certain characters; what is the name of these characters?* *A.* Signs.

Q. *What do two horizontal straight lines signify; thus, 100 cents = 1 dollar?* *A. Equal to:* as, 100 cents = 1 dollar, *read,* 100 *cents are equal to* 1 *dollar.*

Q. *What does a horizontal line crossing a perpendicular tell you to do; thus,* 6+10=16? *A. To add:* thus, 6+10=16, *read,* 6 *and* 10 *are* 16.

Q. *What else does this sign denote?* *A.* A remainder after dividing.

Q. What does one horizontal straight line tell you to do, thus, 8—6=2? *A. To subtract:* thus, 8—6=2, *read,* 6 *from* 8 *leaves* 2.

Q What do two lines, crossing each other in the form of the Roman letter X, *tell you to do; thus,* 6×8=48? *A. To multiply:* thus, 6×8=48, *read,* 6 *times* 8 *are* 48.

Q. What does a horizontal line, with a dot above and below it, tell you to do; thus, 8÷2=4? *A. To divide;* thus, 8÷2=4, *read,* 2 *in* 8, 4 *times.*

Q. By consulting ¶ XVII. *you will perceive that division may be represented in a different manner; how is this done?* *A.* By writing the divisor under the dividend, with a line between them; thus, $\frac{8}{4}=2$, read, 4 in 8, 2 times.

Q. What does $\frac{15}{5}$ *signify, then?* $\frac{20}{5}$ *signify?* $\frac{36}{6}$? $\frac{42}{7}$? $\frac{108}{9}$? $\frac{144}{12}$? $\frac{35}{7}$?

Let me see you write down on the slate the signs of Addition, Subtraction, Multiplication, and Division.

Perform the following examples on the slate, as the signs indicate.

1. 87834+284+65+32+100=88315, *Ans.*
2. 876345723—267001345=609344378, *Ans.*
3. 692784578×27839421=19286721529249338, *Ans.*
4. 202884150÷4025=50406, *Ans.*
5. 2600—600=2000+1828=3828, *Ans.*
6. 3600—400=3200×4=12800, *Ans.*
7. $\frac{260000000}{13000}=20000$, *Ans.*
8. $\frac{108}{9}+\frac{36}{6}=18$, *Ans.*
9. $\frac{108}{9}+\frac{36}{6}+\frac{14}{7}+\frac{15}{5}+\frac{20}{4}=28$, *Ans.*

10. What is the whole number of inhabitants in the world, there being, according to Hassel, in each grand division as follows;—in

Europe, one hundred and eighty millions;
Asia, three hundred and eighty millions;
Africa, ninety-nine millions;
America, twenty-one millions;
Australasia, &c. two millions? *A.* 682000000.

11. What was the number of inhabitants in the following New England towns in 1820, there being in

Portland,	8,581;	Boston,	43,298;
Portsmouth,	7,327;	Providence,	11,767;
Salem,	12,731;	New Haven,	8,327?

A. 92,031.

12. What was the number of inhabitants in the following towns, there being in

New York,	123,706;	Norfolk,	8,478;
Philadelphia,	108,116;	Richmond,	12,067;
Baltimore,	62,738;	Charleston,	24,780;
Washington,	13,247;	Savannah,	7,523;
Albany,	12,630;	New Orleans,	27,176?

A. 400,461.

13. How many more inhabitants were there in New York than Philadelphia? Philadelphia than Baltimore? Baltimore than Boston? Boston than New Orleans? New Orleans than Charleston? Charleston than Albany? Albany than Providence? Providence than New Haven? *A.* Total, 115,379.

14. At 73 cents a bushel, what will 42 bushels of salt cost? What will 800 bushels? *A.* 61466 cents. 15. What will 2970 bushels? What 8900 bushels? *A.* 866510 cents.

16. James had 37 cents, William 10 times as many as James, Rufus 15 times as many as William, Thomas 26 times as many as Rufus, Harry 45 times as many as Thomas, and Stephen 24 times as many as Harry; how many did they all have? *A.* 162487757.

17. There are 60 minutes in one hour; how many hours are there in 120 minutes? In 4800 minutes? *A.* 82 hours. 18. In 172800 minutes? In 1036800 minutes? *A.* 20160 hours.

FEDERAL MONEY.

¶ XXI. Repeat the

TABLE.

10 Mills	..... make	1 Cent, marked	c.
10 Cents	....	1 Dime	d.
10 Dimes	....	1 Dollar ...	$
10 Dollars	....	1 Eagle ...	E.

1. At 10 mills a yard, how many cents will 4 yards of cloth cost? Will 6 yards? Will 8?

2. How many mills are 2 cents? 3 cents? 4 cents? 5 cents? 8 cents? 12 cents?

3. How many cents are 2 dimes? 5 dimes? 6 dimes? 7 dimes? 11 dimes?

4. How many dimes are 2 dollars? 5 dollars? 7 dollars? 10 dollars? 12 dollars?

5. How many eagles are 20 dollars? 30 dollars? lars? 60 dollars? 80 dollars? 100 dollars? 120 doll

Q. How many cents are 4½ pence? *A.*
Q. How many cents are 9 pence? *A.*
Q. How many cents are 18 pence, or ¼ of a dollar? *A.*
Q. How many cents is ⅛ of a dollar? *A.*
Q. How many cents is ¾ of a dollar? *A.*
Q. How many cents is one dollar? *A.*
Q. How many cents is a pistareen? *A.*
Q. How many cents is half a pistareen? *A.*
Q. How many 9 pences in a dollar? *A.*
Q. How many 4 pence-halfpennies in a dollar? *A.*

6. You buy 4 yards of cloth for $1, and give the sho two fifty-cent bits; how much change must he give yo
7. You buy some calico to the amount of 17 cents, a the clerk a pistareen; how much change must he give
8. You give a pistareen for 1 fish; how many cen you give for 2? For 3? For 5? For 7? For 9? F
9. What will 2 yards of ribbon come to in cents, at 4 a yard? At 9 pence? At ¼ of a dollar? At ⅛ a doll ¾ of a dollar?

Q. What is the coin of the United States called? *A* ral Money.
Q. When established? *A.* A. D. 1786.
Q. By what authority? *A.* Congress.
Q. Which is the money unit? *A.* Dollars.
Q. What place do dollars occupy, then? *A.* The units.
Q. How are dollars distinguished from dimes, ce mills? *A.* By a comma, or separatrix, at the right of
Q. What are the figures on the left of dollars? *A.*
Q. What is the first figure on the right of doll Dimes.
Q. What is the second figure? *A.* Cents.
Q. What is the third figure? *A.* Mills.
Q. How many of these coins are real? *A.* 4.
Name them. *A.* The eagle, the dollar, the dime, cent.
Q. Which is a gold coin? *A.* The eagle.
Q. Which are the silver coins? *A.* The dollar and t
Q. Which is a copper coin? *A.* The cent.
Q. Which is imaginary? *A.* The mill, as there is of money of that denomination.*

---

* *There are*, however, half eagles, and half dimes, real coi

Q. What are all the denominations of Federal Money? A. Mills, cents, dimes, dollars, and eagles.

Q. How many of these denominations are used in accounts, and what are they? A. Three—dollars, cents, and mills.

Q. What are dollars and eagles called in accounts? A. Dollars.

Q. What are dimes and cents called? A. Cents.

Note.—The names of coins less than a dollar express their value. Mill is contracted from *Mille*, the Latin for *thousand*; Cent from *Centum*, the Latin for *hundred*; and Dime from *Disme*, the French for *tenth*.

Q. What does this character, $, placed before numbers, denote? A. Federal Money.

Q. As 10 mills make 1 cent, 10 cents 1 dime, &c., increasing from right to left like whole numbers, it follows that any question in Federal Money may be performed as in whole numbers; also that dollars, cents, and mills, may be called either all mills, or all cents and mills; thus, 4 dollars, 25 cents, and 5 mills, may be read, 4255 mills, or 425 cents and 5 mills; but, in order for this, it will sometimes be necessary to write ciphers between the different denominations: when, then, the cents are less than 10, where must a cipher be placed in writing cents with dollars? A. Between the cents and dollars.

Q. Why is this? A. Because, as 100 cents make one dollar, cents, of course, occupy two places; hence, when the cents are less than 10, they must occupy the units' place of cents, and a cipher the tens' place of cents.

Q. In writing down mills with dollars, when there are no cents, how many ciphers must you place between them and dollars? A. Two.

Q. Why? A. Because, as there are two places for cents, when there are no cents, these places must be filled with ciphers.

Q. How are 2 dollars and 5 mills written down, then?
A. $ 2,005.

Q. How are 3 dollars and 2 cents written down, then?
A. $ 3,02.

Q. How are 7 dollars and 8 mills written down? A. $ 7,008.

Q. How are 9 dollars and 1 cent written down? A. $ 9,01.

Q. How are 1 dollar, 1 cent and 1 mill written down?
A. $ 1,011.

Q. How are 20 dollars and 50 cents written down?
A. $ 20,50

REDUCTION OF FEDERAL MONEY.

¶ XXII. *Q. What would you call the changing of numbers from one name, or denomination, to another, retaining the same value; as, 200 cents into 2 dollars?* *A.* Reduction.

RULE.

1. How many mills in one cent? In 8 cents? In 9 cents?

I. *What, then, do you multiply by, to bring cents into mills* *A.* Multiply by 10 mills; that is, annex one cipher. (See ¶ XII.

2. How many cents in 20 mills? In 60 mills? In 80 mills?

II. *What, then, would you divide by, to bring mills into cents* *A.* By 10 mills; that is, cut off the right hand figure.

3. How many cents in one dollar? In 2 dollars? In 8 dollars

III. *How many ciphers, then, do you annex to dollars, to bring them into cents?*

4. How many dollars in 200 cents? In 800 cents?

IV. *How many figures, then, would you cut off, to bring cents into dollars?*

V. *As annexing two ciphers to dollars brings them into cents and one to cents brings cents into mills, how many ciphers then, would you annex to dollars in all, to bring them into mills* *A.* Three.

5. How many mills, then, in 2 dollars? In 8 dollars?

VI. *How many figures would you cut off, to bring mills back into dollars?* *A.* Three.

6. How many dollars in 2000 mills? In 5000 mills?

Exercises for the Slate.

1. Reduce $2, 1 c., 1 m.	to mills.	*A.* 2011 mills.	
2. Reduce $3, 75 c.	to mills.	*A.* 3750 mills.	
3. Reduce $20, 6 c.	to mills.	*A.* 20060 mills.	
		——— *A.* 25821 mill	
4. Reduce $8, 25 c. 8 m.	to mills.		
5. Reduce $4, 28 c. 1 m.	to mills.		
6. Reduce $480, 6 c.	to mills.		
7. Reduce $48	to mills.	*A.* 540599 mills.	
8. Reduce 7680 mills	to dollars.		
9. Reduce 1865 mills	to dollars.		
10. Reduce 172 cents	to dollars.		
11. Reduce 1189 cents	to dollars.	*A.* $23, 15 c. 5 m.	
12. Reduce $260	to cents.		
13. Reduce $130	to cents.		
14. Reduce $2, 45 c.	to cents.		
15. Reduce $24,06	to cents.	*A.* 41651 cents.	

16. Reduce 1265½ cts. to dollars.
17. Reduce 137¼ cts. to dollars.
18. Reduce 1212½ cts. to dollars.
19. Reduce 12000 cts. to dollars. *A.* $146,15 c. 5 m.

ADDITION OF FEDERAL MONEY.

¶ XXIII. 1. What will 20 pencils come to, at 5 cents for 10, and 10 dimes for 10.

2. What is the sum of 50 cents and 5 dimes?

3. What is the sum of 6 cents, 12 cents, 20 cents, and 2 dimes?

4. If you give 25 cents for a top, 25 cents for a knife, and 1 dime for a slate, how much do they all come to?

Q. What is this, which you have now been doing, called? *A.* Addition of Federal Money.

1. A man bought a wagon for $32,50, a harness for $15,20, and a whip for $1, 8 c.; what did the whole cost?

OPERATION.

$. cts.
32,50
15,20
1,08
Ans. $ 48,78

How do you perform this operation? *A.* I place dollars under dollars, cents under cents, adding as in Simple Addition.

Why is there a cipher placed between the $1, and 8 cents? *A.* Because the cents are less than 10.

From the preceding remarks we derive the following

RULE.

I. *How do you write down cents, dollars, &c.?* *A.* Cents under cents, dollars under dollars, &c.

II. *How do you add?* *A.* As in Simple Addition.

III. *Where is the separatrix to be placed?* *A.* Directly under the separating points above.

More Exercises for the Slate.

2. What is the amount of 3 dollars 2 cents, 2 dollars 5 cents, 7 dollars 8 cents, 9 dollars 1 cent, 1 dollar 1 cent 1 mill, and 20 dollars 50 cents? *A.* $42,671.

3. Bought a chaise for $126,18, a watch for $280,25, a coach for $850,62, a hat for $6,50, and a whip for $2,98; how much did all these articles come to? *A.* $1266,53.

4. Bought a cap for $7,50, a coat for $12,18, a pair of silk stockings for $1,62, and a cane for $2,87; what was the cost of the whole? *A.* $24,17.

5. If I give ten dollars one cent for a cart, fifty dollars for a yoke of cattle, seven dollars twenty cents for a plough, sixty-five dollars for a horse, thirty-seven dollars fifty cents for some hay, how much will all these come to? *A.* $169,71.

6. If $\frac{1}{3}$ of an orange be worth 2 cents, how much is $\frac{2}{3}$ worth? How much is a whole orange worth?

7. If $\frac{1}{3}$ of a vessel be worth $25000,50, what is the whole vessel worth? *A.* $75001,50.

8. If $\frac{1}{4}$ of a vessel be worth $3700,12, what is a whole vessel worth? *A.* $14800,48.

9. If $\frac{1}{5}$ of a vessel be worth $10000,50, what is $\frac{2}{5}$ worth? $\frac{3}{5}$? $\frac{4}{5}$? $\frac{5}{5}$? and what part is $\frac{5}{5}$? *A.* $\frac{2}{5}$ = $20001, $\frac{3}{5}$ = 30001,50, $\frac{4}{5}$ = $40002, $\frac{5}{5}$ = $50002,50, *or the whole.*

10. If $\frac{1}{16}$ of the stock in a bank be worth $2356,56, what is $\frac{2}{16}$ worth? (471312)* $\frac{4}{16}$ worth? (942624) *A.* $14139,36.

11. Bought a quart of brandy for 62½ cents, a quarter of flour for 1 dollar 37½ cents, a pound of sugar for 12½ cents, 2 yards of cotton cloth for 75 cents, 1 vest pattern for 87½ cents, 1 dozen of buttons for 8 cents, and 2 cotton balls for 6 cents; how much did the whole amount to? *A.* $3,89.

SUBTRACTION OF FEDERAL MONEY.

¶ **XXIV.** 1. If you buy a top for 20 cents, 12 marbles for 20 cents, 6 oranges for 10 cents, and sell them for 5 dimes, shall you make or lose by trading?

2. Your father gave you 15 mills, 4 mills, 1 mill, 2 dimes, 28 cents; and said, that if you would tell him how much more it would take to make a dollar, he would give you as much more; how much did it take?

3. You bought a yard of ribbon for 24 cents, gave $\frac{1}{4}$ to your sister, and sold the rest at 10 cents for a quarter of a yard; did you make, or lose? and how much?

---

* In order to remedy the inconvenience arising from inspecting many operations, and at the same time to secure the object of giving the amount of several results, the result of each operation will be stated in figures, included in a parenthesis, and annexed to each question, without assigning any value to the figures, leaving this particular entirely to the judgment of the pupil. The pupil should, notwithstanding, be required to prove each operation, and assign to each particular result its true value, the latter of which, however, in most cases, he must necessarily do, to obtain the answer, as it is the amount of all these separate results, with the value assigned to each.

4. If 3 dollars buy one yard of cloth, what is ⅓ of a yard worth?

5. A, B, and C, buy a chaise for 90 dollars; A pays ⅓ of it, how many dollars do B and C pay?

RULE.

I. *How do you write the numbers down?* *A.* As in Addition of Federal Money.

II. *How do you subtract?* *A.* As in Simple Subtraction.

III. *How do you place the separatrix?* *A.* As in Addition of Federal Money.

Exercises for the Slate.

1. A man owed $36,465, and paid $27,696; how much did he then owe?

OPERATION.

$36,465
$27,696

$ 8,769, *A.* = 8 *dollars* 76 *cents* 9 *mills.*

2. You borrow $536,15, and pay $236,18; how much remains unpaid? *A.* $299,97.

3. A merchant bought a quantity of coffee for $526,50, which he afterwards sold for $626,255; how much did he make by the sale? *A.* $99,755.

4. My travelling expenses on a journey were as follows, viz.; stage fare, eighteen dollars; board, nine dollars fifty cents; carrying trunk at different times, seventy-five cents; private conveyance at one time, six dollars thirty-seven and a half cents; and at another, seven dollars; how much had I left, on my return home, of two 50 dollar bills, which I took with me? *A.* $58,37½.

5. From two dollars take twenty cents. *A.* $1,80.

6. From five dollars take one mill. *A.* $4,999.

7. From one dime take 1 cent. *A.* $0,09.

8. How much must you add to three dollars twelve and a half cents, to make four dollars? *A.* $,87½.

9. Subtract 37½ cents from 50 dollars. *A.* $49,62½.

10. From 4 dollars take 3 dollars 99 cents and 9 mills. *A.* $0,001.

11. Suppose I owe the following sums; to Y. $60, 1 cent; to G, $356, 7 cents; to P, $23,50; to D, $700; and my whole stock is worth no more than $1000; am I in debt more than I am worth? and how much? *A.* $139,58 *more than I am worth.*

MULTIPLICATION OF FEDERAL MONEY.

¶ XXV. 1. What will 20 yards of cloth come to, at \$2 per yard? At \$4? At \$6? At \$10?

2. What will 10 yards of tape come to in cents, at 3 mills a yard? At 8 mills? At 2 cents? At 9 cents? At 15 cents? At 12 cents?

3. How many 9 pences are there in a dollar? How many 4½ pences?

4. At 4½ pence a yard, what will 2 yards of ribbon come to in cents? What will 3 yards? 4 yds.? 8 yds.? 12 yds.? 16 yds? 20 yds.? 24 yds.? 32 yds.? 48 yds.? 56 yds.? 64 yds?

5. What will 2 oranges come to in cents, at 2 dimes per orange? What will 3? 5? 6? 7? 9? 12? 15?

Q. What is this, which you have now been doing, called?
A. Multiplication of Federal Money.

RULE.

I. How is the multiplicand to be written down? *A.* As in Addition of Federal Money.

II. How do you multiply? *A.* As in the simple rules.

III. How many places do you retain in the product for cents and mills? *A.* As many as are in the multiplicand.

Exercises for the Slate.

1. At 25 dollars 6 cents 5 mills a month, what will 6 months' labour cost?

OPERATION.

```
$25,065
      6
-------
```

Ans. \$150,390 = 150 *dollars* 39 *cents.*

2. If one pair of shoes cost \$2,25, what will 2 pair cost? (450) 3 pair cost? (675) 4 pair cost? (900) 5 pair cost? (1125) 7 pair? (1575) 9 pair? (2025) *A.* \$67,50. 3. What will 12 pair cost? (2700) 15 pair cost? (3375) 18 pair? (4050) 23 pair? (5175) *A.* \$153. 4. What will 26 pair cost? (5850) 37 pair cost? (8325) 53 pair cost? (11925) 75 pair cost? (16875) *A.* \$429,75. 5. What will 85 pair cost? (19125) 150 pair cost? (33750) 183 pair cost? (41175) 165 pair cost? (37125) *A.* \$1311,75. 6. What will 250 pair cost? (56250) 800 pair *cost? A.* (180000) *A.* \$2362,50.

7. If your travelling expenses for one day are 48 c. 5 mills, how much will the travelling expenses for one year, or 365 days, amount to at that rate? *A.* $177,025. 8. How much will the travelling expenses for 2 years? (354050) For 3 years? (531075) For 5 years? (885125) For 6 years? (1062150) *A.* $2832,40. 9. For 7 years? (1239175) For 9 years? (1593225) For 10 years? (1770250) *A.* $4602,65.

10. What will 2 hogsheads of wine come to, at $32,50 for one hogshead? *A.* $65. 11. What will 3 hogsheads come to? (9750) What will 12? (39000) What will 15? (48750) *A.* $975. 12. What will 25 hogsheads come to? (81250) What will 35 hogsheads? (113750) 150 hogsheads? (487500) *A.* $6825.

13. If $\frac{1}{100}$ of a bank be worth $365,155, what is $\frac{2}{100}$ worth? *A.* $730,31. 14. What is $\frac{3}{100}$ worth? (1095465) $\frac{5}{100}$ worth? (1825775) $\frac{6}{100}$ worth? (2190930) *A.* $5112,17. 15. What is $\frac{8}{100}$ worth? (2921240) $\frac{12}{100}$ worth? (4381860) $\frac{15}{100}$ worth? (5477325) *A.* $12780,425. 16. What is $\frac{16}{100}$ worth? (5842480) $\frac{20}{100}$ worth? (7303100) $\frac{25}{100}$ worth? (9128875) *A.* $22274,455. 17. What is $\frac{30}{100}$ worth? (10954650) $\frac{35}{100}$ worth? (12780425) $\frac{40}{100}$ worth? (14606200) *A.* $38341,275. 18. What is $\frac{45}{100}$ worth? (16431975) $\frac{50}{100}$ worth? (18257750) $\frac{55}{100}$ worth? (20083525) *A.* $54773,25. 19. What is $\frac{60}{100}$ worth? (21909300) $\frac{65}{100}$ worth? (23735075) $\frac{67}{100}$ worth? (24465385) *A.* $70109,76. 20. What is $\frac{75}{100}$ worth? (27386625) $\frac{79}{100}$ worth? (28847245) $\frac{90}{100}$ worth? (32863950) $\frac{100}{100}$ worth? (36515500) *A.* $125613,32.

DIVISION OF FEDERAL MONEY.

¶ **XXVI.** 1. You give 12 cents and 5 mills for 2 sticks of twist; how much is that for one stick?

2. If you give 2 dimes and 5 cents for 2 slates, how much are the slates apiece?

3. If 28 pounds of tea cost 28 dollars, what will 1 pound cost? What will 8 pounds? 5 pounds? 22 pounds?

4. If 20 cwt. of hay cost $40, what will 1 cwt. cost? What will 7 cwt. cost? 11 cwt.? 15 cwt.? 1 cwt., or 4 qrs.? 2 qrs.? 1 qr., or 28 pounds? 14 pounds? 7 pounds? $3\frac{1}{2}$ pounds?

1. *If 12 yards of cloth cost $175,50, what will 1 yard cost?*

OPERATION.

12) 175,50,0

Ans. 14 62 5 mills=14, 62, 5.
$ cts. m.

How do you perform this example? *A.* I divide as in whole numbers.

Where do the 5 mills come from? *A.* In dividing the cents, there is a remainder, which, by annexing a cipher, makes 60 mills, in which 12 is contained 5 times, that is, 5 mills.

From this example we derive the following

RULE.

I. *How do you divide?* *A.* As in Simple Division.

II. *What will the quotient be?* *A.* The answer, in the lowest denomination of the dividend, which may then be brought into dollars.

More Exercises for the Slate.

2. If you divide $35001,50 equally among 125 sailors, how many dollars will each have? *A.* $280,012.

3. If a bank be worth $30515,50, what is $\frac{1}{100}$ of it worth? *A.* $305,155.

4. If a vessel and cargo are valued at $20000, what is $\frac{1}{28}$ worth? *A.* $714,285+.

5. 16 men draw $2050,65 in a lottery; how much is each man's part, if it be equally divided among them? *A.* $128,165+.

6. If a man's salary be $3650,40 a year, what is that a day? *A.* 10 dollars, 1 mill+.

7. Bought 36 lb. of sugar for $10,50; what is that a pound? *A.* $,291=29 cents, 1 mill+.

8. If you buy 383 yards of broadcloth for $5036,50, what is that a yard? *A.* $13,15+.

9. If a man's wages be $365,40 a year, or 52 weeks, what is that a week? *A.* $7,026+. 10. Divide $1000,60 equally among 2 men. (50030) Among 3 men. (333533) Among 4 men. (25015) Among 6. (166766) Among 8. (125075) *A.* 1375,824+. 11. Among 10. (10006) Among 15. (66706) Among 18. (55588) Among 25. (40024) *A.* $262,378+. 12. Among 28. (35735) Among 35. (28588) Among 40. (25015) Among 45. (22235) Among 60. (16676) *A.* $128,249+. 13. Among 70. (14294) Among 85. (11771) Among 95. (10532) Among 100. (10006) Among 150. (667) *A.* $53273+. 14. Among 200. (5003) Among 240. (4169) Among 360. (2779)

Among 400. (2501) Among 550. (1819) *A.* \$16,271+. 15 Among 17. (58858) Among 19. (52663) Among 89. (11242) Among 98. (1021) Among 199. (5028) *A.* 138 dollars, 1 mill+.

To multiply by $\frac{1}{2}$, $\frac{1}{3}$, $\frac{3}{8}$, $5\frac{1}{2}$, &c.

¶ **XXVII.** 1. How much is $\frac{1}{2}$ of 10? $\frac{1}{3}$ of 18? $\frac{1}{4}$ of 16? $\frac{1}{5}$ of 20?

RULE.

I. To multiply 4 by 2, we take 4, 2 times; to multiply by 1, we take 4, 1 time; and to multiply 4 by $\frac{1}{2}$, we take 4, $\frac{1}{2}$ a time, that is, *the half* of 4, and to get this, it is plain that we must divide 4 by 2. *From these remarks what appears to be the rule for multiplying by $\frac{1}{2}$, $\frac{1}{3}$, $\frac{1}{5}$, &c.?* *A.* Divide the multiplicand by the figure below the line.

2. How much is $\frac{1}{5}$ of 20? $\frac{2}{5}$ of 20? $\frac{1}{6}$ of 12? $\frac{2}{6}$ of 12? $\frac{4}{6}$ of 12?

II. In these examples we divide by 5 to get $\frac{1}{5}$ of 20: then it is plain, that $\frac{2}{5}$ is 2 times as much as $\frac{1}{5}$, that is, we multiply $\frac{1}{5}$ of 20, which is 4, by 2, making 8: but we might multiply 20 by the 2 first, and divide by the 5 afterwards; thus, $2\times20=40\div5$, 8 times, the same result as before: *how, then, does it appear that we can proceed to multiply by $\frac{2}{5}$, $\frac{7}{8}$, $\frac{4}{15}$, &c.?* *A.* Divide the multiplicand by the figure below the line, and multiply this quotient by the figure above the line; or multiply first, and then divide.

Exercises for the Slate.

1. What will $2\frac{5}{7}$ yards of cloth cost at 28 cents a yard?

OPERATION.

```
  28          7) 28                       Or,
  2 5/7         ----                       28
 ----           4 = 1/7 of 28.              5
  56            5                        ----
  20          ----                   7) 140
 ----          20 = 5/7 of 28.           ----
$,76 Ans.                                  20
```

How do you multiply by the 2, in $2\frac{5}{7}$? *A.* As usual?

How do you proceed next? *A.* I get $\frac{1}{7}$ of 28, making 4, then multiply the 4 by 5, making 20.

What do you do with the 20? *A.* I add it to 56, making 76.

More Exercises for the Slate.

2. At 30 cents a bushel, what will $19\frac{1}{15}$ bushels of oats cost? *A.* \$5,72.

3. Tell me how many yards 50 rods are, or multiply 50 by $5\frac{1}{2}$. *A.* 275.

4. What will $22\frac{3}{11}$ yards of cloth come to, at \$11 a yard? (245) What will $22\frac{4}{11}$? (246) What will $23\frac{7}{11}$? (260) $3\frac{9}{11}$? (42) $31\frac{10}{11}$? (351) $99\frac{6}{11}$? (1095) *A.* 2239. 5. What will $2\frac{3}{4}$ yards come to at \$12 a yard? (33) What will $5\frac{1}{2}$? (66) $6\frac{2}{3}$? (80) $8\frac{9}{12}$? (105) $10\frac{1}{6}$? (122) $15\frac{3}{6}$? (186) $25\frac{4}{8}$? (306) *A.* \$398.

6. If a man's salary be \$1200 a year, what will $2\frac{1}{60}$ years come to? (2420) What will $3\frac{1}{40}$? (3630) $5\frac{20}{600}$? (6040) $8\frac{555}{600}$? (10710) $12\frac{598}{600}$? (15598) *A.* \$38398.

When the price is an ALIQUOT part of a dollar, or, PRACTICE.

¶ **XXVIII.** 1. At 50 cents a bushel, what will 2 bushels of apples cost? What will 4 bushels? How much is $\frac{1}{2}$ of 4? How much is $\frac{1}{2}$ of 8? What will 8 bushels cost, then? How much is $\frac{1}{2}$ of 12? How much will 12 bushels cost, then? How much is $\frac{1}{2}$ of 40? What will 40 bushels cost, then? How much is $\frac{1}{2}$ of 100? What will 100 bushels cost, then?

2. At 25 cents a peck, what will 2 pecks of salt cost? Will 4 pecks cost? Will 8 pecks cost? How much is $\frac{1}{4}$ of 8? How much is $\frac{1}{4}$ of 16? Will 16 pecks cost, then? How much is $\frac{1}{4}$ of 20? Will 20 pecks cost, then? How much is $\frac{1}{4}$ of 400? Will 400 pecks cost, then?

3. What will 2 oranges cost, at $12\frac{1}{2}$ cents apiece? Will 4 oranges? Will 8 oranges? Will 16 oranges? How much is $\frac{1}{8}$ of 16? How much is $\frac{1}{8}$ of 24? What will 24 oranges cost, then? How much is $\frac{1}{8}$ of 80? Will 80 oranges cost, then?

4. At $6\frac{1}{4}$ cents a pint, what will 2 pints of ale cost? What will 4 pints cost? 8 pints? 16 pints? How much is $\frac{1}{16}$ of 16? How much is $\frac{1}{16}$ of 32? Will 32 pints cost, then?

Q. What part of a dollar is 50 cents? A. $\frac{1}{2}$. *Why? A.* Because 2 times 50 are 100 c. = \$1.

Q. What part of a dollar is 25 cents? A. $\frac{1}{4}$. *Why? A.* Because 4 times 25 c. are 100 c. = \$1.

Q. What are these even parts called? A. Aliquot Parts.

Q. When, then, one number is contained in another exactly 2, 3, 4, &c. times, what is it called? A. An Aliquot Part.

Q. What is the method of finding the cost of articles, by taking aliquot parts, usually called? A. Practice.

Repeat the Table of the aliquot parts of a dollar.

Cts.			Cts.	Cts.	$
50	$=\frac{1}{2}$	of a dollar, because	2×50	$=100$	$=1.$
$33\frac{1}{3}$	$=\frac{1}{3}$	of a dollar, because	$3\times33\frac{1}{3}$	$=100$	$=1.$
25	$=\frac{1}{4}$	of a dollar, because	4×25	$=100$	$=1.$
20	$=\frac{1}{5}$	of a dollar, because	5×20	$=100$	$=1.$
$16\frac{2}{3}$	$=\frac{1}{6}$	of a dollar, because	$6\times16\frac{2}{3}$	$=100$	$=1.$
$12\frac{1}{2}$	$=\frac{1}{8}$	of a dollar, because	$8\times12\frac{1}{2}$	$=100$	$=1.$
10	$=\frac{1}{10}$	of a dollar, because	10×10	$=100$	$=1.$
$6\frac{1}{4}$	$=\frac{1}{16}$	of a dollar, because	$16\times6\frac{1}{4}$	$=100$	$=1.$
5	$=\frac{1}{20}$	of a dollar, because	20×5	$=100$	$=1.$

RULE.

I. *From the illustrations now given, what appears to be a concise rule for calculating the cost of articles, when the price is an aliquot part of a dollar? A.* Divide the number of gallons, yards, &c. by the number of aliquot parts which it takes of the price to make a dollar.

II. *What will the quotient be? A.* The answer, in dollars.

Q. For example, what would you divide by when the price is 50 cents? $33\frac{1}{3}$ cents? 25 cents? 20 cents? $16\frac{2}{3}$ cents? $12\frac{1}{2}$ cents? 10 cents? $6\frac{1}{4}$ cents? 5 cents?

Exercises for the Slate.

1. What will 360010 yards of cloth cost, at $12\frac{1}{2}$ cents a yard?

OPERATION.

8) 360010,00

$ 45001,25, *Ans.*

How do you get the divisor, 8? *A.* $12\frac{1}{2}$ cents $=\frac{1}{8}$ of a dollar.

How do you obtain the 25 cents in the answer? *A.* I annex 2 ciphers for cents, and continue dividing.

2. What cost 2640 bushels of rye, at 50 cents a bushel? *A.* $1320.

3. At 25 cents a bushel, what will 4680 bushels of potatoes cost? (1170) Will 3600 bushels cost? (900) Will 4200 bushels cost? (1050) *A.* $3120.

4. At $6\frac{1}{4}$ cents, or $\frac{1}{16}$ of a dollar, what will 6400 yards of

7

tape cost? (400) Will 32000 yards? (2000) Will 128000 yards? (8000). *A.* $10400.

5. What will 2700 yards of ribbon cost at 12½ cents a yard? (33750) Will 5400 yards? (675) Will 1080 yards? (135) *A.* $1147,50.

6. What will 42124 oranges cost at 5 cents, or $\frac{1}{20}$ of a dollar apiece? *A.* $2106,20.

7. What will be the cost of 1200 yards of cloth, at 50 cents a yard? (600) At 33⅓ cents a yard? (400) At 25 cents a yard? (300) At 20 cents a yard? (240) At 12½ cents a yard? (150) At 6¼ cents a yard? (75) *A.* $1765.

8. At 16⅔ cents a pint, what will 2700 pints of brandy cost? (450) At 10 cents a pint, what will 5400 pints cost? (540) *A.* $990.

9. What will 36002 bushels of salt cost, at $1,12½, or $1 ⅛ of a dollar, a bushel? At $2,25 a bushel?

OPERATIONS.

8) $36002=cost at $1 a bushel.
    4500,25=cost at $,12½ a bush.
    $40502,25 *Ans.*

4) 36002
         2
    72004=cost at $2.
    9000,50=cost at 25c.
Ans. $81004,50=cost at $2¼

10. What will 3700 acres of land cost, at $12,25 ($12¼) an acre? (45325) What will 3700 acres, at $15,50 ($15½) an acre? (57350) *A.* $1026,75.

11. What will 27000 yards of broadcloth cost, at $3,50 ($3½) a yard? (94500) At $2,33⅓ ($2⅓) a yard? (63000) At $1,12½ ($1⅛) a yard? (30375) At $4,06¼ ($4$\frac{1}{16}$) a yard? (10968750) At $1,05 ($1$\frac{1}{20}$) a yard? (28350) *A.* $325912,50+.

Questions, involving the Principles of the foregoing Rules.

1. A man bought a farm for thirty-six hundred dollars, and stock for the same to the amount of seven hundred and twenty dollars; what did both cost? and how much did one cost more than the other? *A. Both,* $4320; *the farm,* $2880 *the most.*

2. What is the amount of the following numbers, viz., ten, thirty, one million, twenty-six thousand, one hundred and one? *A.* 1026141.

3. If the minuend be 26002, and the subtrahend 101, what is the remainder? (25901) If the subtrahend be 601, and the remainder 5025, what is the minuend? (5626) *A.* 31527

4. If the multiplicand be 4200, and the multiplier 48, what is the product? (201600) If the product be 201600, and the multiplicand 4200, what is the multiplier? (48) *A.* 201648.

5. If the divisor be 25, and the dividend 5025, what is the quotient? (201) If the quotient be 201, and the divisor 25, what is the dividend? (5025) If the quotient be 201, and the dividend 5025, what is the divisor? (25) *A.* 5251.

6. If the sum of two numbers be 1800, and the greater 1000, what is the less? (800) If the less be 120, and the sum 1320, what is the greater? (1200) *A.* 2000.

7. $2700342 + 72 + 3 + 1 =$ *Ans.* 2700418.

8. $367895437 - 72591 =$ *Ans.* 367822846.

9. $25432 \times 67345 =$ *Ans.* 1712718040.

10. $360042 \div 8542 =$ *Ans.* $42\frac{1278}{8542}$.

11. $\frac{270000001111}{30000} =$ *Ans.* $9000000\frac{1111}{30000}$.

12. Bought 18 barrels of potatoes, each containing 3 bushels, at 25 cents a bushel; what did the whole cost? *Ans.* $13,50.

13. A farmer sold 30 bushels of rye, at 87 cents a bushel; 30 bushels of corn, at 53 cents a bushel; 8 bushels of white beans, at $1,25 a bushel; two yoke of oxen, at $62 a yoke; 10 calves, at $4 apiece; and 15 barrels of cider, at $2,37½ a barrel; what was the amount of the whole? *A.* $251,625.

14. A merchant, failing in trade, has on hand goods to the amount of $4800, and his borrowed money is $2400: the remainder, after paying his borrowed money, is to be equally divided among 121 creditors: how much will each receive? *A.* 19\frac{101}{121}$.

15. A grocer bought 4 loads of wood, at $2,25 a load (9); 8 bushels of rye, at $,50 a bushel (4); 30 pounds of butter, at 12½ cents a pound (375); 340 pounds of cheese, at 6¼ cents a pound (2125); for which he paid 5 barrels of flour, at $6,25 a barrel (3125); 35 pounds of sugar, at 8½ cents a pound (2975); 3 gallons of molasses, at 25 cents a gallon (75); 15 bushels of salt, at $1,12½ a bushel (16875); what is the balance between the articles bought and sold? *A.* $13,85.

16. What will be the price of 4 bales of goods, each bale containing 60 pieces, and each piece 49 yards, at 37½ cents a yard? *A.* $4410.

17. How many oranges, at 2 cents apiece, can be bought for 4 cents? For $2, or 200 cents? *A.* 102. For $8600? For $10000? *A.* 930000.

18. How many yards of cloth, at $2 a yard, may be bought for 4 barrels of cider, at $3 a barrel? For 8 barrels? *A.* 18. For 28 barrels? For 50 barrels? *A.* 117.

19. How many gallons of molasses, at 23 cents a gallon, may be had for 60 bushels of oats, at 46 cents a bushel? At 69 cents a bushel? *A.* 300 gallons.

FARMERS' BILLS.

Mr. George Stimpson

To Rufus Paywell, Dr.

1828.			
June 5.	To 8 barrels Cider, at $2,12½ a barrel,		$17,00
" 8.	To 6 bushels Corn, " ,58 a bushel,		3,48
			$20,48

July 16, 1828. Received payment,

Rufus Paywell.

Mr. Chauncey Ackley

To Charles Thomas, Dr.

1828.				
June 8.	To 20 Merino Sheep,	at $6	a head,	
" 18.	To 25 Calves,	" 2,12½	" "	
July 1.	To 200 pounds Cheese,	" ,06¼	" pound,	
" 12.	To 18 " Veal,	" ,08¾	" "	
Oct. 15.	To 18 " Clover-Seed,	" ,03¾	" "	
" 18.	To 36 bushels Oats,	" ,27¾	" bushel,	
" 20.	To 17 " Corn,	" ,75	" "	
				$210,61,5

Nov. 15, 1828. Received payment,

Charles Thomas.

MERCHANT'S BILL.

Boston, December 18, 1828.

Mr. Peter Careful

Bought of Stephen Notrust

3800 yards Calico,	at $,17¾ a yard,	
360 bales Cotton Cloth, each bale 60 pieces, each piece 24 yds.	" ,19¾ " "	
40 pieces blue Broadcloth, each 37 yards,	" 4,62½ " "	
400 yards Carpeting,	" 1,18 " "	
200 pieces Nankin, each 42 yards,	" ,39 " "	
		$113651,50

Received payment,

For Stephen Notrust,

John Stimpson.

REDUCTION.

¶ XXIX. WINE MEASURE.

Repeat the

TABLE.

4 Gills (gi.)	make	1 Pint,	marked	pt.
2 Pints		1 Quart,		qt.
4 Quarts		1 Gallon,		gal.
31½ Gallons		1 Barrel,		bl.
42 Gallons		1 Tierce,		tier.
63 Gallons*		1 Hogshead,		hhd.
2 Hogsheads		1 Pipe,		p.
2 Pipes		1 Tun,		T.

1. At 2 cents a gill, what will 1 pint of molasses cost? What will 2 pints? 4 pints? 5 pints? 6 pints? 8 pints? 10 pints? 12 pints? 20 pints? 50 pints? 100 pints?

2. At 24 cts. a gallon, what will 1 qt. of ale cost? What will 2 qts.? ½ a gallon? How much will 48 cts. buy? Will 60 cts.? Will 72 cts.? Will 120 cts.? Will 150 cts.? Will 180 cts.?

3. When rum is a dollar a gallon, what is a barrel worth? What is a hogshead worth? What is $\frac{1}{63}$ of a hogshead worth? (A hogshead is divided into 63 parts, or gallons; therefore 1 gallon is $\frac{1}{63}$.) What is $\frac{2}{63}$? $\frac{15}{63}$? $\frac{40}{63}$?

4. At a dollar a gallon, how many quarts will 50 cents buy? Will $2? Will $2,25? Will $3? Will $3,50? Will $4? Will $6? Will $10? Will $20? Will $30?

5. At $100 a hogshead, what will 2 pipes of wine cost? What will 3 pipes? 4 pipes? 6 pipes? 1 tun? 3 tuns? 4 tuns?

6. How many gills in 4 pints? In 2 quarts? In 3 qts.? In 4 qts.? In 6 qts.? In 12 qts.? In 20 qts.?

Q. From what you have now been doing, for what purposes would you infer that this measure is used? *A.* To measure wine, spirits, vinegar, oil, cider, honey, mead, &c.

Q. What are the denominations of this measure? *A.* Gills, pints, quarts, gallons, barrels, tierces, hogsheads, pipes and tuns.

Note.—The wine gallon contains 231 solid inches, and is in the same proportion to the ale gallon of 282 solid inches as the pound Troy is to the pound avoirdupois.

---

* Hogsheads containing liquors molasses, &c. are of various capacities, generally exceeding 100 gallons.

Exercises for the Slate.

1. In 4 pints how many gills? In 20 pints how many gills? *A.* 96 gills.
2. How many pints in 16 gills? In 80 gills? *A.* 24 pints.
3. How many pints in 2 quarts? In 480 quarts? *A.* 964 pints.
4. How many quarts in 4 pints? In 960 pts? *A.* 482 qts.
5. In 2 hogsheads how many gallons? In 4137 hhds.? *A.* 260757 gallons.

ALE OR BEER MEASURE.

Repeat the

TABLE.

2 Pints	. . . . make	. 1 Quart,	. marked .	qt.
4 Quarts	. . . .	1 Gallon,	. . . .	gal.
36 Gallons	. . . .	1 Barrel,	. . . .	bar.
54 Gallons	. . . .	1 Hogshead,	. . . .	hhd.

1. How many pints in 2 quarts? In 6 qts.? In 10 qts.? In 20 qts.? In 200 qts.? In 600 qts.?
2. How many quarts in 8 pints? In 10 pints? In 12 pints? In 20 pints?
3. What will 1 gallon of beer cost at 2 cents a quart? What will 2 gallons? What will 4 gallons? Will 5 gallons?

Q. From these examples, what would you infer is the use of this measure? *A.* To measure ale, beer, and milk.

Q. What are the denominations of this measure? *A.* Pints, quarts, gallons, barrels and hogsheads.

Note.—A gallon, beer measure, contains 282 cubic inches.

Exercises for the Slate.

6. How many pints in 2 quarts? In 3600 qts.? *A.* 7204 pints.
7. How many quarts in 4 pints? In 7200 pts.? *A.* 3602 quarts.
8. How many gallons in 2 barrels? In 620 bar.? *A.* 22392 gallons.

CLOTH MEASURE.

Repeat the

TABLE.

$2\frac{1}{4}$ Inches (in.) make . .	1 Nail,	. . marked . .	na.
4 Nails	1 Quarter of a yd. . . .		qr.
4 Quarters	1 Yard, ...		yd.
3 Quarters	1 Ell Flemish, . . .		E. Fl.
5 Quarters	1 Ell English, . . .		E. E.
6 Quarters	1 Ell French, . . .		E. Fr.

1. How many nails in 2 qrs.? In 4 qrs.? In 8? In 10? In 11? In 12? In 15?

2. How many quarters in 1 yard? In $\frac{1}{2}$ a yard? In $1\frac{1}{2}$ yds.? In 2 yds.? In 5 yds.? In 7 yds.? In 9 yds.? In 12 yds.? In 20 yds.? In 30 yds.?

3. At 2 cents a quarter, what will 1 yd. of cloth cost? $\frac{1}{2}$ a yd.? $1\frac{1}{2}$ yds.? 2 yds.? 5 yds.? 7 yds.? 9 yds.? 12 yds.? 100 yds.?

4. How many quarters in 2 ells Flemish? 3? 5? 7? 9? 11? 12? 15? 20?

5. How many quarters in 3 ells English? 5? 6? 8? 10? 11? 20?

6. How many quarters in 5 ells French? 7? 9? 11? 12? 15? 20?

7. If $\frac{1}{6}$ of a yard of cloth cost 10 cts., what cost $\frac{2}{6}$? $\frac{3}{6}$? $\frac{1}{2}$? $\frac{4}{6}$? $\frac{5}{6}$? $\frac{6}{6}$? $1\frac{1}{3}$? $1\frac{2}{6}$? $1\frac{1}{2}$? $1\frac{4}{6}$? $1\frac{5}{6}$? 2? 3? 6? 8? 10?

8. If $\frac{1}{4}$ of a yard of tape cost 5 cents, what will 1 yd. cost? What will 2 ells Flemish? 3 ells English? 4 ells French?

Q. From what you have now been doing, for what purposes would you infer that this measure is used? *A.* To measure all kinds of cloth.

Q. What are the denominations of this measure? *A.* Inches, nails, quarters, yards, and ells.

Exercises for the Slate.

9. How many quarters in 2 yards? In 26872 yds.?
A. 107496 qrs.

10. How many yards in 8 quarters? In 107488 qrs.?
A. 26874 yds.

11. How many nails in 3 quarters? In 42000 qrs?
A. 168012 na.

12. How many quarters in 12 nails? In 168000 na.?
A. 42003 qrs.

DRY MEASURE.

Repeat the

TABLE.

2 Pints (pt.) make	. . . 1 Quart,	. . marked	. qt.
8 Quarts	. . . . 1 Peck,	. . . . . .	. . pk.
4 Pecks	. . . . 1 Bushel,	. . . . .	. . bu.

Q. How many bushels make a chaldron? *A.* 36.

1. How many pints in 4 qts.? In 6? 7? 9? 10? 50? 100? 200?

2. How many quarts in 2 pecks? In 3? 4? 5? 6? 7? 8? 10? 12?

3. If you give 5 cents for 1 quart of flax-seed, what will 2 pecks cost? 3 pecks? What will $\frac{1}{8}$ of a peck? $\frac{2}{8}$? $\frac{3}{8}$? $\frac{4}{8}$? $\frac{1}{2}$? $\frac{5}{8}$? $\frac{7}{8}$? $\frac{8}{8}$? How much will 10 cents buy? 15 cts.? 20 cts.? 25 cts.? 30 cts.? 40 cts.?

4. How many pecks in 3 bushels? In 4? In 6? In 10? In 12? In 15? In 20? In 30? In 40? In 50? In 60? In 70? In 80? In 90? In 100?

5. At 10 cts. a peck, what will a bushel of salt cost? What will $\frac{1}{2}$ of a bu.? $\frac{3}{4}$ of a bu.? 2 bu.? 3 bu.? 4 bu.? 6 bu.? 10 bu.?

6. At 40 cts. a bushel, how much rye can you buy for 20 cts.? For 30 cts.? For 60 cts? For 70 cts? For 80 cts.? For 100 cts? For 120 cts.? For 160 cts.? For 360 cts.?

Q. From what you have now been doing, for what purposes would you infer that this measure is used? *A.* To measure corn, seeds, roots, fruits, salt, oysters, coals, &c.

Q. What are the denominations of this measure? *A.* Pints, quarts, pecks, bushels, and chaldrons.

Exercises for the Slate.

13. How many pints in 3 quarts? In 321 qts.? *A.* 648 pts.
14. How many quarts in 6 pints? In 1284 pts.? *A.* 645 qts.
15. How many pecks in 2 bushels? In 32 bu.? *A.* 136 pks.

TROY WEIGHT.

Repeat the

TABLE.

24 Grains (gr.) make	. 1 Pennyweight,	marked pwt.
20 Pennyweights	. . . 1 Ounce,	. . . . oz.
12 Ounces	. . . . 1 Pound,	. . . . lb.

1. How many grains are there in two pennyweights?

2. What are 2 pennyweights of gold worth, if 1 grain is worth 2 cents?

3. How many pennyweights in 5 ounces? In 10? In 12?

4. You carry 2 ounces of old gold to a goldsmith, and he agrees to give you 10 cts. a pennyweight for it; how much will the 2 ounces come to? Will 3 oz.? Will 4 oz.? Will 5 oz.?

5. How many ounces in 2 pounds? In 3 pounds? In 4 pounds? In 5 pounds? In 6 pounds? In 7 pounds? In 10 pounds? In 12 pounds?

6. What will a silver cup come to, that weighs two pounds, if you get 1 dollar an ounce for it? If you get 2 dols.? 3 dols.? 4 dols.? 5 dols.?

Q. How is the fineness of gold expressed? *A.* In carats.

Q. How many carats make an ounce? *A.* 22.

Q. How many parts is an ounce of silver divided into? *A.* 20.

Q. What are the parts called? *A.* Pennyweights.

Q. How many carats fine is such gold as will abide the fire without loss accounted? *A.* 24.

Q. How many ounces are there in a pound of silver, that loses nothing in trial? *A.* 12.

Q. What is that base metal, which is sometimes mixed with gold or silver, called? *A.* Alloy.

Q. From what you have now been doing, for what purposes would you infer that this weight is used? *A.* To weigh jewels, gold, silver, coin, bread and liquors.*

Q. What are the denominations of this weight? *A.* Grains, pennyweights, ounces, and pounds.

Exercises for the Slate.

16. In 2 pounds how many ounces? In 4200 pounds?
A. 50424 oz

17. In 24 ounces how many pounds? In 120000 ounces?
A. 10002 lbs.

18. How many grains in 4 pounds? In 3600 pounds?
A. 20759040 gr.

---

*A pound avoirdupois is heavier than a pound Troy, but an ounce Troy is heavier than an ounce avoirdupois.

AVOIRDUPOIS WEIGHT.

Repeat the

TABLE.

16 Drams (dr.) make	1 Ounce,	. . marked . .	oz.
16 Ounces	1 Pound,*	. . . .	lb.
28 Pounds	1 Quarter of a Hundred Weight,		qr.
4 Quarters	1 Hundred Weight,	. .	cwt.
20 Hundred Weight .	1 Ton,	. . . .	T.

1. How many ounces in 2 pounds? In 3? In 4? In 5?

2. At 2 cts. an ounce, what will 2 lbs. of tea cost? Will 3 lbs.? Will 4 lbs.? Will 5 lbs.?

3. How many pounds in 2 qrs.? In 3 qrs.? In 4 qrs.?

4. What will 2 quarters of raisins come to, at 10 cts. a pound? What will 3 qrs.? What will 4 qrs.?

5. If you give 2 dollars for 1 qr. of flour, what must you pay for 14 lbs.? For 7 lbs.? For 3½ lbs?

6. How many quarters in 2 cwt.? In 4 cwt.? In 5 cwt.? In 6 cwt.? In 8 cwt.? In 12 cwt.? In 15 cwt.? In 20 cwt.? In 30 cwt.?

7. At $1 a quarter, what will 2 cwt. of sugar cost? Will 4 cwt.? Will 5? Will 6? Will 8? Will 10? Will 15? Will 20? Will 30?

8. How many hundred weight in 2 tons? How many in 3 tons? In 4 tons? In 5 tons? In 6 tons? In 8 tons? In 10 tons?

9. You buy 2 tons of hay, and pay 1 dollar a cwt. for it; what will it come to? What will 3 tons? What will 4 tons? What will 5 tons? What will 6 tons? What will 8 tons? What will 10 tons? What will ½ a ton? What will ¼ of a ton? What will $\frac{1}{20}$? (A ton is divided into 20 cwt. or 20 parts; therefore $\frac{1}{20}$ is 1 cwt.) What will $\frac{3}{20}$? $\frac{4}{20}$? $\frac{6}{20}$? $\frac{10}{20}$? $\frac{15}{20}$?

10. What will $\frac{1}{28}$ of a quarter of coffee come to, at 11 cts. a pound? What will $\frac{2}{28}$? $\frac{6}{28}$? $\frac{7}{28}$? $\frac{9}{28}$? $\frac{10}{28}$? What will 3 pounds? 5? 7? 9? 10?

Q. *From what you have now been doing, for what purposes would you infer that this weight is used?* A. To weigh all coarse goods, that are subject to waste, such as hay, flour, &c. and all metals, except gold and silver.

Q. *What are the denominations of this weight?* A. Drams, ounces, pounds, quarters, hundreds and tons.

---

* *The pound avoirdupois is equal to 14 oz. 11 pwt. 15 1-2 grains Troy.*

Exercises for the Slate.

19. How many hundred weight in 2 tons? In 2832 tons? *A.* 56680 cwt.

20. How many ounces in 2 pounds? In 104000 pounds? *A.* 1664032 oz.

21. In 8 quarters how many hundred weight? In 240 qrs.? *A.* 62 cwt.

22. In 1 hundred weight how many quarters? *A.* 4 qrs. In 4 qrs. how many pounds? *A.* 112.

Note.—By the last example it appears, that 112 pounds make 1 hundred weight; but in our populous towns, especially sea-ports, traders buy and sell by the 100 pounds.

APOTHECARIES' WEIGHT.

Repeat the

TABLE.

20 Grains (gr.) make . .	1 Scruple, .	marked .	℈.
3 Scruples	1 Dram, ...		ʒ.
8 Drams	1 Ounce, ...		℥.
12 Ounces	1 Pound, ...		℔.

Q. What is the use of this weight? A. By it apothecaries compound their medicines.

Q. Do they buy and sell by this weight? A. They buy and sell by avoirdupois weight.

Q. What are the denominations of this weight? A. Grains, scruples, drams, ounces, and pounds.

Exercises for the Slate.

23. How many drams in 2 ounces? In 4360 ounces? *A.* 34896 drams.

24. How many ounces in 16 drams? In 6464 drams? *A.* 810 ounces.

LONG MEASURE.

Repeat the

TABLE.

3 Barley Corns (b.c.) make	1 Inch, . . marked	in.
12 Inches	1 Foot, ...	ft.
3 Feet	1 Yard, ...	yd.
$5\frac{1}{2}$ Yards	1 Rod, Pole or Perch,	rd.
40 Rods	1 Furlong, ...	fur.
8 Furlongs	1 Mile, ...	m.
3 Miles	1 League, ...	lea.
$69\frac{1}{2}$ Statute Miles	1 Degree on the earth,	°

360 Degrees the circumference of the earth.

1. If a man travel one furlong in 5 minutes, how far can he go in 10 minutes? In 15? In 20? In 30?

2. How many barley corns in 2 inches? In 3? In 4? In 5? In 6? In 7? In 8? In 9? In 20?

3. How many inches are 6 barley corns? Are 12? Are 15? Are 18? Are 21? Are 24? Are 27? Are 60?

4. How many inches in 3 feet? In 5 feet? In 12 feet?

5. How many rods is one furlong? Are 2? Are 4? Are 6?

6. How many furlongs in 80 rods? In 160? In 240?

7. How many furlongs are there in 2 miles? In 3? In 4? In 6? In 7? In 8? In 9?

8. How many miles in 16 furlongs? In 24? In 32? In 48? In 56?

9. How much is $\frac{1}{8}$ of a mile? Is $\frac{2}{8}$? Is $\frac{3}{8}$? Is $\frac{4}{8}$? Is $\frac{1}{2}$? Is $\frac{7}{8}$? Is $\frac{8}{8}$?

10. If you travel a mile in 16 minutes, how much time will it take to travel $\frac{1}{8}$ of a mile? How much $\frac{4}{8}$? $\frac{6}{8}$? $\frac{7}{8}$? $\frac{8}{8}$? 1 furlong? 4 furlongs? 16 furlongs?

Q. What is the use of long measure? *A.* To measure length only.

Q. What is the use of the league? *A.* Distances at sea are measured by it.

Q. What are the denominations of this measure? *A.* Barley corns, inches, feet, yards, rods, furlongs, miles, leagues and degrees.

Exercises for the Slate.

25. How many furlongs in 2 miles? In 26784 miles? *A. 214288* fur.

26. How many furlongs in 80 rods? In 627360 rods? *A.* 15686 fur.

27. In 2 rods how many yards? In 11010 rods? (To multiply by $5\frac{1}{2}$, $8\frac{1}{4}$, &c., consult ¶ XXVII.) *A.* 60566.

28. In 11 yards how many rods? *A.* 2 rods. In 66 yards how many rods? We cannot easily divide 66 by $5\frac{1}{2}$, but we may multiply 66 by 2, making 132 half yards, which we can divide by 11, the half yards in $5\frac{1}{2}$; thus, $66\times2=132\div11=12$ rods, *Ans.* *Hence, to divide by $5\frac{1}{2}$, $30\frac{1}{4}$, &c., we need only bring the divisor into halves, quarters, &c., also the dividend into the same, and the quotient will be the answer.*

29. In 132 yards how many rods? In 4224 yards? *A.* 792 rods.

30. How many barley corns in 2 inches? In 278365 inches? *A.* 835101.

LAND OR SQUARE MEASURE.

Repeat the

TABLE.

144 Square Inches	. . . make . . .	1 Square Foot.
9 Square Feet	. . . .	1 Square Yard.
$30\frac{1}{4}$ Square Yards, or $272\frac{1}{4}$ Square Feet,	. . . .	1 Square Rod.
40 Square Rods	. . . .	1 Square Rood.
4 Square Roods	. . . .	1 Square Acre.
640 Square Acres	. . . .	1 Square Mile.

1. How many square inches in 2 square feet?

2. How many square feet in 3 square yards? In 4? In 5? In 6? In 7? In 8? In 9? In 10? In 11? In 12? In 20? In 30? In 40? In 50? In 60?

3. How many square yards in 36 square feet? In 45? In 54? In 63? In 72? In 81? In 90? In 180? In 270? In 360? In 450? In 540?

4. How many square rods in 2 square roods? In 3? In 4? In 6? In 8?

5. How many square roods in 80 square rods? In 120? In 160? In 240?

6. At a dollar a square rod, what will 2 square roods come to? What will 3? 4? How many rods can you buy for 120 dollars? For 160 dollars? For 240 dollars? For 320 dollars? For 400 dollars? For 40 dollars? For 20 dollars? For 10 dollars? What would $\frac{1}{40}$ of a rood come to? $\frac{4}{40}$? $\frac{6}{40}$? $\frac{20}{40}$? $\frac{1}{40}$? $\frac{30}{40}$? $\frac{3}{4}$? $\frac{10}{40}$? $\frac{1}{4}$?

Q. What is the use of this measure? *A.* To measu[r]
and breadth only.

Q. What are the denominations of this measure? *A*
feet, yards, rods, roods, acres, and miles.

Exercises for the Slate.

31. In 2 square roods how many square rods? In 45
roods? *A.* 18080 rods.

32. How many square acres in 2 square miles?
square miles? *A.* 1500160 sq. a.

33. How many square yards in 4 square rods? I[n]
square rods? *A.* 5856521.

SOLID OR CUBIC MEASURE.

Repeat the

TABLE.

1728 Solid Inches . . make . . . 1 Solid F[oot]
40 Feet of Round Timber, or }
50 Feet of Hewn Timber, } . . 1 Ton, or

Also,

27 Solid Feet . . . . make . . . . 1 Solid Y[ard]
128 Solid Feet, or 8 feet long, }
4 wide, and 4 high, } . 1 Cord of

1. What will a parcel of wood come to, which is 8 fe[et]
4 feet high, and 4 wide, at the rate of 6 dollars a cord?

2. If you pay 80 dollars for a ton of round timber, w[hat]
80 feet come to? What will 120 feet come to?

Q. What is the use of this measure? *A.* To measure
breadth, and depth.

Q. What are the denominations of this measure? *A.*
feet, yards, tons, and cords.

Exercises for the Slate.

34. In 2 cords of wood, how many solid feet? In 2[8]
A. 3840 feet.

35. How many solid inches in 2 solid feet? In 28[0]
feet? *A.* 4841856 solid inches.

36. How many solid feet in 345600 solid inches? In
solid inches. *A.* 600 solid feet.

TIME.

Repeat the

TABLE.

60 Seconds (s.)	make	1 Minute, marked	m.
60 Minutes		1 Hour,	h.
24 Hours		1 Day,	d.
$365\frac{1}{4}$ Days		1 Year,	yr.
100 Years		1 Century,	cen.

Also,

7 Days	make	1 Week, marked	w.
4 Weeks		1 Month,	mo.
13 Months, 1 Day, 6 Hours,		1 Julian Year.	
12 Calendar Months		1 Year,	yr.

1. If a man earn a dollar a day, how much will he earn in 2 weeks? In 5 weeks? In 7? In 9? In 20? In 40? In $\frac{1}{7}$ of a week? In $\frac{5}{7}$? In $\frac{4}{7}$? In $\frac{2}{7}$? How many weeks could you hire him for $7? For 14? For 28?

2. A boy is to have $2 a week in a store; how much must the merchant pay him for one month's time? For 2? For 5? For 10? How long must he stay to come to 8 dollars? To 32? To 40? To 80?

3. If a man has $12 a month, what will two months' work come to? What will 4? What will 5? What will 7? What will 9? What will 1 year? 2 years? $\frac{1}{12}$ of a month? $\frac{6}{12}$? $\frac{3}{12}$? $\frac{12}{12}$?

4. How many seconds in 2 hours? In 3? In 4? In 5? In 6?

5. If a watch click once in one second of time, how many times will it click in 2 minutes? In 3? In 4? In 5? In 6? How many times in $\frac{1}{60}$ of an hour? In $\frac{30}{60}$? In $\frac{1}{2}$? In $\frac{15}{60}$?

6. How many minutes are there in 1 hour? In $\frac{1}{2}$ an hour? In $\frac{1}{4}$? In $\frac{3}{4}$? In $\frac{45}{60}$? In $\frac{15}{60}$? In $1\frac{1}{2}$? In 2? In 4? In 10?

7. How many hours in 2 days? In 3? In $\frac{1}{2}$? In $\frac{1}{4}$? In $\frac{5}{24}$? In $\frac{6}{24}$? In $\frac{15}{24}$?

8. How many days in 2 weeks? In 4 weeks? In 6? In 8? In 12? In 20? In 30?

9. How many weeks in 2 months? In 3? In 4? In 5? In 6? In 7? In 8? In 9? In 10? In 20? In 50? In 100?

10. How many months in 8 weeks? In 12? In 20? In 40?

11. How many calendar months in 2 years? In 4? In 7? In 12?

12. How many years are there in 24 calendar months? In 48? In 84? In 144?

13. If a man have 1 dollar for $\frac{1}{12}$ of a year, what will $\frac{2}{12}$ come to? What will 1 year come to? How long could you have him for 6 dollars? For 12? For 24? For 36? For 60? For 144?

14. If work be a dollar a day, what would a year's wages come to? What would $\frac{2}{365}$ of a year? What $\frac{3}{365}$? What $\frac{7}{365}$? What $\frac{65}{365}$? What $\frac{365}{365}$?

Q. How many days are there in each month?

A. Thirty days hath September,
April, June, and November;
February hath twenty-eight,
And thirty-one the others rate.

Q. How many days are there in January? In February? In March? In April? In May? In June? In July? In August? In September? In October? In November? In December?

Q. How many days has February in Bissextile, or Leap Year? *A.* 29.

Q. When the year of our Lord can be divided by 4 without a remainder, what is the year called? *A.* Bissextile, or Leap Year.

Q. What year was 1824? *A.* Bissextile.

Q. Why? *A.* Because 1824 can be divided by 4 without a remainder?

Q. When will the next Leap Year be?

Q. How many days are there in Bissextile or Leap Year? *A.* 366.

Q. What are the denominations of this measure? *A.* Seconds, minutes, hours, days, years, centuries; also weeks and months.

Exercises for the Slate.

37. How many hours in 120 minutes? In 960360 minutes? *A.* 16008 hours.

38. How many days in 4 years? ($4 \times 365\frac{1}{4}$) In 44 years? *A.* 17532 days.

39. Reduce 2 weeks to days; 318 weeks to days. *A.* 2240 days.

CIRCULAR MOTION.

Q. What is meant by Circular Motion? *A.* The motion of *the earth and* other planets round the sun.

Q. What is it used for? *A.* For reckoning latitude and longitude?

Repeat the

TABLE.

60 Seconds, (″) . . make . . 1 Minute, marked ′.
60 Minutes . . . . 1 Degree, ... °.
30 Degrees . . . . 1 Sign, ... S.
12 Signs, or 360 Degrees, the whole Great Circle of the Zodiac.

1. How many seconds are in 3 minutes? In 6? In 10?
2. How many minutes are in 2 degrees? In 5? In 6? In 7? In 9?
3. How many degrees in 2 signs? In 3? In 4? In 6?
4. How many degrees does the sun travel over in one day, or 24 hours? *A.* 360.

Q. How many degrees does the sun pass over in one day, not reckoning the night, when the days and nights are of an equal length? *A.* 180.

Q. How many degrees in one hour? *A.* 15.

Q. Why? *A.* Because the sun travels over the circumference of the earth in 24 hours, he must travel over 15 degrees in one hour, for 15 times 24 are 360.

5. How many degrees in 2 hours? In 4? In 6? In 10? In 20?
6. What is the difference of time, then, between London, (through which the meridian runs, and from which longitude is generally computed,) and 15 degrees east of London? *A.* One hour.

Q. Why? *A.* Because the sun travels 15 degrees in one hour.

7. What is the difference of time between London and 30 degrees west of London? What 60 degrees? What 75 degrees?
8. The difference in distance between Washington in the District of Columbia, and Missouriopolis in Missouri, is 15 degrees; what is the difference of time between those places?

Q. What are the denominations of Circular Motion? *A.* Seconds, minutes, degrees, signs, and circles.

Exercises for the Slate.

40. How many degrees in 2 signs? In 3602 signs? *A.* 108120 degrees.

8*

41. In 120 minutes how many degrees? In 192060 minutes? *A.* 3202 degrees.

TABLES OF PARTICULARS.

Q. How many single things make a dozen? . . *A.* 12.
Q. How many dozen a gross? . . . . *A.* 12.
Q. How many dozen a great gross? . . . . . *A.* 144.
Q. How many single things a score? . . . . *A.* 20.
Q. How many score a hundred? . . . . *A.* 5.
Q. How many sheets make a quire of paper? *A.* 24.
Q. How many quires make a ream? . . . . . *A.* 20.

1. How many single things are in 2 dozen? In 3? In 5? In 12?
2. What will 2 dozen of pearl buttons come to, at 2 cents apiece? What will 3 dozen? What 4?
3. What will a great gross of buttons come to, at $1 a dozen?
4. What will 2 quires of paper come to, if you give 1 cent for a sheet? What will 3 quires? What will 4 quires?
5. If you give 20 cents for 1 quire of paper, how many cents must you pay for 1 ream? For 2 reams?

Q. How many pounds in a barrel of pork? *A.* 200.
Q. How many pounds in a barrel of beef? *A.* 200.

6. What will a barrel of pork come to, at 4 cents a pound?
7. What will 2 barrels of beef come to at 4 cents a pound?

BOOKS

Q. When a sheet is folded into 2 leaves, what is it called? *A.* Folio.
Q. When folded into 4 leaves, what is it called? *A.* Quarto, or 4to.
Q. When folded into 8 leaves, what is it called? *A.* Octavo, or 8vo.
Q. When folded into 12 leaves, what is it called? *A.* Duodecimo, or 12mo.
Q. When folded into 18 leaves, what is it called? *A.* 18mo.

STERLING MONEY.

Repeat the

TABLE.

4 Farthings, (qrs.) make	. .	1 Penny, marked d.
12 Pence	. . . .	1 Shilling, ... s.
20 Shillings	. . . .	1 Pound, ... £.

1. How many farthings are there in 2 pence? In 3? In 4? In 6? In 8? In 10? In 12? In 20?

2. How many pence are there in 2 shillings? In 3? In 5? In 7?

3. How many shillings are there in 2 pounds? In 3? In 6?

4. If 1 pair of gloves is worth 8 pence, how many pence are 2 pair worth? 3 pair? 4 pair? 5 pair? 6 pair? 10 pair?

5. When a bushel of wheat cost 10 shillings, what will 2 bushels cost? What will 4? What will 6? What will 9? What will 12?

6. If 1 cart cost 1 pound, how many shillings will buy 2? How many 3? How many 4? How many 5? How many 9? How many 12?

7. How many farthings will buy 2 inkstands, if they cost a penny apiece? How many will buy 3? How many will buy 4?

Note.—The characters used for English money are £. for *Libra*, the Latin for pounds; *s.* for *Solidi*, the Latin for shillings; *d.* for *Denarii*, the Latin for pence; and *qrs.* for *Quadrantes*, the Latin for farthings.

A Pound Sterling is equal to \$ 4,44$\frac{4}{9}$ *cents, Fed. Money.*
An English Guinea, " 4,66$\frac{2}{3}$ *cents,* " "
An English Shilling, " 22$\frac{2}{9}$ *cents,* " "
\$ 1, *Federal Money, is equal to* 4 *s.* 6 *d. Sterling.*

Q. How is 1 farthing sometimes written? *A.* $\frac{1}{4}$ d.
Q. How are 2 farthings sometimes written? *A.* $\frac{1}{2}$ d.
Q. How are 3 farthings sometimes written? *A.* $\frac{3}{4}$ d.

Exercises for the Slate.

42. How many shillings in 3 pounds? In 4200 pounds? *A.* 84060 shillings.

43. How many pence in 3 shillings? In 2600 shillings? *A.* 31236 pence.

44. How many farthings in 4 pence? In 2700 pence? *A.* 10816 farthings.

45. How many pence in 8 farthings? In 6200 farthings? *A.* 1552 pence.

PRACTICAL APPLICATION OF REDUCTION,

Involving the Rule, with Miscellaneous Examples.

1. At 20 cents a quart, what will 2 gallons of rum cost? What will 4 gallons? What will 5 gallons?

2. At $60 a hogshead, what will 2 pipes of wine come to? What will 4? What will 6?

3. How many minutes in 2 hours? In 4? In 6?

4. How many weeks in 14 days? In 21? In 49?

Q. What are such questions as these in? A. Reduction.

Q. What, then, is the changing numbers from one denomination to another called? A. Reduction.

Q. In 2 bushels of corn, how many pecks?

Q. Are 8 pecks of corn as much as 2 bushels?

Q. Is the value altered, then?

Q. Do you multiply or divide, to find how many furlongs there are in 2 miles?

Q. When, then, the reduction is performed by multiplication, what is it called? A. Reduction Descending.

Q. Do you multiply or divide, to find how many gallons there are in 8 quarts?

Q. When, then, the reduction is performed by division, what is it called? A. Reduction Ascending.

Q. From the preceding remarks, how many kinds of Reduction do there appear to be? and what are they?

A. Two;—Reduction Ascending and Descending.

RULE.

Q. What do you multiply 2 furlongs by, to bring them into rods?

I. *As a general rule, then, what do you multiply furlongs by? A.* By what makes a furlong.

II. *What do you multiply days, bushels, &c. by?* *A.* Bushels by what makes a bushel, days by what makes a day, &c.

Q. What do you divide by to bring 8 gills into pints?

III. *As a general rule, then, what must you divide gills, minutes, quarters, &c. by?* *A.* Gills by gills, minutes by minutes, quarters by quarters, &c.

Q. Why do you divide thus? *A.* Because 4 gills are equal to 1 pint, 60 minutes to 1 hour, &c.

Q. What do you multiply pounds (money) by? *A.* Shillings.

Q. What do you multiply shillings by? *A.* Pence.

Q. What do you multiply days by? *A.* Hours.

Q. What do you multiply by, to bring 25 pounds into farthings? *A.* By 20, 12, and 4.

Q. How do you bring 40 cwt. into tons? 1 cwt. into drams? ton into drams? Drams into tons? Drams into cwt.? 1 lb. into grains? Grains into pounds? Ells Flemish into nails? Nails into ells Flemish? Quarters into ells English? Ells English into quarters? 5 bushels into pints? 200 pints into bushels? 360 degrees into inches? Feet into furlongs? 5 weeks into seconds? Seconds into weeks? Years into seconds? Seconds into years? How do you tell what 1 tun of wine will cost, at 6 cents a gill?

Exercises for the Slate.

46. At 6 cents a pound, what will 2 qrs. 8 lbs. of sugar cost?

OPERATION.	
qrs. lbs. 2 8 28 — 56 lbs. 8 lbs. — 64 lbs. 6 cts. — $ 3,84 *Ans.*	*Why do you multiply the 2 qrs. by 28 lbs.?* *A.* Because, since it takes 28 pounds to make 1 qr., there will of course be 28 times as many pounds as quarters; that is, 28 times the quarters. *What do you do with the 8 lbs.?* *A.* I add it to 56 lbs., making 64 lbs. *Why do you multiply the 64 lbs. by 6 cents?* *A.* Because every pound of sugar cost 6 cents; that is, 6 times the pounds.

Proof of the foregoing Example.

47. How many quarters of sugar can you buy for 384 cents, at 6 cents a pound?

OPERATION.

6) 384

28) 64 (2 qrs.
56

8 lbs. *Ans.* 2 qrs. 8 lbs.

Why do you divide 384 cents by 6 cents? A. Because, since there are 6 times as many cents as pounds, as often as 6 is contained in 384, so many pounds there will be.

Why do you divide 64 pounds by 28 pounds? A. Because every 28 pounds make 1 quarter.

What, then, appears to be the method of proof? A. Reverse the operation; that is, make the divisors in the operation the multipliers in the proof, and the multipliers the divisors.

More Exercises for the Slate.

48. At 5 cents a gill, what will 8 pints of rum cost? *A.* $1,60.
49. How many gills in 20 pints? In 40 pints? *A.* 240 gills.
50. How many pints in 80 gills? In 120? *A.* 50 pints.
51. A merchant sold 5 hhds. of brandy at 3 dollars a gallon; how much did it amount to? *A.* $945.
52. How many hogsheads in 126 gallons? In 945? *A.* 17.
53. At 6 cents a quart, what will 1 hhd. of molasses come to? *A.* $15,12.
54. How many quarts are there in 3 hhds.? In 100? *A.* 25956.
55. How many hogsheads are there in 756 qts.? In 252 qts.? *A.* 4.
56. Sold 1 tun of wine at 5 cents a gill; what did I get for it? *A.* $403,20.
57. How many gills in 10 tuns? *A.* 80640.
58. At 8 cents a pound, what will 3 cwt. 2 qrs. of raisins cost? *A.* $31,36.
59. How many pounds, in 15 cwt. 1 qr. *A.* 1708.
60. At 8 pence a peck, how many pence will 3 bushels of salt cost? *A.* 96.
61. How many pence are there in 20 shillings? *A.* 240.
62. How many pence in 5£? In 10? *A.* 3600.
63. How many pounds in 480 d.? In 1440 d.? *A.* 8.
64. How many farthings in 4£? *A.* 3840.
65. At 6 pence per lb., what will 2 qrs. of rice cost? *A.* 336 pence.
66. How many qrs. of rice, at 6 pence per lb., may be bought for 336 pence? *A.* 2 qrs.

67. At 6 cents a gill, what will 1 tun of wine cost? *A.* $483,84.

68. How many tuns of wine, at 6 cents a gill, may be bought for $483,84? *A.* 1 tun.

69. At 9 pence per quart, what will 16 gals. 2 qts. of molasses come to in pence? *A.* 594 pence.

70. How many gallons of molasses, at 9 pence per quart, may be bought for 594 pence? *A.* 16 gals. 2 qts.

71. What is the value of a silver cup, weighing 10 oz. 5 pwts. 18 grs., at 5 mills per grain? *A.* $24,69.

72. At 5 mills per grain, what will be the weight of a silver cup that $24,69 will purchase? *A.* 10 oz. 5 pwts. 18 grs.

73. At 12 cents a pound, what cost 5 cwt. 2 qrs. 18 lbs. of sugar? *A.* $76,08 cts.

74. How many hundred weight of sugar, at 12 cents per lb., may be bought for $76,08? *A.* 5 cwt. 2 qrs. 18 lbs.

75. At 9 pence an hour, what will 2 yrs. 6 mo. 3 weeks, 6 da. 12 h. labour come to in pence? *A.* 187380.

76. How many years' work, at 9 pence per hour, may be obtained for 187380 pence? *A.* 2 yrs. 6 mo. 3 weeks, 6 da. 12 h.

77. At 20 cents a nail, what is the price of 4 yds. of cloth? *A.* $12,80.

78. At 320 cents a yard, what will 64 nails of cloth cost? *A.* $12,80.

79. At 1£ 2 s. 6½ d. per yard, what will 20 yds. of broadcloth cost in farthings? *A.* 21640.

80. At 6 cents a pint, what will 20 bu. 0 pks. 3 qts. 1 pt. of flax-seed cost? *A.* $77,22.

81. How many shillings, at 2 farthings a gill, will 5 T. 1 p. 1 hhd. 2 gals. 2 qts. 1 pt. 3 gills, cost? *A.* 1935 s. 7 d. 2 qrs.

82. At 2 shillings a quarter, how many dollars will 8 yds. 1 qr. of broadcloth cost? *A.* 11.

83. At 2 pence a gill, how many dollars will buy 50 gals. 2 qts. 1 pt. 2 gills of ale? *A.* $45,055.

84. In 3600 dollars, how many farthings? *A.* 1036800.

85. In 1036800 farthings, how many dollars? *A.* 3600.

86. In 25 guineas, of 28 s. each, how many pence? *A.* 8400.

87. In 8400 pence, how many guineas? *A.* 25.

88. In 15 lbs. how many ounces, drams, scruples, and grains? *A.* 180 ounces, 1440 drams, 4320 scruples, 86400 grains.

89. In 86400 grains, how many scruples, drams, and ounces? *A.* 4320 scruples, 1440 drams, 180 ounces.

90. In 256 miles, 30 rods, how many rods? *A.* 81950.

91. In 81950 rods, how many miles? *A.* 256 m. 30 rods.

92. In 15 lea. 1 m. 6 fur. 28 rods, 4 yds. how many barley corns? *A.* 8903304.

93. In 8903304 barley corns, how many leagues? *A.* 1 1 m. 6 fur. 28 rods, 4 yds.

94. In 360 ells English, 4 qrs. 2 n. how many nails 1804 qrs. 7218 na.

95. In 7218 na. how many ells English? *A.* 360 E. E. 2 na.

96. In 3000 E. Flemish, how many nails? *A.* 36000.

97. In 36000 nails, how many E. Flemish? *A.* 3000.

98. In 500 acres, how many roods? *A.* 2000.

99. In 2000 roods, how many acres? *A.* 500.

100. In 15 tons of hewn timber, how many solid in *A.* 1296000.

101. In 1296000 solid inches of hewn timber, how tons? *A.* 15.

102. In 20 cords of wood, how many solid inches 4423680.

103. In 4423680 solid inches, how many cords of w *A.* 20.

104. In 500 bushels, 3 pecks, 7 qts. 1 pt. how many p *A.* 32063.

105. In 32063 pints, how many bushels? *A.* 500 bu. 7 qts. 1 pt.

106. In 20 tuns of wine, how many gills? *A.* 161280.

107. In 161280 gills of wine, how many tuns? *A.* 20.

108. In 366 years, 300 days, 20 hours, 50 minutes, a seconds, how many seconds? *A.* 11378955037.

109. In 11378955037 seconds, how many years? *A.* 36 300 days, 20 hours, 50 minutes, and 37 seconds.

110. In 8 signs of the Zodiac, how many seconds 864000.

111. In 20 purses, each containing 21 guineas, how shillings, pence, and farthings? *A.* 11760 s. 141120 d. 5 far.

112. In 5 ingots of silver, each 5 lbs. 6 oz. 20 grs., how grains? *A.* 158500.

113. In 239130 grains, how many ingots, each 6 lbs. 15 grs.? *A.* 6.

114. A lady sent a tankard to a silversmith, that we 5 lbs. 3 oz. and ordered it to be made into spoons, each to 2 oz. 2 pwts.; how many spoons did it make? *A.* 30.

115. A goldsmith, having 15 ingots of silver, each wei 2 lbs. 7 oz. 3 pwts. which he wishes to make into bowls of 8 oz., tankards of 1 lb. 10 oz., salts of 11 oz., and spoons o 15 pwts., and of each an equal number, how many will th of each sort?

Bring 2 lbs. 8 oz., 1 lb. 10 oz., 11 oz., and 1 oz. 15 into pennyweights; add them up for a divisor; then

2 lbs. 7 oz. 3 pwts. into pennyweights, multiply by 15 for a dividend, divide, and the quotient will be the answer. *A.* 7.

116. In 26880 lbs. of sugar, how many hhds., each 12 cwt.? *A.* 20.

117. How many barley corns will reach round the globe, it being 360 degrees? *A.* 4755801600.

118. In running 300 miles, how many times will a wheel 9 feet 2 inches in circumference, turn round? *A.* 172800.

119. In 172800 turns of a wheel measuring 9 feet 2 inches, how many miles? *A.* 300.

120. How many times will a wheel, which is 15 feet 9 inches in circumference, turn round in going from Providence to Norwich, it being 45 miles? *A.* 15085+.

121. A farmer rents a plantation of 400 acres, of which no more than 200 are to be tilled; how many poles are there in the remainder? *A.* 32000.

122. In a lunar month, of 27 days, 7 hours, 43 minutes, 5 seconds, how many seconds? *A.* 2360585.

123. How many seconds is it from the birth of our Saviour to Christmas, 1828, allowing the year to contain 365¼ days, or 365 days, 6 hours? *A.* 57687292800.

124. When a person is 21 years old, how many seconds old is he? *A.* 662709600.

125. It is supposed the wars of Bonaparte, in 20 years, caused the death of 2000000 of persons; how many was this per hour, allowing the year to contain 365 days 6 hours? *A.* $11\frac{71480}{175320}$.

COMPOUND ADDITION.

¶ XXX. 1. William bought an arithmetic for 2 s. 6 d., and an inkstand for 6 d.; how many shillings did both cost?

2. Harry purchased a vest; the cloth and making cost 5 s., the buttons 9 d., and the thread 3 d.; how much did the vest cost?

3. William, Harry, and Thomas gathered some nuts; and when they measured them, it was found that William had 2 qts. and 1 pt., Harry 3 qts. 1 pt., and Thomas 2 qts.; how many pecks did they gather in all?

4. How many pecks are 3 qts. + 1 qt. + 4 qts.?
5. How many yards are 2 qrs. + 3 qrs. + 3 qrs.?
6. How many gallons are 1 qt. + 2 qts. + 1 qt.?
7. How many pence are 1 qr. + 3 qrs. + 1 qr.?
8. How many shillings are 3 d. + 8 d. + 1 d.?

9. How many pounds are 10 s. + 18 s. + 12 s.?
10. How many hours are 50 m. + 20 m. + 10 m.?
11. How many feet are 4 in. + 10 in. + 11 in.?
12. How many minutes in 45 sec. + 15 sec. + 10 sec.?
13. How many pounds in 8 oz. + 12 oz. + 12 oz.?
14. How many bushels in 1 pk. + 3 pks. + 2 pks.?
15. Sold a Virgil, that cost me 12 s. 6 d., so as to gain 1 s. 6 d. how much did I get for it?

Q. What is this, which you have now been doing, called?
A. Compound Addition.
Q. Why do you call it Compound? why not Simple Addition?
A. Because there are more denominations than one.
Q. What do you mean by more denominations than one?
A. Shillings, pence, &c. in one sum; pecks, quarts, pints, &c in another sum.
Q. What, then, is the collecting numbers of different denominations into one sum called? *A.* Compound Addition.

Operation by Slate illustrated.

1. A man bought a cart for 6£ 12 s. 3 d., a load of hay for 3£ 9 s. 7 d., and a cow for 4£ 4 s. 1 d.; what did he pay for the whole?

OPERATION.

	20	12
£	s.	d.
6	12	3
3	9	7
4	4	1
Ans. £14	5	11

How do you write the numbers down? *A.* Pounds under pounds, shillings under shillings, &c.

How do you get the 11 *d. in the answer?* *A.* I find, by adding up the column of pence, that it makes 11 d., which I write under the column of pence.

How do you get the 5 *shillings?* *A.* Adding up the column of shillings, I find it makes 25 s. = 1£ 5 s. (for 20 s. in 25, 1 time, and 5 over), writing the 5 s. under the column of shillings.

What is to be done with the 1£? *A.* I must, of course, add pounds to pounds, and, to do this, I join, or carry it to the next column, which is pounds.

How do you get the 14£? *A.* Adding up the column of pounds makes 13 pounds, and 1£ (to carry) makes 14£.

From these illustrations we derive the following

RULE.

I. *How do you place the numbers to be added?* *A.* Pounds under pounds, shillings under shillings, drams under drams, &c.

II. *At which hand do you begin to add?* *A.* At the right.

III. *How do you add up the first column?* *A.* As in Simple Addition.

IV. *What do you divide the amount by?* *A.* By as many of this denomination as make one of the next higher, as in Reduction.

V. *What do you do with the remainder?* *A.* Write it underneath.

VI. *What do you carry to the next column?* *A.* The quotient.

VII. *How long do you proceed in this way?* *A.* Till I come to the last column.

VIII. *How do you proceed with this?* *A.* Add it up, and set the whole amount down, as in Simple Addition.

Proof.—Q. *What is the proof?* *A.* The same as in Simple Addition.

More Exercises for the Slate.

2. Bought a cart for 2 £ 15 s., a plough for 18 s.; how much did both cost? *A.* 3 £ 13 s.

3. Bought a coat for 5£ 6 s., a watch for 1£ 19 s.; how much did they come to? *A.* 7£ 5 s.

4. A man bought one load of hay for 6£ 3 s. and another for 7£ 15 s. 6 d.; how much did he give for both? *A.* 13£ 18 s. 6 d.

5. Sold an ox for 10£ 15 s. 6 d., a cow for 6£ 19 s. 11 d., a horse for 12£ 6 s. 4 d.; how much money did I receive? *A.* 30£ 1 s. 9 d.

6. Bought of a grocer 3 gals. 2 qts. of rum, 5 gals. 3 qts. of gin, and 4 gals. 1 qt. of molasses; how many gallons did I buy in all? *A.* 13 gals. 2 qts.

7. Sold 4 hhds. of molasses, the first of which contained 42 gals. 2 qts. 1 pt., the second 65 gals. 0 qt. 1 pt., the third 50 gals. 3 qts., and the fourth 55 gals. 1 qt. 1 pt.; how much was sold in all? *A.* 213 gals. 3 qts. 1 pt.

8. A grocer sold 4 hhds. of sugar, weighing as follows, the first 7 cwt. 1 qr. 14 lbs., the second 5 cwt. 2 qrs. 10 lbs., the

third 9 cwt. 1 qr. 15 lbs., the fourth 7 cwt. 1 qr. 10 lbs.; what did the whole weigh? *A.* 29 cwt. 2 qrs. 21 lbs.

9. A merchant bought 4 pieces of cloth; the first containing 20 yds. 3 qrs. 1 na., the second 15 yds. 3 qrs. 1 na., the third 26 yds., and the fourth 10 yds. 1 qr. 3 na.; how many yards did he buy in all? *A.* 73 yds. 0 qr. 1 na.

10. A man bought 3 bu. 3 pks. of wheat at one time, 6 bu. at another time, 7 bu. 2 pks. 7 qts. at a third, and 4 bu. 1 pk. 6 qts. at a fourth; how many bushels did he buy in all? *A.* 21 bu. 3 pks. 5 qts.

11. A man bought two loads of hay, one weighing 19 cwt. 1 qr., and the other 18 cwt. 2 qrs.; how much did both weigh? *A.* 37 cwt. 3 qrs.

12. A man travelled in one day 27 miles, 3 fur., in another day 30 m. 2 fur. 25 rods; how far did he travel in all? *A.* 57 m. 5 fur. 25 rods.

13. A merchant bought 3 bales of cotton; the first contained 4 cwt. 3 qrs. 18 lbs., the second 3 cwt. 1 qr. 5 lbs., and the third 5 cwt. 0 qr. 24 lbs.; what was the weight of the whole? *A.* 13 cwt. 1 qr. 19 lbs.

14. A man has 3 farms; the first containing 150 acres, 2 roods, 25 rods; the second, 200 acres, 1 rood, 15 rods; and the third, 100 acres, 1 rood, 10 rods: how many acres has he in all? *A.* 451 acres, 1 rood, 10 rods.

15. William resided in Providence, his native place, till he was 15 yrs. 6 m. 4 days old; he then went to Boston, where he resided 7 yrs. 2 m. 2 da.; from Boston he emigrated to Salem, where he remained 4 yrs. 3 da.; from Salem he went to Portsmouth, and resided there two years precisely: now, how much time did he spend in these places in all? *A.* 28 yrs. 8 m. 1 w. 2 d.

16. A man brings to market 3 loads of wood; the first containing 1 cord, 64 feet, 864 in.; the second, 2 cords, 63 ft. 64 in.; and the third, 1 cord, 60 ft. 931 in.; how much did he bring in all? *A.* 5 cords, 60 ft. 131 inches.

17. A goldsmith bought 4 ingots of silver, the first of which weighed 8 lbs. 2 oz. 12 pwts., the second 5 lbs. 4 oz. 5 pwts., the third 6 lbs. 10 oz. 11 pwts., and the fourth 6 lbs. 11 oz. 15 pwts.; what was the weight of the whole? *A.* 27 lbs. 5 oz. 3 pwts.

18. James is 10 yrs. 2 mo. 3 wks. 4 da. old, Thomas is 11 yrs. 11 mo. 5 da. old, Rufus is 9 yrs. 10 mo. old, Harry is 14 yrs. old; what is the sum of all their ages? *A.* 46 yrs. 2 da.

By multiplying the answers to the following sums by 2, the true answers may be obtained.

Note.—It will be well for the learner, not only in this, but in all rules, to prove his results, when practicable.

19. Add together 17£ 13 s. 11 d. 1 qr., 13£ 10 s. 2 d. 2 qrs.,

10£ 17 s. 3 d. 1 qr., 7£ 7 s. 6 d. 2 qrs., 2£ 2 s. 3d. 2 qrs. 18£ 17 s. 10 d. 2 qrs. *A.* 35£ 4 s. 6 d. 3 qrs. × 2.

20. Add together 46£ 16 s. 5 d. 1 qr., 2£ 8 s. 9 d. 2 qrs., 58£ 16 s. 10 d. 1 qr., 316£ 15 s. 8 d. 2 qrs., 651£ 18 s. 9 d. 2 qrs., 405£ 16 s. 5 d. *A.* 741£ 6 s. 6 d. × 2.

21. Add together 30£ 10 s. 3 d. 2 qrs., 14£ 9 s. 8 d. 0 qr. 1£ 0 s. 1 d. 2 qrs., 2£ 8 s. 7 d. 2 qrs., 42£ 9 s. 6 d. 2 qrs., 28£ 5 s 4 d. 2 qrs. *A.* 59£ 11 s. 9 d. 3 qrs. × 2.

22. Add together 15 lbs. 10 oz. 18 pwts. 22 grs., 3 lbs. 3 oz. 15 pwts. 20 grs., 7 lbs. 7 oz. 18 pwts. 13 grs., 5 lbs. 8 oz. 13 pwts. 16 grs., 3 lbs. 6 oz. 9 pwts. 6 grs., 6 oz. 10 pwts. 11 grs. *A.* 18 lbs. 4 oz. 3 pwts. 8 grs. × 2.

23. Add together 2 cwt. 3 qrs. 27 lbs., 1 cwt. 2 qrs. 16 lbs., 3 cwt. 1 qr. 25 lbs., 5 cwt. 2 qrs. 12 lbs., 2 cwt. 2 qrs. 14 lbs., 5 cwt. 1 qr. 15 lbs. *A.* 10 cwt. 3 qrs. 12 lbs. 8 oz. × 2.

24. Add together 70 yds. 2 qrs. 1 na., 12 yds. 1 qr. 1 na., 9 yds. 0 qr. 1 na., 40 yds. 2 qrs. 1 na., 56 yds. 1 qr. 1 na., 48 yds. 1 qr. 1 na. *A.* 118 yds. 2 qrs. 1 na. × 2.

25. Add together 1 pk. 6 qts. 1 pt., 2 pks. 5 qts., 1 pk. 4 qts., 1 pk. 3 qts. 1 pt., 2 pks. 5 qts., 3 pks. 4 qts. 0 pt. *A.* 6 pks. 6 qts. 0 pt. × 2.

26. Add together 38 gals. 2 qts. 1 pt. 2 gi., 16 gals. 1 qt. 3 gi., 20 gals. 2 qts. 1 pt. 1 gi., 18 gals. 1 qt. 1 pt., 7 gals. 1 qt. 2 gi., 30 gals. 2 qts. 1 pt. *A.* 66 gals. × 2.

27. Add together 80 lea. 1 m. 5 fur. 30 po., 50 lea. 2 m. 6 fur. 20 po., 40 lea. 1 m. 7 fur. 15 po., 30 lea. 2 m. 4 fur. 25 po., 70 lea. 1 m. 3 fur. 10 po., 60 lea. 2 m. 2 fur. 4 po. *A.* 167 lea. 0 m. 2 fur. 32 po. × 2.

28. Add together 367 acres, 2 roods, 30 rods; 815 acres, 1 rood, 16 rods; 40 acres, 2 roods, 20 rods; 60 acres, 2 roods, 30 rods. *A.* 642 acres, 0 roods, 28 rods. × 2.

29. *Solid measure.*—Add together 12 feet, 1335 inches; 15 feet, 1615 inches; 2 feet, 755 inches; 13 feet, 1283 inches. *A.* 22 feet, 766 inches × 2.

30. Add together 20 yrs. 363 da. 20 h. 50 m. 30 sec., 20 yrs. 40 da. 10 h. 30 m. 20 sec., 12 yrs. 110 da. 13 h. 16 sec., 13 yrs. 8 da. 10 h. 20 m. 14. sec., 7 yrs. 20 da. 8 h. 10 m. 12 sec. *A.* 36 yrs. 271 da. 19 h. 25 m. 46 sec. × 2.

31. Add together 11S. 29°. 16′. 59″., 20° 45′. 11″., 8S. 3°. 10′. 50″., 3 S. 10°. 6′. 10″. *A.* 12S. 1°. 39′. 35″. × 2.

COMPOUND SUBTRACTION.

¶ **XXXI.** 1. William had 2 qts. of walnuts, and gave *away 1 pt.*; how many had he left?

2. James, owing Rufus 1 s. 6 d., paid him 6 d.; how much did he then owe?

3. Thomas bought a knife for 9 d., and sold it for 1 s. 6 d.; how much did he make by the trade?

4. Harry and Rufus purchased 3 qts. of walnuts; Harry paid so much towards them, that he is entitled to 2 qts. 1 pt.; now, what is Rufus' part?

5. A servant, returning with a two gallon jug of molasses, perceived that it had leaked out some considerable, and, wishing to know how much, by emptying it into a 6 quart and 1 pint measure, found it exactly filled the measure; how much had leaked out?

6. From 1 gallon take 3 quarts.
7. From 8 gills take 2 pints.
8. From 1 ounce take 12 drams.
9. From 2 quarters take 3 nails.
10. From 1 pound take 11 shillings.
11. From 2 shillings take 10 pence.
12. From 2 quarters take 20 pounds.
13. From 3 weeks take 7 days.

Q. *What is this, which you have now been doing, called?* A. Compound Subtraction.

Q. *What, then, is the taking one number from another of different denominations called?* A. Compound Subtraction.

Q. *Wherein does Compound differ from Simple Subtraction?* A. Simple consists of only one denomination; Compound of more than one.

Operation by Slate illustrated.

1. A merchant bought a piece of cloth containing 10 yds. 2 qrs. 3 na., and sold 7 yds. 3 qrs. 2 na.; how much had he left?

OPERATION.

	yds.	qrs.	na.
		4	4
	10	2	3
	7	3	2
A.	2	3	1

In this example, how are the numbers written down? A. The less under the greater, with nails under nails, quarters under quarters, &c., as in Compound Addition.

How do you get the 1 na. in the answer? A. I begin with nails, the least denomination, and say, 2 na. from 3 na. leaves 1 na.

How do you proceed to get the 3 qrs.? A. I cannot take 3 qrs. from 2 qrs., but I can borrow, as in Simple Subtraction, 1 yd. = 4 qrs. from the yards; then say, 4 qrs. joined or added to the 2 qrs. (the top figure) makes 6 qrs., from which taking 3 qrs. leaves 3 qrs.

But suppose that, instead of adding first, we subtract first, how would you proceed? *A.* Taking 3 qrs. from 4 qrs. (borrowed) leaves 1 qr., and 2 qrs. makes 3 qrs.

How do you get the 2 yds.? *A.* I must carry 1 yd. (for the yard which I borrowed) to 7 yds., making 8 yds., which, subtracted from 10 yds., leaves 2 yds.

From this example we derive the following

RULE.

I. *How do you write the numbers down?* *A.* The less under the greater, placing each denomination as in Compound Addition.

II. *With which denomination do you begin to subtract?* *A.* The least denomination.

III. *How do you subtract each denomination?* *A.* From the denomination above it, as in Simple Subtraction.

IV. *If the lower number in any denomination be greater than the upper, how do you proceed?* *A.* Borrow as many units as make *one* in the next higher denomination, from which subtract the lower number.

V. *What must the remainder be added to?* *A.* The upper number.

VI. *How many do you carry in such cases?* *A.* One.

VII. *How do you subtract the last denomination?* *A.* As in Simple Subtraction.

PROOF. *How do you prove the operation.* *A.* By adding the remainder and subtrahend together, as in Simple Subtraction, the amount of which must be equal to the minuend.

More Exercises for the Slate.

2. If, from a piece of cloth containing 10 yds. 2 qrs., you cut off 2 yds. 2 qrs., how much will there be left? *A.* 8 yds.

3. A bought of B a bushel of barley for 8 s. 6 d.; he gave B 1 bu. of rye, worth 4 s. 3 d., and paid the rest in money; how much did he pay? *A.* 4 s. 3 d.

4. A bought of B a bale of cotton for 20£ 4 s., and B bought of A 4 tierces of rice for 15£ 18 s.; A paid B the rest in money; how much did he pay? *A.* 4£ 6 s.

5. A man bought a wagon for 6£ 10 s., and sold it for 12£ 18 s.; how much did he make by the trade? *A.* 6£ 8 s.

6. A man bought one load of hay for 4£ 10 s., and another for 5£ 15 s.; how much more did he give for one than the other? *A.* 1£ 5 s.

7. A man bought two loads of hay, one weighing 18 cwt. 3 qrs. 25 lbs., and the other 17 cwt. 0 qr. 26 lbs; how much did one weigh more than the other? *A.* 1 cwt. 2 qrs. 27 lbs.

8. A merchant bought a piece of broadcloth, containing 40 yds., from which he sold 36 yds. 1 qr. 2 na.; how much did he have left? *A.* 3 yds. 2 qrs. 2 na.

9. A grocer bought a hhd. of rum, containing 65 gals., and by accident 2 gals. 2 qts. 1 pt. leaked out; how many gallons did he have left? *A.* 62 gals. 1 qt. 1 pt.

10. A merchant bought a quantity of corn, weighing 20 cwt. 2 qrs. 15 lbs., of which he sold 10 cwt. 3 qrs. 12 lbs.; how much had he left? *A.* 9 cwt. 3 qrs. 3 lbs.

11. A grocer retailed 10 gals. 3 qts. 1 pt. 1 gi. of rum from a hhd. containing 54 gals. 2 qts. 1 pt. 2 gi., how much had he left? *A.* 43 gals. 3 qts. 0 pt. 1 gi.

12. If, from a box of butter, containing 20 lbs., there be sold 10 lbs. 8 oz., how much will there be left? *A.* 9 lbs. 8 oz.

13. If, from a field, containing 40 acres, 2 roods, 20 poles, there be taken 19 acres, 3 roods, 30 poles, how much will there be left? *A.* 20 acres, 2 roods, 30 poles.

14. William engaged himself in a store for 3 yrs.; after having stayed 2 yrs. 2 mo. 2 w. 2 da., how much longer had he to stay? *A.* 9 mo. 1 w. 5 da.

15. A farmer, having raised 40 bu. of corn, kept 23 bu. 2 pks. for his own use, and sold the rest; how much did he sell? *A.* 16 bu. 2 pks.

16. A farmer made in one year, from his orchard, 200 bbls. 14 gals. of cider, of which he sold precisely 118 bbls. 3 qts. 1 pt.; how much had he left for his own use? *A.* 82 bbls. 13 gals. 1 pt.

17. If, from a parcel of wood, containing 40 cords and 64 feet, there be sold 39 cords and 32 feet, how much will there be left? *A.* 1 cord 32 feet.

18. The distance from Providence to Norwich is 45 miles; now, when a man has travelled 30 m. 7 fur. 20 rods of the distance, how much farther has he to travel? *A.* 14 m. 20 rods.

19. From 14£ 15 s. 6 d. 2 qrs. take 12£ 15 s. 6 d. 3 qrs. *A.* 1£ 19 s. 11 d. 3 qrs.

20. From 1£ take 2 s. *A.* 18 s.
21. From 1£ take 2 d. *A.* 19 s. 10 d.
22. From 1£ take 2 qrs. *A.* 19 s. 11 d. 2 qrs.
23. From 1 lb. take 19 grs. *A.* 11 oz. 19 pwts. 5 grs.
24. From 1 ton take 10 oz. *A.* 19 cwt. 3 qrs. 27 lbs. 6 oz.
25. From 1 lb. take 15 grs. *A.* 11 oz. 19 pwts. 9 grs.
26. From 1 yd. take 2 qrs. *A.* 2 qrs.
27. From 1 bu. take 1 pt *A.* 3 pks. 7 qts. 1 pt.
28. From 1 yd. take 1 b.c. *A.* 2 ft. 11 in. 2 b.c.
29. From 1 yd. take 1 in. *A.* 2 ft. 11 in.
30. From 1 sq. yd. take 3 sq ft. *A.* 6 sq. ft.
31 From 1 ton r. timber take 50 cu in. *A.* 39 sol. ft. 1678 sol. in.

32. From 1 yr. take 12 h. *A.* 11 mo. 3 w. 6 da. 12 h.
33. From 12£ 2 qrs. take 6 d. *A.* 11£ 19 s. 6 d. 2 qrs.
34. From 10 cwt. 10 oz. take 5 drs. *A.* 10 cwt. 9 oz. 11 drs.
35. From 1 E. E. 2 qrs. take 3 na. *A.* 1 E. E. 1 qr. 1 na.
36. From 8 gals. 3 gills take 1 pt. *A.* 7 gals. 3 qts. 1 pt. 3 gi.
37. From 12 m. 15 rods take 3 fur. *A.* 11 m. 5 fur. 15 rods.
38. From 1 mo. 2 h. take 45 m. *A.* 1 mo. 1 h. 15 m.

COMPOUND MULTIPLICATION.

¶ XXXII. 1. If one knife cost 9 d., how many shillings will buy 2 knives? Will buy 4? Will buy 6? Will buy 8? Will buy 12?

2. William, having a basket that would hold 1 qt. 1 pt., filled it with nuts; how many qts. can be put in a basket that will hold twice as much? 3 times as much? 4 times as much?

3. At 1 s. 6 d. a bushel, how many shillings will 2 bushels of apples cost? Will 4 bu.? Will 6 bu.? Will 8 bu.?

4. Multiply 2 s. 6 d. by 2.

5. Multiply 2 pwts. 12 grs. by 2.

6. Multiply 2 bu. 4 qts. by 3.

7. Multiply 3 gals. 2 qts. by 2.

8. Multiply 20 m. 20 sec. by 3.

9. How many pence in 2 times 2 farthings? 4 × 2 farthings, or 2 qrs.? 4 × 3 qrs.? How many shillings in 2 times 6 d.? 2 × 12 d.? 4 × 6 d.? 5 × 12 d.? 8 × 3 d.? 4 × 7 d.? 3 × 5 d.? 8 × 12 d.?

10. How many pounds in 4 times 10 s.? 3 × 10 s.? 4 × 6 s.?

Operation by Slate illustrated.

1. A merchant bought 5 yards of cloth for 2£ 6 s. 1 d. 3 qrs. per yard; what did the whole cost?

OPERATION.

	20	12	4
£	s.	d.	qrs.
2	6	1	3
			5
Ans. £11	10	8	3

How do you get the 3 qrs. in the answer?

A. 5 times 3 qrs. are 15 qrs. = 3 d. 3 qrs., writing down the 3 qrs. and carrying 3 d. as in Compound Addition.

How do you get the 8 d.? *A.* 5 times 1 d. are 5 d., and 3 d. (to carry) makes 8 d.

How do you get the 10 s.? *A.* 5 times 6 s. are 30 s. = 1£. 10 s., writing down the 10 s. and carrying the 1£.

How do you get the 11£? *A.* 5 times 2£ are 10£, an (to carry) makes 11£.

From these illustrations we derive the followi

RULE.

I. *With which denomination do you begin to multiply* With the lowest.

II. *How do you multiply that, and each denomination* Separately, as in Simple Multiplication.

III. *How do you divide each product and carry? A.* . Compound Addition.

PROOF. *What is the proof? A.* As in Simple Multiplic

More Exercises for the Slate.

2. At 5 s. 6 d. a gallon, what will 2 gals. of rum cost 11 s.

3. At 2 s. 6 d. 1 qr. a quart, what will 2 qts. of brandy (5-0-2) What will 3 qts.? (7-6-3) What will 4 qts.? (10-1) will 5 qts.? (12-7-1) What will 6 qts. (15-1-2). *A.* 2£. 10

4. How much wine in 7 bottles, each containing 2 qts 2 gills? (4-3-0-2) How much in 8 bottles? (5-2) In 9? (6- In 10? (6-3-1) In 11? (7-2-0-2). *A.* 30 gals. 3 qts. 1 pt. 2

5. What is the weight of 3 doz. silver spoons, each doz. w ing 2 lbs. 6 oz. 12 pwts. 3 grs.? (7-7-16-9) What will 4 doz. w (10-2-8-12) What will 5 doz.? (12-9-0-15) What will 6 (15-3-12-18). *A.* 45 lbs. 10 oz. 18 pwts. 6 grs.

6. Bought 4 loads of hay, each load weighing 1 T. 10 qrs. 20 lbs. 5 oz. 15 drs.; what was the weight of the w (6-2-2-25-7-12) What would be the weight of 5 l (7-13-1-17-13-11) Of 11 loads? (16-17-2-0-1-5) Of 12 l (18-8-0-20-7-4). *A.* 49 T. 1 cwt. 3 qrs. 7 lbs. 14 oz.

7. At the rate of 36 lea. 2 m. 3 fur. a day, how far will sel sail in 6 days? (220-2-2) In 15 days? (551-2-5) In 10 (367-2-6) In 9 days? (331-0-3.) *A.* 1471 lea. 2 m.

8. In 8 bales of cloth, each bale containing 12 pieces piece 27 yds. 1 qr. 2 na., how many yards? *A.* 2628 yds.

COMPOUND DIVISION.

¶ XXXIII. 1. William had 2 qts. 1 pt. of walnuts, *he wished* to divide equally among his two little brc *how many* must he give each?

2. James bought 2 books for 2 s. 6 d.; how much did he pay apiece?

3. If you pay 1 s. 6 d. for 2 inkstands, how many shillings would that be apiece?

4. A man bought 4 lambs for 6 shillings; how many pence did he pay apiece?

5. William has 3 pks. 4 qts. of walnuts, which he wishes to put into 4 little baskets, each of which will hold 7 qts; will his baskets hold all his walnuts, or not? and inform me how you do it.

6. 3 men have 4 gals. 2 qts. of cider allowed them every day; how much is that apiece?

7. How many pence is $\frac{1}{3}$ of 1 s.? $\frac{1}{3}$ of 2 s.? $\frac{1}{3}$ of 3 s.? $\frac{1}{3}$ of 1 s. 6 d.? $\frac{1}{3}$ of 1 s. 3 d.? $\frac{1}{6}$ of 1 s. 6 d.? $\frac{1}{6}$ of 2 s. 6 d.? $\frac{1}{2}$ of 1 d.? $\frac{1}{2}$ of 1 d. 2 qrs.

Q. What is this, which you have now been doing, called? A. Compound Division.

Q. Wherein does it differ from Simple? A. Simple consists of only one denomination; Compound, of more than one.

Q. What, then, is the process called, by which we find how many times one number is contained in another of different denominations? A. Compound Division.

Operation by Slate illustrated.

1. A man bought 2 loads of hay for 15£ 3 s. 8 d.; how much was that a load?

OPERATION.

		20	12
	£	s.	d.
2)	15	3	8
Ans.	£7	11	10

How do you get the 7£ in the answer? A. I begin as in Short Division of whole numbers, and say, 2 is contained in 15£, 7 (£) times, and 1£ over, writing down the 7 times.

What do you do with the 1£ over? A. 1£ = 20 s., which I join or carry to the 3 s., making 23 s.

How do you proceed, then, to get the 11 s.? A. I say, 2 in 23, 11 times, and 1 s. over, writing down the 11 s. underneath.

How do you get the 10 d.? A. The 1 s. over being equal to 12 d., I join or carry it to 8 d., making 20 d.; then, 2 in 20, 10 times.

From these illustrations we derive the following

RULE.

I. *At which hand do you begin to divide, and how do you proceed? A.* With the highest denomination, and divide as in Simple Division.

II. *If you have a remainder, how do you proceed?* *A.* Find how many of the next lower denomination this remainder is equal to, which add to the next denomination; after which divide as in whole numbers.

Proof. *What is the proof?* *A.* The same as in Simple Division.

More Exercises for the Slate.

2. If 8 tons of hay cost 40£ 14 s. 8 d., what will 1 ton cost? *A.* 5£ 1 s. 10 d.
3. If 11 gals. of brandy cost 5£ 16 s. 5 d., what will 1 gallon cost? *A.* 10 s. 7 d.
4. If a man spend 60£ 13 s. 4 d. a week, how much is that a day? *A.* 8£ 13 s. 4 d.
5. If 1 cwt. of rice cost 2£ 6 s. 8 d., what will 1 lb. cost? *A.* 0£ 0 s. 5 d.
6. You have 31£ 9 s. 6 d. to be divided equally among 2 men; how much would it be apiece? (15-14-9) How much would it be apiece to be divided among 3? (10-9-10) Among 6? (5-4-11). *A.* 31£ 9 s. 6 d.
7. Divide 2 gals. 2 qts. by 4; (0-2-1) by 5; (0-2) by 10; (0-1) by 2; (1-1). *A.* 2 gals. 2 qts. 1 pt.
8. Divide 96 acres 2 roods, 16 rods, by 7; (13-3-8) by 8; (12-0-12) by 12; (8-0-8). *A.* 33 acres, 3 roods, 28 rods.

Questions to exercise the foregoing Rules.

1. What is the sum of the following numbers, viz. one, two thousand, thirty thousand, four millions, twenty thousand, nineteen, four hundred millions? *A.* 404052020.
2. Bought a coat for 15 dollars, a vest for 1 dollar 37½ cents, a pair of boots for 6 dollars 12½; what did the whole cost me? *A.* $22,50.
3. Bought a horse for $75, and sold him for 37½ cents less than he cost me; what did I get for him? *A.* $74,62,5.
4. What will 3200 yards of tape come to at 6¼ cents, or $\frac{1}{16}$ of a dollar, a yard? (200) At 12½ cents, or ⅛ of a dollar? (400) At 25 cents, or ¼ of a dollar? (800). *A.* $1400.
5. How many yards in 31557600 rods? *A.* 173566800.
6. How many years in 31557600 seconds, allowing the year to contain 365¼ days? *A.* 1 year.
7. At 4 cents a gill, what will 1 tun of wine cost? *A.* $322,56.
8. How much wine can be bought for $322,56, at 4 cents a gill? *A.* 1 tun.
9. How many rods in 1100 yds.? In 3300 yds.? *A.* 800 rods.

10. How many dollars in 300£? In 900£? *A.* $4000.
11. Reduce 5£ 17 s. 6 d. to farthings. *A.* 5640 farthings.
12. How many pounds in 5640 farthings? *A.* 5£ 17 s. 6 d.
13. Multiply 3600 by $25\frac{1}{6}$. *A.* 90600.
14. What will 1 ton of clover-seed cost, at 5 mills an ounce? *A.* $179,20.
15. At 2 cents an inch, what will 1 yard of cloth cost? *A.* 72 cents.
16. Reduce 1 tun to gills. *A.* 8064 gills.
17. Reduce 20 bushels to pints. *A.* 1280 pints.
18. Reduce 4 tons to drams. *A.* 2293760.
19. How many barley-corns will reach across the Atlantic Ocean, allowing it to be 3000 miles? *A.* 570240000.
20. How many times will a watch click in 20 years, if it click at the usual rate of 60 times in a minute? *A.* 631152000.
21. A father left legacies to his children as follows: to Thomas, 75£ 14 s. 6 d., to William 3 times as much as Thomas, to his daughter Mary $\frac{1}{6}$ as much as Thomas, and to Susan, his youngest child, as much as all the rest, lacking 20£ 13 s. 8 d., how much did each receive? *A.* William 227£ 3 s. 6 d., Mary 12£ 12 s. 5 d., Susan 294£ 16 s. 9 d.

Mr. Charles Testy

To Lewis P. Child, Dr.

1827.		
Jan. 1.	To 3 yds. Linen Cloth, at 1 s. 6 d. a yard,	
" 15.	" 1 ton of Hay, at 4 s. 6 d. a hundred,	
Feb. 28.	" 25 bushels of Rye, at 3 s. 9 d. a bushel,	
Mar. 9.	" 3 Cows, at 5£ 10 s. a head, - - -	
		£25 18 s. 3 d.

Halifax, April 1, 1827.

Received payment,

Lewis P. Child.

FRACTIONS.

¶ **XXXIV.** 1. If one third ($\frac{1}{3}$) of an apple cost 2 cents, what will a whole apple cost?

2. If one third cost 3 cents, what will a whole one cost? If one third cost 4 cents, what will one whole apple cost? If one third cost 6 cents? 8 cents? 9 cents? 20 cents? 50 cents? 100 cents?

3. If you pay 3 cents for one fifth ($\frac{1}{5}$) of an orange, what will a whole orange cost?

4. If you pay 2 dollars for one eighth ($\frac{1}{8}$) of a ticket, what will a whole ticket cost?

Q. How many halves to an apple, or any thing?

Q. How many thirds? Fifths? Eighths? Sixteenths?

Q. When an apple, or any thing, is divided into two equal parts, would you call one of these parts a half or a third? Into 3 equal parts, what is one part called?

Q. Into 4 parts, what is one part called?

Q. Into 5 parts, what is one part called?

Q. Into 8 parts, what is 1 part called?

Q. Into 8 parts, what are 2 parts called?

Q. Into 8 parts, what are 5 parts called?

Q. When an apple, or any thing, is divided into two equal parts, how would you express one part, on the slate, in figures?

A. I set the 1 *down, and draw a line under it; then write the* 2 *under the line.*

Let me see you write down, in this manner, on the slate, one half. One third. One fourth. One fifth. One sixth. Two sixths. Three sixths. Three eighths. Eight twelfths.

Q. *What are such expressions as these called?* *A.* Fractions.

Q. *When, then, any whole thing, as an apple, a unit, &c. is broken or divided into equal parts, what are these parts called?* *A.* Fractions.

Q. *Why called fractions?* *A.* Because *fraction* signifies *broken.*

Q. *You have seen, that, when any whole thing is divided into* 3 *parts, these parts are called thirds; into* 4 *parts, called fourths: what, then, does the fraction take its name or denomination from?* *A.* From the number of parts into which any thing is divided.

Q. *When an apple is divided into* 6 *parts, and you are desirous of giving away* 5 *parts, how would you express these parts?* *A.* $\frac{5}{6}$.

Q. *What is the* 6 (*in* $\frac{5}{6}$) *called?* *A.* The denominator.

Q. *Why so called?* *A.* Because it gives the name or denomination to the parts.

Q. *What is the* 5 (*in* $\frac{5}{6}$) *called?* *A.* Numerator.

Q. *Why so called?* *A.* Because it numerates or numbers the parts.

Q. *Which is the numerator, then?* *A.* The number above the line.

Q. *Which is the denominator?* *A.* The number below the line.

Q. *What, then, does the denominator show?* *A.* The number of parts a unit, or any thing, is divided into.

Q. *What does the numerator show?* A. How many parts are taken, or used.

Q. *In the expressions* $\frac{2}{5}$, $\frac{7}{16}$, $\frac{9}{12}$, $\frac{4}{50}$, *which are the numerators, and which are the denominators?*

Q. *If you own* $\frac{28}{40}$ *of a vessel, how many parts is the vessel supposed to be divided into? and how many parts do you own?* A. 40 parts, and I own 28 parts.

Q. *Is* $\frac{1}{3}$ *of an apple more than* $\frac{1}{4}$ *of it?*

Q. *What fraction, then, is greater than* $\frac{1}{3}$? *Than* $\frac{1}{4}$? *Than* $\frac{1}{5}$? *Than* $\frac{2}{3}$? *Than* $\frac{4}{5}$? *What fraction is less than* $\frac{3}{5}$? *Than* $\frac{2}{5}$? *Than* $\frac{4}{5}$? *Than* $\frac{2}{3}$?

Q. *From these remarks, what appears to be a correct definition of fractions?* A. They are broken parts of a whole number.

Q. *How are they represented?* A. By one number placed above another, with a line drawn between them.

Q. *In Simple Division, you recollect, that the remainder was represented in like manner; what, then, may justly be considered the origin of fractions?* A. Division.

Q. *What may the numerator be considered?* A. The dividend.

Q. *What may the denominator be considered?* A. The divisor.

Q. *What, then, is the value of a fraction?* A. The quotient of the numerator divided by the denominator.

Q. *What is the quotient of* 1 *dollar divided among* 2 *men?* A. $\frac{1}{2}$.

Q. *What is the quotient of* 7 *divided by* 8? A. $\frac{7}{8}$.

Q. *How, then, are fractions represented?* A. By the sign of division.

Q. *What does* $\frac{2}{3}$ *express?*

A. The quotient, of which { 2 is the dividend. 3 is the divisor.

---

1. If 3 apples be divided equally among 8 boys, what part of one apple will each boy receive? 1 *apple among* 8 *boys would be* $\frac{1}{8}$ *of an apple apiece, and* 3 *apples would be* 3 *times as much; that is,* $\frac{3}{8}$ *of an apple apiece.* *Ans.* $\frac{3}{8}$.

2. If 4 oranges be divided equally among 8 boys, what part of an orange is each boy's part? 1 *orange among* 8 *boys* $=\frac{1}{8}$, *and* 4 *oranges are* 4 *times as much; that is,* $\frac{4}{8}$, *Ans.* If 2 oranges among 7 boys? A. $\frac{2}{7}$. 9 oranges among 13 boys? 20 oranges among 37 boys?

3. One orange among 2 boys is $\frac{1}{2}$ of an orange apiece; how

much is 1 divided by 2, then? *Ans.* $\frac{1}{2}$. How much is 1 divided by 3? *A.* $\frac{1}{3}$. The quotient of 5 divided by 6? *A.* $\frac{5}{6}$. Of 3 by 5? Of 7 by 9? Of 8 by 13? Of 11 by 15?

4. What part of one apple is a third part of 2 apples? *A third part of one apple is* $\frac{1}{3}$, *and a third part of 2 apples must be twice as much; that is,* $\frac{2}{3}$ *of* 1 *apple.* *A.* $\frac{2}{3}$.

5. What part of 1 apple is one fourth ($\frac{1}{4}$) part of 3 apples? $\frac{1}{4}$ *of* 3 *apples is* 3 *times as much as* $\frac{1}{4}$ *of* 1 *apple; that is,* $\frac{3}{4}$ *of* 1 *apple.* *A.* $\frac{3}{4}$.

6. What part of one apple is $\frac{1}{5}$ of 3 apples? *A.* $\frac{3}{5}$. What part of 1 apple is $\frac{1}{5}$ of 4 apples? *A.* $\frac{4}{5}$. $\frac{1}{6}$ of 4 apples is what part of 1 apple? *Ans.* $\frac{4}{6}$.

A Proper Fraction. Q. *We have seen that the denominator shows how many parts it takes to make a whole or unit; when, then, the numerator is less than the denominator, is the fraction greater, or less, than a whole thing or unit?* *A.* It must be less.

Q. *What is such a fraction called?* *A.* A Proper Fraction.

Q. *How may it always be known?* *A.* The numerator is less than the denominator.

Q. *What kind of fractions are* $\frac{4}{5}$, $\frac{3}{4}$, $\frac{7}{8}$, &c.?

An Improper Fraction. Q. *When the numerator is as large, or larger than the denominator, as,* $\frac{8}{8}$, $\frac{16}{11}$, $\frac{4}{3}$, *it is plain, that the fraction expresses* 1 *whole, or more than* 1 *whole; what is such a fraction called?* *A.* An Improper Fraction.

Q. *How may it be known?* *A.* The numerator is greater than the denominator.

Q. *What kind of fractions are* $\frac{5}{3}$, $\frac{12}{7}$, $\frac{8}{8}$, &c.

A Mixed Number. Q. *What is a mixed number?* *A.* A fraction joined with a whole number.

Q. *What kind of fractions are* $15\frac{3}{4}$, $16\frac{1}{5}$, &c.

Q. *What kind of fractions are each of the following expressions, viz.* $15\frac{3}{8}$, $\frac{5}{6}$, $\frac{21}{5}$, $8\frac{4}{8}$, $\frac{18}{20}$, $7\frac{1}{3}$, $\frac{50}{9}$?

¶ XXXV. To change an Improper Fraction to a Whole or Mixed Number.

1. How many whole apples are there in 6 thirds ($\frac{6}{3}$) of an apple? In 8 quarters ($\frac{8}{4}$)? In $\frac{12}{3}$? In $\frac{16}{8}$? In $\frac{24}{3}$? In $\frac{50}{25}$? In $\frac{200}{100}$?

2 How many weeks in $\frac{14}{7}$ of a week? In $\frac{28}{7}$? In $\frac{42}{7}$? In $\frac{56}{7}$? In $\frac{84}{7}$?

3. How many pints in $\frac{8}{4}$ gills? In $\frac{32}{4}$ gills? In $\frac{48}{8}$ gills? In $\frac{120}{8}$ gills?

4. How much is $\frac{8}{8}$ of a dollar? *A.* \$1. Is $\frac{9}{8}$? *A.* 1 and $\frac{1}{8}$ = $1\frac{1}{8}$. Is $\frac{10}{8}$? Is $\frac{16}{8}$? Is $\frac{17}{8}$? Is $\frac{24}{8}$? Is $\frac{25}{8}$?

Q. What is the finding how many whole things are contained in an improper fraction called? A. Reducing an improper fraction to a whole or mixed number.

---

1. James, by saving $\frac{1}{16}$ of a dollar a day, would save in 33 days $\frac{33}{16}$; how many dollars would that be?

OPERATION.

16)33

Ans. $2\frac{1}{16}$ dollars.

In this example, as $\frac{16}{16}$ make 1 dollar, it is plain, that as many times as 16 is contained in 33, so many dollars it is, 16 is contained 2 times and 1 over; that is, $2\frac{1}{16}$ dollars.

RULE I. *What, then, is the rule for reducing an improper fraction to a whole or mixed number? A.* Divide the numerator by the denominator.

More Exercises for the Slate.

2. A regiment of soldiers, consuming $\frac{1}{5}$ of a barrel of pork a day, would consume in 28 days $\frac{28}{5}$ of a barrel; how many barrels would that be? *A.* $5\frac{3}{5}$ barrels.

3. A man, saving $\frac{1}{5}$ of a dollar a day, would save in 365 days $\frac{365}{5}$; how many dollars would that be? *A.* \$73.

4. Reduce $\frac{1201}{60}$ to a mixed number. *A.* $20\frac{1}{60}$.

5. Reduce $\frac{874}{12}$ to a mixed number. *A.* $72\frac{10}{12}$.

6. Reduce $\frac{38}{9}$ to a mixed number. *A.* $4\frac{2}{9}$.

7. Reduce $\frac{134}{11}$ to a mixed number. *A.* $12\frac{2}{11}$.

8. Reduce $\frac{167}{12}$ to a mixed number. *A.* $13\frac{11}{12}$.

9. Reduce $\frac{6384}{272}$ to a mixed number. *A.* $23\frac{128}{272}$.

10. Reduce $\frac{1728}{12}$ to a whole number. *A.* 144.

¶ XXXVI. To reduce a Whole or Mixed Number to an Improper Fraction.

1. How many halves will 2 whole apples make? Will 3? Will 4? Will 6? Will 20? Will 100?

2. How many thirds in 2 whole oranges? In $2\frac{1}{3}$? In $2\frac{2}{3}$? In 3? In $3\frac{1}{3}$? In 8? In 12?

3. A father, dividing one whole apple among his children, gave them $\frac{1}{5}$ of an apple apiece; how many children were there?

4. James, by saving $\frac{1}{8}$ of a dollar a day, found, after several days, that he had saved $1\frac{3}{8}$ of a dollar; how many 8ths did he save? and how many days was he in saving them?

5. How many 7ths in 2 whole oranges? In $2\frac{4}{7}$? In $2\frac{5}{7}$? In $3\frac{4}{7}$?

This rule, it will be perceived, is exactly the reverse of the last, and proves the operations of it.

---

1. In $30\frac{3}{8}$ of a dollar, how many 8ths?

OPERATION.

$$
\begin{array}{r l}
30\frac{3}{8} & \\
8 & \\
\hline
240 = & \text{the 8ths in 30 dollars.} \\
3 = & \text{the 8ths in } \frac{3}{8}. \\
\hline
243 = & \frac{243}{8}. \ \textit{Ans.}
\end{array}
$$

RULE I. *What, then, is the rule for reducing a mixed or whole number to an improper fraction?* *A.* Multiply the whole number by the denominator of the fraction.

II. *What do you add to the product?* *A.* The numerator.

III. *What is to be written under this result?* *A.* The denominator.

More Exercises for the Slate.

2. What improper fraction is equal to $20\frac{1}{60}$? *A.* $\frac{1201}{60}$.
3. What improper fraction is equal to $72\frac{10}{12}$? *A.* $\frac{874}{12}$.
4. What improper fraction is equal to $4\frac{2}{9}$? *A.* $\frac{38}{9}$.
5. What improper fraction is equal to $12\frac{2}{3}$? *A.* $\frac{38}{3}$.
6. What improper fraction is equal to $16\frac{5}{12}$? *A.* $\frac{197}{12}$.
7. What improper fraction is equal to $17\frac{2}{11}$? *A.* $\frac{189}{11}$.
8. What improper fraction is equal to $144\frac{1}{12}$? *A.* $\frac{1729}{12}$.
9. Reduce $30\frac{5}{20}$ pounds to 20ths. *As $\frac{1}{20}$ of a pound* = 1 s., $\frac{2}{20}$ = 2 s., *the question is the same as if it had been stated thus:* In 30£ 5 s. how many shillings? *A.* $\frac{605}{20}$ = 605 shillings.
10. In $14\frac{3}{7}$ weeks, how many 7ths? *A.* $\frac{101}{7}$ = 101 days.
11. *In* $26\frac{3}{8}$ pecks, how many 8ths? *A.* $\frac{211}{8}$ = 211 quarts.

¶ XXXVII. To reduce a Fraction to its lowest Terms.

Q. When an apple is divided into 4 parts, 2 parts, or $\frac{2}{4}$, are evidently $\frac{1}{2}$ of the apple: now, if we take $\frac{1}{2}$, and multiply the 1 and 2 both by 2, we shall have $\frac{2}{4}$ again; why does not this multiplying alter the value? *A.* Because, when the apple is divided into 4 parts, or quarters, it takes 2 times as many parts, or quarters, to make one whole apple, as it will take parts, when the apple is divided into only 2 parts, or halves: hence, multiplying only increases the number of parts of a whole, without altering the value of the fraction.

Q. Now, if we take $\frac{2}{4}$, and multiply both the 2 and 4 by 2, we obtain $\frac{4}{8} = \frac{1}{2}$; what, then, is $\frac{1}{2}$ equal to? *A.* $\frac{2}{4}$, or $\frac{4}{8}$.

Q. Now it is plain that the reverse of this must be true; for, if we divide both the 4 and 8 in $\frac{4}{8}$ by 2, we obtain $\frac{2}{4}$, and, dividing the 2 and 4 in $\frac{2}{4}$ by 2, we have $\frac{1}{2}$; what, then, may be inferred from these remarks respecting multiplying or dividing both the numerator and denominator of the same fraction? *A.* That they may both be multiplied, or divided, by the same number, without altering the value of the fraction.

Q. What are the numerator and denominator of the same fraction called? A. The terms of the fraction.

Q. What is the process of changing $\frac{4}{8}$ into its equal $\frac{1}{2}$ called? A. Reducing the fraction to its lowest terms.

Mental Exercises.

1. Reduce $\frac{2}{4}$ to its lowest terms.
2. Reduce $\frac{4}{8}$ to its lowest terms.
3. Reduce $\frac{3}{6}$ to its lowest terms.
4. Reduce $\frac{9}{12}$ to its lowest terms.
5. Reduce $\frac{10}{20}$ to its lowest terms.
6. Reduce $\frac{4}{20}$ to its lowest terms.
7. Reduce $\frac{50}{200}$ to its lowest terms.

Operation by Slate illustrated.

1. One minute is $\frac{1}{60}$ of an hour, and 15 minutes are $\frac{15}{60}$; what part of an hour will $\frac{15}{60}$ make, reduced to its lowest terms?

OPERATION.

$$5)\frac{15}{60} = \frac{3}{12} \;(3) = \frac{1}{4}$$

How do you get the $\frac{3}{12}$ in this example? A. By dividing 15 and 60, each, by 5. *How do you get the $\frac{1}{4}$? A.* By dividing 3 and 12, each, by 3. How do you know that $\frac{1}{4}$

is reduced to its lowest terms? *A.* Because there is no number greater than 1 that will divide both the terms of $\frac{1}{4}$ without a remainder.

From these illustrations we derive the following

RULE.

I. *How do you proceed to reduce a fraction to its lowest terms?* *A.* Divide both the terms of the fraction by any number that will divide them without a remainder, and the quotients again in the same manner.

II. *When is the fraction said to be reduced to its lowest terms?* *A.* When there is no number greater than 1 that will divide the terms without a remainder.

More Exercises for the Slate.

2. Reduce $\frac{72}{84}$ of a barrel to its lowest terms. *A.* $\frac{6}{7}$.
3. Reduce $\frac{28}{36}$ of a hogshead to its lowest terms. *A.* $\frac{7}{9}$.
4. Reduce $\frac{80}{320}$ of a tun to its lowest terms. *A.* $\frac{1}{4}$.
5. Reduce $\frac{144}{1728}$ of a foot to its lowest terms. *A.* $\frac{1}{12}$.
6. Reduce $\frac{224}{448}$ of a gallon to its lowest terms. *A.* $\frac{1}{2}$.
7. Reduce $\frac{1428}{2856}$ of an inch to its lowest terms. *A.* $\frac{1}{2}$.

¶ XXXVIII. To multiply a Fraction by a Whole Number.

1. If 1 apple cost $\frac{1}{3}$ of a cent, what will 2 apples cost? How much is 2 times $\frac{1}{3}$?

2. If a horse eat $\frac{1}{4}$ of a bushel of oats in one day, how many bushels will he eat in 2 days? In 3 days? How much is 2 times $\frac{1}{4}$? 3 times $\frac{1}{4}$?

3. William has $\frac{3}{8}$ of a melon, and Thomas 2 times as much; what is Thomas' part? How much is 2 times $\frac{3}{8}$? 2 times $\frac{4}{8}$? 2 times $\frac{1}{3}$? 3 times $\frac{2}{8}$? 6 times $\frac{2}{15}$?

Q. From these examples, what effect does multiplying the numerator by any number appear to have on the value of the fraction, if the denominator remain the same? *A.* It multiplies the value by that number.

Q. 2 times $\frac{1}{4}$ is $\frac{2}{4} = \frac{1}{2}$: but, if we divide the denominator 4 (*in* $\frac{1}{4}$) by 2, we obtain $\frac{1}{2}$; *what effect, then, does dividing the denominator by any number have on the value of a fraction,* if the numerator remain the same? *A.* It multiplies the value by that number.

Q. What is the reason of this? *A.* Dividing the denominator makes the parts of a whole so many times *larger;* and, if as many are taken, as before, (which will be the case if the numerator remain the same,) the value of the fraction is evidently increased so many times.

Again, as the numerator shows how many parts of a whole are taken, multiplying the numerator by any number, if the denominator remain the same, increases the number of parts taken; consequently, it increases the value of the fraction.

4. At $\frac{3}{16}$ of a dollar a yard, what will 4 yards of cloth cost? 4 times $\frac{3}{16}$ are $\frac{12}{16}=\frac{3}{4}$ of a dollar, *Ans. But, by dividing the denominator of* $\frac{3}{16}$ *by 4, as above shown, we immediately have* $\frac{3}{4}$ *in its lowest terms.*

From these illustrations we derive the following

RULE.

I. *How can you multiply a fraction by a whole number?* *A.* Multiply the numerator by it, without changing its denominator.

II. *How can you shorten this process?* *A.* Divide the denominator by the whole number, when it can be done without a remainder.

Exercises for the Slate.

1. If a horse consume $\frac{3}{15}$ of a bushel of oats in one day, how many bushels will he consume in 30 days? *A.* $\frac{90}{15}=6$ bushels.

2. If 1 pound of butter cost $\frac{3}{20}$ of a dollar, what will 205 pounds cost? *A.* $\frac{615}{20}=30\frac{15}{20}=30\frac{3}{4}$ dollars.

3. Bought 400 yards of calico, at $\frac{3}{8}$ of a dollar a yard; what did it come to? *A.* $\frac{1200}{8}=\$150$.

4. How much is 6 times $\frac{3}{17}$? *A.* $\frac{18}{17}=1\frac{1}{17}$.

5. How much is 8 times $\frac{15}{42}$? *A.* $\frac{120}{42}=2\frac{36}{42}=2\frac{6}{7}$.

6. How much is 12 times $\frac{8}{15}$? *A.* $\frac{96}{15}=6\frac{6}{15}=6\frac{2}{5}$.

7. How much is 13 times $\frac{211}{850}$? *A.* $\frac{2743}{850}=3\frac{193}{850}$.

8. How much is 314 times $\frac{3}{4}$? *A.* $\frac{942}{4}=235\frac{2}{4}=235\frac{1}{2}$.

9. How much is 513 times $\frac{7}{11}$? *A.* $\frac{3591}{11}=326\frac{5}{11}$.

10. How much is 530 times $\frac{21}{23}$? *A.* $\frac{11130}{23}=483\frac{21}{23}$.

Divide the denominator in the following.

11. How much is 42 times $\frac{11}{42}$? *A.* 11.

12. How much is 13 times $\frac{19}{20}$? *A.* $\frac{247}{20}=12\frac{7}{20}$.

13. How much is 60 times $\frac{5}{120}$? *A.* $\frac{5}{2}=2\frac{1}{2}$.

14. At $2\frac{1}{8}$ dollars a yard, what will 9 yards of cloth cost? *9 times 2 are 18, and 9 times $\frac{1}{8}$ are $\frac{9}{8}=1\frac{1}{8}$, which, added to 18, makes $19\frac{1}{8}$ dollars. A. This process is substantially the same as ¶ XXVII., by which the remaining examples in this rule may be performed.*

15. Multiply $3\frac{1}{4}$ by 367. *A.* $1192\frac{3}{4}$.

16. Multiply $6\frac{7}{8}$ by 211. *A.* $1450\frac{5}{8}$.

17. Multiply $3\frac{5}{60}$ by 42. *A.* $129\frac{30}{60}=129\frac{1}{2}$.

¶ XXXIX. To multiply a Whole Number by a Fraction.

Q. When a number is added to itself several times, this repeated addition has been called multiplication; but the term has a more extensive application. It often happens that not a whole number only, but a certain portion of it, is to be repeated several times, as, for instance, If you pay 12 cents for a melon, what will $\frac{3}{4}$ of one cost? $\frac{1}{4}$ of 12 cents is 3 cents; and to get $\frac{3}{4}$, it is plain that we must repeat the 3, 3 times, making 9 cents, the answer; when, then, a certain portion of the multiplicand is repeated several times, or as many times as the numerator shows, what is it called? *A.* Multiplying by a fraction. How much is $\frac{1}{3}$ of 12? $\frac{2}{3}$ of 12? $\frac{1}{5}$ of 20? $\frac{3}{5}$ of 20? $\frac{1}{4}$ of 8? $\frac{3}{4}$ of 8? $\frac{1}{5}$ of 40? $\frac{2}{5}$ of 40? $\frac{3}{5}$ of 40? $\frac{4}{5}$ of 40?

Q. We found in Multiplication, ¶ X., that when two numbers are to be multiplied together, either may be the multiplier; hence, to multiply a whole number by a fraction, is the same as a fraction by a whole number; consequently, the operations of both are the same as that described in ¶ XXVII.; what, then, is the rule for multiplying a whole number by a fraction? (For answer, see ¶ XXVII.)

Exercises for the Slate.

1. What will 600 bushels of oats cost, at $\frac{3}{16}$ of a dollar a bushel? *A.* $\$112\frac{1}{2}$.

2. What will 2700 yards of tape cost, at $\frac{1}{8}$ of a dollar a yard? *A.* $\$337\frac{1}{2}$.

3. *Multiply* 425 by $5\frac{1}{5}$. *A.* 2210.

4. Multiply 272 by $15\frac{3}{4}$. *A.* 4284.
5. Multiply 999 by $21\frac{2}{9}$. *A.* 21201.
6. Multiply 20 by $5\frac{3}{40}$. *A.* $101\frac{1}{2}$.

¶ XL. To divide a Fraction by a Whole Number.

1. If 3 apples cost $\frac{3}{4}$ of a cent, what will 1 apple cost? How much is $\frac{3}{4} \div 3$?

2. If a horse eat $\frac{1}{2}$ or $\frac{2}{4}$ of a bushel of meal in 2 days, how much will he eat in one day? How much is $\frac{2}{4} \div 2$?

3. A rich man divided $\frac{6}{8}$ of a barrel of flour among 6 poor men; how much did each receive? How much is $\frac{6}{8} \div 6$?

4. If 3 yards of calico cost $\frac{3}{8}$ of a dollar, how much is it a yard? How much is $\frac{3}{8} \div 3$?

5. If 3 yards of cloth cost $\frac{9}{16}$ of a dollar, how much is it a yard?

The foregoing examples have been performed by simply dividing their numerators, and retaining the same denominator, for the following reason, that the numerator tells how many parts any thing is divided into; as, $\frac{4}{8}$ are 4 parts, and, to divide 4 parts by 2, we have only to say, 2 in 4, 2 times, as in whole numbers. But it will often happen, that the numerator cannot be *exactly* divided by the whole number, as in the following examples.

6. William divided $\frac{3}{4}$ of an orange among his 2 little brothers; what was each brother's part?

We have seen, ¶ XXXVII., *that the value of the fraction is not altered by multiplying both of its terms by the same number;* hence, $\frac{3}{4} \times 2 = \frac{6}{8}$. *Now,* $\frac{6}{8}$ *are* 6 *parts, and William can give* 3 *parts to each of his two brothers; for* 2 *in* 6, 3 *times.* *A.* $\frac{3}{8}$ of an orange apiece.

Q. In this last example, if (in $\frac{3}{4}$) we multiply the denominator 4 by 2, (the whole number,) we have $\frac{3}{8}$, the same result as before; why is this? *A.* Multiplying the denominator makes the parts so many times smaller; and, if the numerator remain the same, no more are taken than before; consequently, the value is lessened so many times.

From these illustrations we derive the following

RULE.

I. When the numerator can be divided by the whole number without a remainder, how do you proceed? *A.* Divide the nu merator by the whole number, writing the denominator under the quotient.

II. *When the numerator cannot be thus divided, how do you proceed?* *A.* Multiply the denominator by the whole number, writing the result under the numerator.

Exercises for the Slate.

1. If 8 yards of tape cost $\frac{8}{15}$ of a dollar, how much is it a yard? How much is $\frac{8}{15} \div 8$?

2. Divide $\frac{6}{8}$ by 8. *A.* $\frac{6}{64} = \frac{3}{32}$.

3. Divide $\frac{4}{15}$ by 6. *A.* $\frac{4}{90} = \frac{2}{45}$.

4. Divide $\frac{7}{23}$ by 8. *A.* $\frac{7}{184}$.

5. Divide $\frac{8}{80}$ by 8. (*Divide the numerator.*) *A.* $\frac{1}{80}$.

6. Divide $\frac{16}{250}$ by 4. *A.* $\frac{4}{250} = \frac{2}{125}$.

Note. When a mixed number occurs, reduce it to an improper fraction, then divide as before.

7. Divide \$$6\frac{3}{4}$ among five men. *A.* $6\frac{3}{4} = \frac{27}{4} \div 5 = \frac{27}{20} = 1\frac{7}{20}$.

8. Divide $2\frac{11}{13}$ by 4. *A.* $\frac{37}{52}$.

9. Divide $16\frac{3}{8}$ by 5. *A.* $\frac{131}{40} = 3\frac{11}{40}$.

10. Divide $25\frac{4}{21}$ by 20. *A.* $\frac{529}{420} = 1\frac{109}{420}$

11. Divide $8\frac{3}{7}$ by 6. *A.* $\frac{59}{42} = 1\frac{17}{42}$.

12. Divide $114\frac{1}{5}$ by 280. *A.* $\frac{571}{1400}$.

¶ XLI. To multiply one Fraction by another.

1. A man, owning $\frac{5}{8}$ of a packet, sells $\frac{3}{4}$ of his part; what part of the whole packet did he sell? How much is $\frac{3}{4}$ of $\frac{5}{8}$?

$\frac{3 \times 5}{4 \times 8} = \frac{15}{32}$ *Ans.* The reason of this operation will appear from the following illustration.

Once $\frac{1}{8}$ *is* $\frac{1}{8}$, *and* $\frac{1}{4}$ *of* $\frac{1}{8}$ *is evidently* $\frac{1}{8}$ *divided by* 4, *which is done,* ¶ XL., *by multiplying the denominator* 8 *by the* 4, *making* 32; *that is,* $\frac{1}{4}$ *of* $\frac{1}{8} = \frac{1}{32}$.

Again, *if* $\frac{1}{4}$ *of* $\frac{1}{8}$ *be* $\frac{1}{32}$, *then* $\frac{1}{4}$ *of* $\frac{5}{8}$ *will be* 5 *times as much, that is,* $\frac{5}{32}$.

Again, *if* $\frac{1}{4}$ *of* $\frac{5}{8}$ *be* $\frac{5}{32}$, *then* $\frac{3}{4}$ *will be* 3 *times* $\frac{5}{32} = \frac{15}{32}$. *Ans., as before.*

The above process, by close inspection, will be found to consist in multiplying together the two numerators for a new numerator, and the two denominators for a new denominator.

Should a whole number occur in any example, it may be reduced to an improper fraction, by placing the figure 1 under it: thus 7 becomes $\frac{7}{1}$; for, since the value of a fraction (¶ XXXIV.) is the numerator divided by the denominator, the value of $\frac{7}{1}$ is 7; for, 1 in 7, 7 times.

From these illustrations we derive the following

RULE.

Q. How do you proceed to multiply one fraction by another?
A. Multiply the numerators together for a new numerator; and the denominators together for a new denominator.

Note. If the fraction be a mixed number, reduce it to an improper fraction, then proceed as before.

Mental Exercises.

2. How much is $\frac{1}{2}$ of $\frac{1}{2}$?
3. How much is $\frac{1}{2}$ of $\frac{3}{4}$?
4. How much is $\frac{3}{4}$ of $\frac{5}{7}$?
5. How much is $\frac{5}{6}$ of $\frac{6}{7}$?
6. How much is $\frac{1}{3}$ of $\frac{1}{4}$?
7. How much is $\frac{2}{3}$ of $\frac{1}{8}$?
8. How much is $\frac{1}{10}$ of $\frac{1}{10}$?
9. How much is $\frac{3}{4}$ of $\frac{1}{100}$?

Q. What are such fractions as these sometimes called?
A. Compound Fractions.

Q. What does the word of *denote?* *A.* Their continual multiplication into each other.

Exercises for the Slate.

1. A man, having $\frac{3}{50}$ of a factory, sold $\frac{2}{3}$ of his part; what part of the whole did he sell? How much is $\frac{2}{3}$ of $\frac{3}{50}$? $\frac{2}{3}\times\frac{3}{50}=\frac{6}{150}=\frac{1}{25}$, *Ans.*

2. At $\frac{7}{25}$ of a dollar a yard, what will $\frac{5}{8}$ of a yard of cloth cost? How much is $\frac{7}{25}$ of $\frac{5}{8}$? *A.* $\frac{7}{40}$.

3. Multiply $\frac{3}{8}$ of $\frac{5}{7}$ by $\frac{3}{7}$. *A.* $\frac{3\times5\times3}{8\times7\times7}=\frac{45}{392}$.

4. Multiply $\frac{3}{4}$ of $\frac{2}{3}$ by $\frac{4}{9}$. *A.* $\frac{24}{108}=\frac{2}{9}$.

5. Multiply $\frac{5}{11}$ of $\frac{9}{13}$ by $\frac{7}{8}$. *A.* $\frac{315}{1144}$.

6. Multiply $\frac{267}{300}$ by $\frac{7}{5}$. *A.* $\frac{1869}{1500}=\frac{623}{500}=1\frac{123}{500}$.

Note. If the denominator of any fraction be equal to the numerator of any other fraction, they may both be dropped on the principle explained in ¶ XXXVII.; thus $\frac{2}{3}$ of $\frac{3}{4}$ of $\frac{5}{6}$ may be shortened, by dropping the numerator 3, and denominator 3; the remaining terms, being multiplied together, will produce the fraction required in lower terms, thus: 2 of $\frac{1}{4}$ of $\frac{5}{6}=\frac{2}{4}$ of $\frac{5}{6}$ $=\frac{10}{24}=\frac{5}{12}$, *Ans.*

The answers to the following examples express the fraction in its lowest terms.

7. How much is $\frac{1}{2}$ of $\frac{1}{3}$ of $\frac{3}{4}$ of $\frac{5}{6}$? *A.* $\frac{5}{48}$.
8. How much is $\frac{3}{4}$ of $\frac{1}{2}$ of $\frac{4}{5}$? *A.* $\frac{3}{10}$.
9. How much is $5\frac{1}{2}$ times $5\frac{1}{2}$? *A.* $30\frac{1}{4}$.
10. How much is $16\frac{1}{2}$ times $16\frac{1}{2}$? *A.* $272\frac{1}{4}$.
11. How much is $20\frac{3}{11}$ times $\frac{1}{2}$ of $\frac{1}{3}$? *A.* $\frac{223}{66}=3\frac{25}{66}$.

¶ XLIX. To find the Least Common Multiple of two or more numbers.

Q. 12 is a number produced by multiplying 2 (a factor) by some other factor; thus $2\times6=12$; what, then, may the 12 be called? *A.* The multiple of 2.

Q. 12 is also produced by multiplying not only 2, but 3 and 6, likewise, each by some other number; thus, $2\times6=12$; $3\times4=12$; $6\times2=12$; when, then, a number is a multiple of several factors or numbers, what is it called? *A.* The common multiple of these factors.

Q. As the common multiple is a product consisting of two or more factors, it follows that it may be divided by each of these factors without a remainder; how, then, may it be determined, whether one number is a common multiple of two or more numbers, or not? *A.* It is a common multiple of these numbers, when it can be divided by each without a remainder.

Q. What is the common multiple of 2, 3, and 4, then? *A.* 24.

Q. Why? *A.* Because 24 can be divided by 2, 3, and 4, without a remainder.

Q. We can divide 12, also, by 2, 3, and 4, without a remainder; what, then, is the least number, that can be divided by 2 or more numbers, called? *A.* The least common multiple of these numbers.

Q. It sometimes happens, that one number will *divide* several other numbers, without a remainder; as, for instance, 3 will divide 12, 18, and 24, without a remainder; when, then, several numbers can be thus divided by one number, what is the number called? *A.* The common divisor of these numbers.

Q. 12, 18, and 24, may be divided also, each, by 6, *even*, what, then, is the greatest number called, which will divide 2 or more numbers without a remainder? *A.* The greatest common divisor.*

---

* In ¶ XXXVII., in reducing fractions to their lowest terms, we were sometimes obliged, in order to do it, to perform several operations in dividing; but, had we only known the greatest common divisor of both terms of the fraction, we might have reduced them by simply dividing once; hence it may sometimes be convenient to have a rule

To find the greatest common divisor of two or more numbers.

1. What is the greatest common divisor of 72 and 84?

OPERATION.

```
72)84(1
   72
   --
   12)72(6
      72
      --
```

A. 12, *common divisor.*

In this example, 72 is contained in 84, 1 time, and 12 remaining; 72, then, is not a factor of 84. Again, if 12 be a factor of 72, it must also be a factor of 84; for, 72+12=84. By dividing 72 by 12, we do find it to be a factor of 72, (for 72÷12=6 with no remainder); therefore 12 is a common factor or divisor of 72 and 84; and, as the greatest common divisor of two or more numbers never exceeds their difference; so 12, the difference between 84 and 72, must be the greatest common divisor.

Hence, the following RULE. *Divide the greater number by the less, and, if there be no remainder, the less number itself is the common divisor; but, if there be a remainder, divide the divisor by the remainder, always dividing the last divisor by the last remainder, till nothing remain: the last divisor is the divisor sought.*

Note. If there be more numbers than two, of which the greatest common divisor is to be found, find the common divisor of two of them first, and then of that common divisor, and one of the other numbers, and so on.

2. Find the greatest common divisor of 144 and 132. *A.* 12.
3. Find the greatest common divisor of 168 and 84. *A.* 84.
4. Find the greatest common divisor of 24, 48, and 96. *A.* 24.

Let us apply this rule to reducing fractions to their lowest terms. See ¶ XXXVII.

5. Reduce $\frac{132}{144}$ to its lowest terms.

$12)\frac{132}{144}=\frac{11}{12}$, *Ans.*

In this example, by using the common divisor, 12, found in the answer to sum No. 2, we have a number that will reduce the fraction to its lowest terms, by simply dividing both terms but once.

After the same manner perform the following examples.

6 Find the common divisor of 750 and 1000; also reduce $\frac{750}{1000}$ to its lowest terms. *A.* 250, and $\frac{3}{4}$.

7. Reduce $\frac{420}{630}$ to its lowest terms. *A.* $\frac{2}{3}$.

8. Reduce $\frac{660}{1000}$ to its lowest terms. *A.* $\frac{33}{50}$.

Should it be preferred to reduce fractions to their lowest terms by ¶ XXXVII., the following rules may be found serviceable.

Any number ending with an even number or cipher is divisible by 2.

Any number ending with 5 or 0 is divisible by 5; also if it end in 0, it is divisible by 10.

1. What is the least common multiple of 6 and 8?

OPERATION.	
2)6 . 8	
3 . 4	

In this example, it will be perceived that the divisor 2 is a factor, both of 6 and 8, and that dividing 6 by 2 gives its other factor 3 (for 6÷2=3); likewise dividing the 8 by 2 gives its other factor 4 (for 8÷2=4); consequently, if the divisors and quotients be multiplied together, their product must contain all the factors of the numbers 6 and 8; hence this product is the common multiple of 6 and 8, and, as there is no other number greater than 1, that will divide 6 and 8, 4×3×2 =24 will be the least common multiple of 6 and 8.

Note. When there are several numbers to be divided, should the divisor not be contained in any one number, without a remainder, it is evident, that the divisor is not a factor of that number; consequently, it may be omitted, and reserved to be divided by the next divisor.

2. What is the common multiple of 6, 3 and 4?

OPERATION.

3)6 . 3 . 4

2)2 . 1 . 4

1 . 1 . 2

Ans. 3×2×2=12

In dividing 6, 3 and 4 by 3, I find that 3 is not contained in 4 even; therefore, I write the 4 down with the quotients, after which I divide by 2, as before. Then, the divisors and quotients, multiplied together, thus, 2×2×3=12, *Ans.*

From these illustrations we derive the following

RULE.

I. ***How do you proceed first to find the least common multiple of two or more numbers?*** *A.* Divide by any number that will divide two or more of the given numbers without a remainder, and set the quotients, together with the undivided numbers, in a line beneath.

II. ***How do you proceed with this result?*** *A.* Continue dividing, as before, till there is no number greater than 1 that will divide two or more numbers without a remainder; then multiplying the divisors and numbers in the last line together, will give the least common multiple required.

More Exercises for the Slate.

3. Find the least common multiple of 4 and 16. *A.* 16.
4. Find the least common multiple of 10 and 15. *A.* 30.
5. Find the least common multiple of 30, 35 and 6. *A.* 210.
6. Find the least common multiple of 27 and 51. *A.* 459
7. Find the least common multiple of 3, 12 and 8. *A.* 24.

8. Find the least common multiple of 4, 12 and 20. *A.* 60.
9. Find the least common multiple of 2, 7, 14 and 49. *A.* 98.

¶ XLIX. To reduce Fractions of different Denominators to a Common Denominator.

Q. When fractions have their denominators alike, they may be added, subtracted, &c. as easily as whole numbers; for example, $\frac{1}{8}$ and $\frac{2}{8}$ are $\frac{3}{8}$; but in the course of calculations by numbers, we shall meet with fractions whose denominators are unlike; as, for instance, we cannot add, as above, $\frac{2}{5}$ and $\frac{3}{4}$ together; what, then, may be considered the object of reducing fractions of different denominators to a common denominator? *A.* To prepare fractions for the operations of addition, subtraction, &c., of fractions.

Q. What do you mean by a common denominator? *A.* When the denominators are alike.

1. Reduce $\frac{2}{3}$ and $\frac{5}{6}$ to a common denominator.

OPERATION.

Numer. $2\times6=\underline{12}$, new numer.
Denom. $3\times6=18$, com. denom.

Numer. $5\times3=\underline{15}$, new numer.
Denom. $6\times3=18$, com. denom.

In performing this example, we take $\frac{2}{3}$, and multiply both its terms by the denominator of $\frac{5}{6}$; also, we multiply both the terms of $\frac{5}{6}$ by 3, the denominator of $\frac{2}{3}$; and, as both the terms of each fraction are multiplied by the same number, consequently the value of the fractions is not altered, ¶ XXXVII.

From these illustrations we derive the following

RULE.

I. *What do you multiply each denominator by for a new denominator?* *A.* By all the other denominators.

II. *What do you multiply each numerator by for a new numerator?* *A.* By the same numbers (denominators) that I multiply its denominator by.

Note. As, by multiplying in this manner, the same denominators are continually multiplied into each other, the process may be shortened; for, having found one denominator, it may be written under each new numerator. This, however, the intelligent pupil will soon discover of himself; and, perhaps, it is best he should.

More Exercises for the Slate.

2. Reduce $\frac{3}{4}$ and $\frac{7}{8}$ to a common denominator. *A.* $\frac{24}{32}$, $\frac{28}{32}$.

3. Reduce $\frac{4}{5}$ and $\frac{2}{3}$ to a common denominator. *A.* $\frac{12}{15}$, $\frac{10}{15}$.

4. Reduce $\frac{3}{4}$ and $\frac{10}{11}$ to a common denominator. *A.* $\frac{33}{44}$, $\frac{40}{44}$.

5. Reduce $\frac{8}{9}$, $\frac{6}{7}$ and $\frac{1}{2}$ to a common denominator.

A. $\frac{112}{126}$, $\frac{108}{126}$, $\frac{63}{126}$.

6. Reduce $\frac{7}{9}$, $\frac{4}{5}$ and $\frac{3}{7}$ to a common denominator.

A. $\frac{245}{315}$, $\frac{252}{315}$, $\frac{135}{315}$.

Compound fractions must be reduced to simple fractions before finding the common denominator; also the fractional parts of mixed numbers may first be reduced to a common denominator, and then annexed to the whole numbers.

7. Reduce $\frac{1}{2}$ of $\frac{2}{3}$ and $\frac{5}{7}$ to a common denominator.

A. $\frac{14}{42}$, $\frac{30}{42}$.

8. Reduce $14\frac{3}{4}$ and $\frac{5}{6}$ to a common denominator.

A. $14\frac{18}{24}$, $\frac{20}{24}$.

9. Reduce $10\frac{3}{8}$ and $\frac{1}{2}$ of $\frac{5}{6}$ to a common denominator.

A. $10\frac{36}{96}$, $\frac{40}{96}$.

10. Reduce $8\frac{11}{21}$ and $14\frac{1}{7}$ to a common denominator.

A. $8\frac{77}{147}$, $14\frac{21}{147}$.

Notwithstanding the preceding rule finds a common denominator, it does not always find the least common denominator. But, since the common denominator is the product of all the given denominators into each other, it is plain, that this product (¶ XLII.) is a common multiple of all these several denominators; consequently, the least common multiple found by ¶ XLII. will be the least common denominator.

11. What is the least common denominator of $\frac{2}{3}$, $\frac{5}{6}$ and $\frac{1}{2}$?

OPERATION.

```
3)3 . 6 . 2
  ---------
2)1 . 2 . 2
  ---------
  1 . 1 . 1
Ans. 2×3=6
```

Now, as the denominator of each fraction is 6ths., it is evident that the numerator must be proportionably increased; that is, we must find how many 6ths each fraction is; and, to do this, we can take $\frac{2}{3}$, $\frac{5}{6}$, and $\frac{1}{2}$ of the 6ths., thus:

$\frac{2}{3}$ of 6 = 4, *the new numerator, written over the* 6, = $\frac{4}{6}$.

$\frac{5}{6}$ of 6 = 5, *the new numerator, written over the* 6, = $\frac{5}{6}$.

$\frac{1}{2}$ of 6 = 3, *the new numerator, written over the* 6, = $\frac{3}{6}$.

Ans. $\frac{4}{6}$. $\frac{5}{6}$. $\frac{3}{6}$.

Hence, to find the least common denominator of several

fractions, find the least common multiple of the deno[minators] for the common denominator, which, multiplied by each fraction, will give the new numerator for said fraction.

12. Reduce $\frac{3}{4}$ and $\frac{5}{8}$ to the least common denominator.

A. $\frac{6}{8}$, $\frac{5}{8}$.

13. Reduce $\frac{3}{5}$ and $\frac{1}{10}$ to the least common denominator.

A. $\frac{6}{10}$, $\frac{1}{10}$.

14. Reduce $14\frac{5}{6}$ and $13\frac{3}{4}$ to the least common denominator.

A. $14\frac{10}{12}$, $13\frac{9}{12}$.

Fractions may be reduced to a common, and even to the least common denominator, by a method much shorter than either of the preceding, by multiplying both the terms of a fraction by any number, that will make its denominator like the other denominators, for a common denominator; or by dividing both the terms of a fraction by any numbers that will make the denominators alike, for a common denominator. This method oftentimes will be found a very convenient one in practice.

Reduce $\frac{3}{4}$ and $\frac{2}{8}$ to a common, and to a least common, denominator.

$\frac{3}{4}\times2=\frac{6}{8}$; then $\frac{6}{8}$ and $\frac{2}{8}$=*common denominator*, *A.*

$2)\frac{2}{8}=\frac{1}{4}$; then $\frac{3}{4}$ and $\frac{1}{4}$=*least common denominator*, *A.*

In this example both the terms of one fraction are multiplied, and both the terms of the other divided, by the same number; consequently, (¶ XXXVII.,) the value is not altered.

Reduce $\frac{9}{12}$ and $\frac{1}{3}$ to the least common denominator.

A. $\frac{9}{12}$, $\frac{4}{12}$.

Reduce $\frac{500}{1000}$ and $\frac{1}{4}$ to the least common denominator.

A. $\frac{2}{4}$, $\frac{1}{4}$.

Reduce $\frac{80}{240}$ and $\frac{5}{24}$ to the least common denominator.

A. $\frac{8}{24}$, $\frac{5}{24}$.

Reduce $\frac{4}{80}$ and $\frac{3}{20}$ to the least common denominator.

A. $\frac{1}{20}$, $\frac{3}{20}$.

ADDITION OF FRACTIONS.

¶ **XLIV.** 1. A father gave money to his sons as follows, to William $\frac{1}{8}$ of a dollar, to Thomas $\frac{2}{8}$, and to Rufus $\frac{3}{8}$; how much is the amount of the whole? How much are $\frac{1}{8}$, $\frac{2}{8}$, and $\frac{3}{8}$, added together?

2. A mother divides a pie into 6 equal pieces, or parts, and

gives $\frac{2}{6}$ to her son, and $\frac{3}{6}$ to her daughter; how much did she give away in all? How much are $\frac{2}{6}$ and $\frac{3}{6}$ added together?

3. How much are $\frac{1}{9}+\frac{3}{9}+\frac{2}{9}$?

4. How much are $\frac{3}{11}+\frac{4}{11}+\frac{2}{11}$?

5. How much are $\frac{2}{13}+\frac{7}{13}+\frac{1}{13}$?

6. How much are $\frac{3}{20}+\frac{6}{20}+\frac{2}{20}$?

When fractions like the above have a common denominator expressing parts of a whole of the same size, or value, it is plain, that their numerators, being like parts of the same whole, may be added as in whole numbers; but sometimes we shall meet with fractions, whose denominators are unlike, as, for example, to add $\frac{1}{2}$ and $\frac{1}{4}$ together. These we cannot add as they stand; but, by reducing their denominators to a common denominator, by ¶ XLIII., they make $\frac{4}{8}$ and $\frac{2}{8}$, which, added together as before, make $\frac{6}{8}$, *Ans.*

1. Bought 3 loads of hay, the first weighing $19\frac{3}{4}$ cwt., the second $20\frac{1}{5}$ cwt. and the third $22\frac{2}{3}$ cwt.: what was the weight of the whole?

$\frac{3}{4}$, $\frac{1}{5}$, $\frac{2}{3}$, reduced to a common denominator, are equal to $\frac{45}{60}$, $\frac{12}{60}$ and $\frac{40}{60}$: these, joined to their respective whole numbers, give the following expressions, viz.

OPERATION.

Cwt.		Cwt.
$19\frac{3}{4}$	=	$19\frac{45}{60}$
$20\frac{1}{5}$	=	$20\frac{12}{60}$
$22\frac{2}{3}$	=	$22\frac{40}{60}$

Ans. $62\frac{37}{60}$ *cwt.*

By adding together all the 60ths, viz. 45, 12 *and* 40, *we have* $\frac{97}{60}=1\frac{37}{60}$; *then writing the* $\frac{37}{60}$ *down, and carrying the whole number,* 1, *to the amount of the column of whole numbers, makes* 62, *which, joined with* $\frac{37}{60}$, *makes* $62\frac{37}{60}$, *Ans.*

2. How much is $\frac{1}{2}$ of $\frac{3}{4}$, and $\frac{2}{3}$, added together? *$\frac{1}{2}$ of $\frac{3}{4}=\frac{3}{8}$; then $\frac{3}{8}$ and $\frac{2}{3}$, reduced to a common denominator, give $\frac{9}{24}$ and $\frac{16}{24}$, which, added together as before, give $\frac{25}{24}=1\frac{1}{24}$, Ans.*

From these illustrations we derive the following

RULE.

I. *How do you prepare fractions to add them?* *A.* Reduce compound fractions to simple ones, then all the fractions to a common or least common denominator.

II. *How do you proceed to add?* *A.* Add their numerators.

More Exercises for the Slate.

3. What is the amount of $16\frac{3}{4}$ yards, $17\frac{1}{5}$ yds. and $3\frac{1}{3}$ yards? *A.* $37\frac{17}{60}$.

4. Add together $\frac{3}{5}$ and $\frac{4}{7}$. *A.* $1\frac{6}{35}$.

5. Add together $\frac{5}{7}$, $\frac{8}{9}$ and $\frac{8}{11}$. *A.* $2\frac{229}{693}$.

6. Add together $\frac{2}{13}$, $\frac{8}{9}$ and $\frac{1}{4}$. *A.* $1\frac{137}{468}$.

7. Add together $14\frac{3}{4}$ and $15\frac{5}{6}$. *A.* $30\frac{7}{12}$.

8. Add together $\frac{1}{2}$ of $\frac{5}{6}$ and $\frac{3}{8}$ of $\frac{1}{3}$. *A.* $\frac{156}{288}$.

9. Add together $3\frac{1}{5}$, $\frac{1}{2}$ of $\frac{5}{6}$, and $\frac{7}{8}$. *A.* $4\frac{59}{120}$.

SUBTRACTION OF FRACTIONS.

¶ **XLV.** 1. William, having $\frac{3}{4}$ of an orange, gave $\frac{1}{4}$ to Thomas; how much had he left? How much does $\frac{1}{4}$ from $\frac{3}{4}$ leave?

2. Harry had $\frac{3}{8}$ of a dollar, and Rufus $\frac{5}{8}$; what part of a dollar has Rufus more than Harry? How much does $\frac{3}{8}$ from $\frac{5}{8}$ leave?

3. How much does $\frac{10}{16}$ from $\frac{12}{16}$ leave?

4. How much does $\frac{10}{25}$ from $\frac{12}{25}$ leave?

5. How much does $\frac{6}{15}$ from $\frac{9}{15}$ leave?

6. How much does $\frac{50}{100}$ from $\frac{72}{100}$ leave?

From the foregoing examples, it appears that fractions may be subtracted by subtracting their numerators, as well as added, and for the same reason.

1. Bought $20\frac{2}{3}$ yards of cloth, and sold $15\frac{3}{4}$ yards; how much remained unsold?

OPERATION.

$\frac{2}{3}$ and $\frac{3}{4}$, reduced to a common denominator, make $\frac{8}{12}$ and $\frac{9}{12}$; then,

$20\frac{2}{3} = 20\frac{8}{12}$
$15\frac{3}{4} = 15\frac{9}{12}$

$4\frac{11}{12}$ *yds., Ans.*

In this example, we cannot take $\frac{9}{12}$ *from* $\frac{8}{12}$, *but, by borrowing* 1 (unit), *which is* $\frac{12}{12}$, *we can proceed thus:* $\frac{12}{12}$ *and* $\frac{8}{12}$ *are* $\frac{20}{12}$, *from which taking* $\frac{9}{12}$, *or* 9 *parts from* 20 *parts, leaves* 11 *parts, that is,* $\frac{11}{12}$; *then, carrying* 1 (*unit, for that which I borrowed*) *to* 15, *makes* 16; *then,* 16 *from* 20 *leaves* 4, *which, joined with* $\frac{11}{12}$, *makes* $4\frac{11}{12}$, *Ans.*

2. From $\frac{2}{5}$ take $\frac{1}{6}$. $\frac{2}{5}$ and $\frac{1}{6}$, *reduced to a common denominator, give* $\frac{12}{30}$ *and* $\frac{5}{30}$; *then,* $\frac{5}{30}$ *from* $\frac{12}{30}$ *leaves* $\frac{7}{30}$, *Ans.*

From these illustrations we derive the following

RULE.

1. *What is the rule?* *A.* Prepare the fractions as in addition; then, the difference of the numerators, written over the denominator, will give the difference required.

More Exercises for the Slate.

2. From $\frac{40}{41}$ take $\frac{5}{6}$. *A.* $\frac{35}{246}$.
3. From $\frac{11}{15}$ take $\frac{4}{11}$. *A.* $\frac{61}{165}$.
4. From $\frac{15}{16}$ take $\frac{4}{7}$. *A.* $\frac{41}{112}$.
5. From $\frac{7}{8}$ take $\frac{2}{5}$. *A.* $\frac{19}{40}$.
6. From $1\frac{15}{16}$ take $\frac{5}{9}$. *A.* $1\frac{55}{144}$.
7. From $\frac{3}{4}$ of $\frac{4}{8}$ take $\frac{1}{9}$. *A.* $\frac{19}{72}$.
8. From $\frac{1}{2}$ of $\frac{6}{10}$ take $\frac{1}{2}$ of $\frac{9}{26}$. *A.* $\frac{66}{520}$.
9. From $19\frac{1}{2}$ take $\frac{2}{3}$ of 19. *A.* $\frac{41}{6}=6\frac{5}{6}$.

DIVISION OF FRACTIONS.

¶ XLVI. To divide a Whole Number by a Fraction.

Lest you may be surprised, sometimes, to find in the following examples a quotient very considerably larger than the dividend, it may here be remarked, by way of illustration, that 4 is contained in 12, 3 times, 2 in 12, 6 times, 1 in 12, 12 times; and a half ($\frac{1}{2}$) is evidently contained twice as many times as 1 whole; that is, 24 times. Hence, *when the divisor is* 1 (unit), *the quotient will be the same as the dividend; when the divisor is more than* 1 (unit), *the quotient will be less than the dividend; and when the divisor is less than* 1 (unit), *the quotient will be more than the dividend.*

1. At $\frac{3}{4}$ of a dollar a yard, how many yards of cloth can you buy for 6 dollars? 1 dollar is $\frac{4}{4}$, and 6 dollars are 6 times $\frac{4}{4}$, that is, $\frac{24}{4}$; then $\frac{3}{4}$ or 3 parts are contained in $\frac{24}{4}$, or 24 parts, as *many times as* 3 is contained in 24; that is, 8 times. *A.* 8 yards.

In the foregoing example, the 6 was first brought in 4ths, or quarters, by multiplying it by the denominator of the divisor, thereby reducing it to parts of equal size with the divisor; hence we derive the following

RULE.

I. *How do you proceed to divide a whole number by a fraction?* *A.* Multiply the dividend by the denominator of the dividing fraction, and divide the product by the numerator.

Exercises for the Slate.

2. At $\frac{5}{16}$ of a dollar a bushel, how many bushels of rye can I have for 80 dollars?

OPERATION.

```
         80 dividend.
         16 denominator.
        ----
        480
        80
        ----
Numer. 5)1280
        ----
Quotient,  256 bushels, Ans.
```

In this example, we see more fully illustrated the fact that division is the opposite of multiplication; for, to *multiply* 80 by $\frac{5}{16}$, we should *multiply* by the numerator, and *divide* by the denominator, ¶ XXXIX.

3. If a family consume $\frac{3}{8}$ of a quarter of flour in one week, how many weeks will 48 quarters last the same family?
A. 128 weeks.

4. If you borrow of your neighbour $\frac{1}{10}$ of a bushel of meal at one time, how many times would it take you to borrow 96 bushels? *A.* 960 times.

5. How many yards of cloth, at $\frac{1}{5}$ of a dollar a yard, may be bought for 200 dollars? *A.* 1000 yards.

6. How many times is $\frac{36}{7}$ contained in 720? *A.* 140.

7. How many times is $8\frac{1}{3}$ contained in 300? *Reduce $8\frac{1}{3}$ to an improper fraction.* *A.* 36.

8. Divide 620 by $8\frac{2}{11}$. *A.* $75\frac{7}{9}$.

9. Divide 84 by $\frac{168}{320}$. *A.* 160.

10. Divide 92 by $4\frac{1}{2}$. *A.* $20\frac{4}{9}$.

11. Divide 100 by $2\frac{3}{4}$. *A.* $36\frac{4}{11}$.

12. Divide 86 by $15\frac{7}{8}$. *A.* $5\frac{53}{127}$.

13. How many rods in 220 yards? *A.* 40 rods.

14. How many sq. rods in 1210 sq. yards? *A.* 40 sq. rods.
15. How many barrels in 1260 gallons? *A.* 40 barrels.

¶ XLVII. To divide one Fraction by another.

1. At $\frac{1}{4}$ of a cent an apple, how many apples may be bought for $\frac{3}{4}$ of a cent? How many times $\frac{1}{4}$ in $\frac{3}{4}$? How many times $\frac{2}{5}$ in $\frac{6}{5}$?

2. William gave $\frac{3}{8}$ of a dollar for one orange; how many oranges, at that rate, can he buy for $\frac{6}{8}$ of a dollar? How many for $\frac{9}{8}$ of a dollar? For $\frac{12}{8}$? For $\frac{24}{8}$? For $\frac{27}{8}$? For $\frac{60}{8}$?

Hence we see that fractions, having a common denominator, may be divided by dividing their numerators, as well as subtracted and added, and for the same reason.

1. At $\frac{1}{3}$ of a dollar a yard, how many yards of cloth may be bought for $\frac{3}{4}$ of a dollar?

OPERATION.

Reducing the fractions $\frac{1}{3}$ *and* $\frac{3}{4}$ *to a common denominator, thus:—*

$$\frac{1}{3} \quad \frac{3}{4}$$

$$\frac{4}{12}, \quad \frac{9}{12}$$

Then $\frac{4}{12}$ *is contained in* $\frac{9}{12}$ *as many times as 4 is contained in* $9, = 2\frac{1}{4}$.

A. $2\frac{1}{4}$ yards.

In this example, as the common denominator is not used, it is plain that we need not find it, but only multiply the numerators by the same numbers as before. This will be found to consist in multiplying the numerator of the divisor into the denominator of the dividend, and the denominator of the divisor into the numerator of the dividend. But it will be found to be more convenient, in practice, to invert the divisor, then multiply the upper terms together for a numerator, and the lower terms for a denominator; thus, taking the last example;

$\frac{1}{3}$ *and* $\frac{3}{4}$, *by inverting the divisor, become* $\frac{3}{1}$ *and* $\frac{3}{4}$; *then,* $\frac{3}{1} \times \frac{3}{4} = \frac{9}{4} = 2\frac{1}{4}$ *yards, as before, Ans.*

PROOF. $\frac{9}{4}$, *the quotient, multiplied by* $\frac{1}{3}$, *the divisor, thus,* $\frac{9}{4} \times \frac{1}{3}$, *gives* $\frac{9}{12} = \frac{3}{4}$ *the divisor*

From these illustrations we derive the following

RULE.

1. *How do you proceed to divide one fraction by another?*
A. I invert the divisor, then multiply the upper terms together for a new numerator, and the lower for a new denominator.

Note. Mixed numbers must be reduced to improper fractions, and compound to simple terms.

PROOF. It would be well for the pupil to prove each result, as in Simple Multiplication, by multiplying the divisor and quotient together, to obtain the dividend.

More Exercises for the Slate.

2. At $\frac{1}{5}$ of a dollar a peck, how many pecks of salt may be bought for $\frac{7}{8}$ of a dollar? *A.* $4\frac{3}{8}$ pecks.
3. Divide $\frac{3}{7}$ by $\frac{3}{14}$. *A.* $\frac{42}{21}=2$
4. Divide $\frac{6}{7}$ by $\frac{1}{25}$. *A.* $\frac{150}{7}=21\frac{3}{7}$.
5. Divide $1\frac{4}{5}$ by $\frac{6}{11}$. *A.* $3\frac{3}{10}$.
6. Divide $9\frac{1}{4}$ by $\frac{1}{2}$ of $\frac{4}{8}$. *A.* 37.
7. How many times is $\frac{1}{3}$ contained in $\frac{1}{2}$? *A.* $1\frac{1}{2}$.
8. How many times is $\frac{5}{16}$ contained in $\frac{29}{31}$? *A.* $2\frac{154}{155}$.
9. What number multiplied by $\frac{2}{7}$ will make $\frac{8}{13}$? *A.* $2\frac{1}{26}$.

REDUCTION OF FRACTIONS.

It will be recollected, that in Reduction (¶ XXIX.) whole numbers were brought from higher to lower denominations by multiplication, and from lower to higher denominations by division; hence, fractions of one denomination may be reduced to another after the same manner, and by the same rules.

¶ XLVIII. To reduce Whole Numbers to the Fraction of a greater Denomination.

1. What part of 2 miles is 1 mile?
2. What part of 4 miles is 1 mile? Is 2 miles? Is 3 miles?
3. What part of 1 yd. is 1 qr.? Is 2 qrs.? Is 3 qrs.?
4. What part of 8 gallons is 1 gallon? Is 3 gallons?
5. What part of 9 oz. is 1 oz.? Is 2 oz.? Is 5 oz.?
6. What part of 7 yds. is 1 yd.? Is 6 yds.? Is 7 yds.?
7. What part of $21 is $17? Is $11? Is $13?
8. What part of 271 inches is 11 in.? Is 251 in.?
9. What part of 1 month is 1 day? Is 2 days?
10. What part of 1 hour is 11 minutes? Is 21 minutes?
11. What part of 19 cents is 11 cents? Is 3 cents?
12. What part of 1 d. is 1 farthing? Is 2 qrs.? Is 3 qrs.?
13. What part of 1 s. is 1 d.? Is 2 d.? Is 3d.?

1. What part of a bushel is 3 pks. 4 qts.?

OPERATION.

1 bu.	3 pks. 4 qts.
4	8
4	24
8	4
32 *Denom.*	28 *Numer.*

Then $\frac{28}{32}=\frac{7}{8}$, *Ans.*

1 bu. = 32 *qts.; and* 3 *pks* 4 *qts.* = 28 *qts.* Then, *as* 32 *qts. make* 1 *bushel,* 1 *qt. is* $\frac{1}{32}$ *of a bushel, and* 28 *qts* 28 *times* $\frac{1}{32}=\frac{28}{32}=\frac{7}{8}$ *bu. Ans., the same as before.*

From these illustrations we derive the following

RULE.

I. *How do you obtain the numerator?* *A.* Bring the given denominations to the lowest denomination mentioned for a numerator.

II. *How do you obtain the denominator?* *A.* Bring 1 (or an integer) of that higher denomination into the same denomination for a denominator.

More Exercises for the Slate.

2. What part of 1£ is 2s. 6d.? *A.* $\frac{30}{240}=\frac{1}{8}$.
3. What part of 1 cwt. is 3 qrs. 15 lbs. 14 oz.? *A.* $\frac{799}{896}$.
4. What part of 1 yd. is 3 qrs. 3 na.? *A.* $\frac{15}{16}$.
5. What part of 1 bu. is 3 pecks, 7 qts. 1 pt.? *A.* $\frac{63}{64}$.
6. What part of 1 tun is 1 gal. 0 qts. 2 pts. 1 gil.? *A.* $\frac{41}{8064}$.
7. What part of 15 pipes is 25 gals.? *A.* $\frac{5}{378}$.
8. What part of 2 m. is 7 fur. 11 in. 2 b. c.? *A.* $\frac{33271}{76032}$.
9. What part of 1 mo. is 19 days? *A.* $\frac{19}{30}$.
10. What part of 1 mo. is 25 days, 13 hours? *A.* $\frac{613}{720}$.
11. What part of 1 mo. is 22 days, 15 h. 1 min. *A.* $\frac{32581}{43200}$.

¶ XLIX. To reduce a Fraction to Whole Numbers of less Denominations,

OR,

To find the Value of a Fraction.

1. How much is $\frac{1}{4}$ of a shilling? How much $\frac{1}{16}$ of a lb. $\frac{5}{16}$ *of a lb.?* $\frac{9}{16}$ of a lb.? $\frac{11}{16}$ of a lb.? $\frac{14}{16}$ of a lb.? $\frac{1}{28}$ of 1 *of a cwt.?* $\frac{2}{28}$? $\frac{5}{28}$? $\frac{15}{28}$? $\frac{27}{28}$? $\frac{1}{2}$ an hour? $\frac{3}{4}$? $\frac{3}{5}$?

Operation by Slate illustrated.

1. What is the value of $\frac{5}{6}$ of a pound?

OPERATION.

```
Numer. 5
       20 s.
       ———
Denom. 6)100(16 s.
         6
         —
         40
         36
         —
16 s. 8 d. Ans.   4
         12
         ——
       6)48(8 d.
         48
         ——
```

How do you proceed in this example? and why? *A.* As 1£ = 20 s., $\frac{5}{6}$ of one pound is the same as $\frac{5}{6}$ of 20 s., and, to get $\frac{5}{6}$ of 20, we multiply the numerator 5 and 20 together, making 100; which, divided by the denominator 6, gives 16 s. and $\frac{4}{6}$ of another shilling remaining. This $\frac{4}{6} = \frac{4}{6}$ of 12 d.; then, $\frac{4}{6}$ of 12 d. = 8 d.

From these illustrations we derive the following

RULE.

I. *What do you multiply the numerator by?* *A.* By as many of the next denomination as make one of that; that is, pounds by what makes a pound, ounces by what makes an ounce, as in reduction of whole numbers.

II. *What do you divide the product by?* *A.* By the denominator.

III. *If there be a remainder, how do you proceed?* *A.* Multiply and divide as before.

More Exercises for the Slate.

2. What is the value of $\frac{3}{4}$ of a cwt.? *A.* 3 *qrs.*

3. What is the value of $\frac{1}{3}$ of an acre? *A.* 1 *rood,* $13\frac{1}{3}$ *rds*

4. What is the value of $\frac{20}{23}$ of a pound Troy? *A.* 10 *oz.* 8 *pwts.* $16\frac{16}{23}$.

5. What is the value of $\frac{213}{272}$ of a hhd.? *A.* 49 *gallons* $1\frac{92}{272}$ *qts.*

6. What is the value of $\frac{713}{371}$ of a pound avoirdupois? *A.* 1 *lb.* $14\frac{278}{371}$ *oz.*

7. What is the value of $\frac{500}{630}$ of a hogshead? *A.* 50 *gallons.*

8. What is the value of $\frac{9}{13}$ of a day? *A.* 16 *h.* 36 *m.* $55\frac{5}{13}$ *sec.*

¶ L. To reduce Fractions of a higher Denomination into lower.

We have seen (¶ XXXVIII.) that fractions are multiplied by multiplying their numerators, or dividing their denominators.

1. Reduce $\frac{1}{480}$ £ to the fraction of a penny.

OPERATION.

Numer.	1
	20 s.
	20
	12 d.
New numer.	240

Then, $\frac{240}{480}$ (*Numer.* 240 over *Denom.* 480) $= \frac{1}{2}$d. *Ans.*,

In this example, we multiply the 1, in $\frac{1}{480}$ *as in Reduction of whole numbers, viz., pounds by what makes a pound, shillings by what makes a shilling, &c. But this operation may be expressed differently, thus;* $\frac{1}{480} \times 20 \times 12 = \frac{240}{480} = \frac{1}{2}$*d.; or, by dividing the denominators thus;* $\frac{1}{480} \div 20 = \frac{1}{24} \div 12 = \frac{1}{2}$ d., *Ans., as before, in its lowest terms.*

RULE.

How, then, would you proceed?

A. Multiply the fraction as in Reduction of whole numbers

More Exercises for the Slate.

2. Reduce $\frac{1}{240}$ of a pound to the fraction of a shilling. *A.* $\frac{1}{12}$ s.

3. Reduce $\frac{1}{1920}$ of a pound to the fraction of a farthing. *A.* $\frac{1}{2}$ qr.

4. Reduce $\frac{1}{1008}$ of a hogshead to the fraction of a gallon. *A.* $\frac{1}{16}$ gal.

5. Reduce $\frac{5}{171}$ of a bushel to the fraction of a quart. *A.* $\frac{160}{171}$ qt.

6. Reduce $\frac{1}{1441}$ of a day to the fraction of a minute. *A.* $\frac{1440}{1441}$ m.

7. Reduce $\frac{5}{1008}$ of a cwt. to the fraction of a pound. *A.* $\frac{5}{9}$ lb.

8. Reduce $\frac{4}{2520}$ of a hhd. to the fraction of a pint. *A.* $\frac{4}{5}$ pt.

9. Reduce $\frac{3}{180}$ of a pound to the fraction of a shilling. *A.* $\frac{1}{3}$ s.

¶ LI. To reduce Fractions of a lower Denomination into a higher.

We have seen, that, to divide a fraction, (¶ XL.) we must multiply the denominator, or divide the numerator.

This rule is the reverse of the last, (¶ L.), and proves it.

1. Reduce $\frac{1}{2}$ of a penny to the fraction of a pound.

OPERATION.

$$\begin{array}{lr} \textit{Denom.} & 2 \\ & 12 \\ \hline & 24 \\ & 20 \\ \hline \textit{New denom.} & 480 \end{array}$$

Then, $\frac{1}{480}$, *Ans.*

In this example, we divide as in Reduction, (¶ XXIX), viz. pence by pence, shillings by shillings; but, in order for this, we must either multiply the denominator or divide the numerator by the same numbers that we should divide by in Reduction of whole numbers. The same result will be obtained if performed thus: $\frac{1}{2} \times \overset{d.}{12} \times \overset{s.}{20} = \frac{1}{480}$ £, *Ans.*

Hence the following

RULE.

1. *How do you proceed?* *A.* Divide as in Reduction of whole numbers.

More Exercises for the Slate.

2. Reduce $\frac{1}{12}$ of a shilling to the fraction of a pound. *A.* $\frac{1}{240}$ £.

3. Reduce $\frac{1}{2}$ of a farthing to the fraction of a pound. *A.* $\frac{1}{1920}$ £.

4. Reduce $\frac{1}{16}$ of a gallon to the fraction of a hogshead. *A.* $\frac{1}{1008}$ *hhd.*

5. Reduce $\frac{160}{171}$ of a quart to the fraction of a bushel. *A.* $\frac{5}{171}$ *bu.*

6. Reduce $\frac{1440}{1441}$ of a minute to the fraction of a day. *A.* $\frac{1}{1441}$.

7. Reduce $\frac{5}{9}$ of a pound to the fraction of a cwt. *A.* $\frac{5}{1008}$

8. Reduce $\frac{4}{5}$ of a pint to the fraction of a hhd. *A.* $\frac{4}{2520} = \frac{1}{630}$.

9. Reduce $\frac{3}{9}$ of a shilling to the fraction of a pound. *A.* $\frac{3}{180} = \frac{1}{60}$.

DECIMAL FRACTIONS.

¶ LII. Q. When such fractions as these occur, viz. $\frac{5}{100}$, $\frac{250}{1000}$, how is a unit supposed to be divided? A. 10 equal parts, called tenths; and each tenth into 10 other parts, called hundredths; and each hundredth into 10 equal parts, called thousandths, &c.

Q. How is it customary to write such expressions? A taking away the denominator, and placing a comma befor numerator.

Let me see you write down, in this manner, $\frac{5}{10}$, $\frac{25}{100}$, $\frac{525}{1000}$.

Q. What name do you give to fractions written in this ner? A. Decimal Fractions.

Q. Why called decimal? A. From the Latin word de signifying *ten;* because they increase and decrease in a fold proportion, like whole numbers.

Q. What are all other fractions called? A. Vulgar, or mon fractions.

Q. In whole numbers, we are accustomed to call the r hand figure, units, from which we begin to reckon, or nu rate; hence it was found convenient to make the same pla starting point in decimals; and, to do this, we make use comma; what, then, is the use of this comma? A. It me shows where the units' place is.

Q. What are the figures on the left of the comma call A. Whole numbers.

Q. What are the figures on the right of the comma call A. Decimals.

Q. What, then, may the comma properly be called? A. Se ratrix.

Q. Why? A. Because it *separates* the decimals from whole numbers.

Q. What is the first figure at the right of the separa called? A. 10ths.

Q. What is the second, third, fourth, &c.? A. The sec is hundredths, the third thousandths, the fourth ten thousand and so on, as in the numeration of whole numbers.

Let me see you write down again $\frac{5}{10}$ *in the form of a de mal.*

Q. As the first figure at the right of the separatrix is ten in writing down $\frac{5}{100}$, then, where must a cipher be plac A. In the tenths' place.

Let me see you write down in the form of a decimal $\frac{5}{1}$ A. ,05.

Write down $\frac{7}{1000}$, $\frac{8}{100}$, $\frac{1}{100}$.

Q. How would you write down in decimals $\frac{7}{1000}$? *A.* By placing 2 ciphers at the right of the separatrix, that is, before the 7.

Let me see you write it down? *A.* ,007.

Let me see you write down $\frac{2}{1000}$? *A.* ,002.

Q. Why do you write 2 down with 2 ciphers before it? *A.* Because in $\frac{2}{1000}$, the 2 is thousandths; consequently, the 2 must be thousandths when written down in decimals.

Q. What does ,5 signify? *A.* $\frac{5}{10}$.

Q. What does ,05 signify? *A.* $\frac{5}{100}$.

Q. Now, as $\frac{5}{10} = \frac{1}{2}$, and as multiplying $\frac{5}{100}$ by 10 produces $\frac{50}{100}$, which is also equal to $\frac{1}{2}$, how much less in value is ,05 than ,5? *A.* Ten times.

Q. Why? *A.* Because the parts in $\frac{5}{100}$ are ten times *smaller* than in $\frac{5}{10}$; and, as the numerator is the same in both expressions, consequently, the value is lessened 10 times.

Q. How, then, do decimal figures decrease in value from the left towards the right? *A.* In a tenfold proportion.

Q. What does ,50 mean. *A.* 5 tenths, and no hundredths.

Q. What, then, is the value of a cipher at the right of decimals? *A.* No value.

Q. We have seen that ,5 is 10 times as much in value as ,05, or $\frac{5}{100}$; what effect, then, does a cipher have placed at the left of decimals? *A.* It decreases their value in a tenfold proportion.

Q. Since decimals decrease from the left to the right in a tenfold proportion, how, then, must they increase from the right to the left? *A.* In the same proportion.

Q. Since it was shown, that ,5 $= \frac{5}{10}$; ,25 $= \frac{25}{100}$, what, then, will always be the denominator of any decimal expression? *A.* The figure 1, with as many ciphers placed at the right of it as there are decimal places.

Let me see you write down the following decimals on your slate, and change them into a common, or vulgar fraction, by placing their proper denominators under each, viz. ,5 ,05 ,005 ,62 ,0225 ,37.

Q. ,25 is $\frac{25}{100} = \frac{1}{4}$, and ,5 is $\frac{5}{10} = \frac{1}{2}$; which, then, is the most in value, ,25 or ,5?

Q. By what, then, is the value of any decimal figures determined? *A.* By their distance from the units' place, or separatrix.

Q. When a whole number and decimal are joined together, thus; 2,5, what is the expression called? *A.* A mixed number.

Q. As any whole number may be reduced to tenths, hundredths, thousandths, &c. by annexing ciphers, (for multiplying by 10, 100, &c.) thus, 5 is 50 tenths, 500 hundredths, &c.; how, then, may any mixed number be read, as 25,4? *A*. 254 tenths, giving the name of the decimal to all the figures.

Q. How is 25,36 read? *A*. 2536 hundredths.

Q. How is 5,125 read? *A*. 5125 thousandths.

Q. What would 5125 thousandths be, written in the form of a vulgar or common fraction? *A*. $\frac{5125}{1000}$.

This is evident from the fact, that $\frac{5125}{1000}$ (an improper fraction), reduced to a mixed number again, is equal to 5,125.

The pupil may learn the names of any decimal expression, as far as ten-millionths, also how to read or write decimals, from the following

TABLE.

	Hundreds.	Tens.	Units.		Tenths.	Hundredths.	Thousandths.	Ten-Thousandths.	Hundred-Thousandths.	Millionths.	Ten-Millionths.	
$\frac{5}{10}$=	.	.	.	,	5	.	.	.	.	.	.	*read* 5 Tenths.
$\frac{6}{100}$=	.	.	.	,	0	6	.	.	.	.	.	*read* 6 Hundredths.
$\frac{25}{1000}$=	.	.	.	,	0	2	5	.	.	.	.	*read* 25 Thousandths.
$\frac{1328}{10000}$=	.	.	.	,	1	3	2	8	.	.	.	*read* 1328 Ten-Thousandths
$7\frac{8}{10}$=	.	.	7	,	8	.	.	.	.	.	.	*read* 7, and 8 Tenths.
$6\frac{9}{1000000}$=	.	.	6	,	0	0	0	0	0	9	.	*read* 6, and 9 Millionths.
$26\frac{25}{100}$=	.	2	6	,	2	5	.	.	.	.	.	*read* 26, and 25 Hundredths.
$3\frac{8}{10000000}$=	.	.	3	,	0	0	0	0	0	0	8	*read* 3, and 8 Ten-Millionths.
365=	3	6	5	,	0	0	0	0	0	0	0	*read* 365.

Exercises for the Slate.

Write in decimal form 7 tenths, 42 hundredths, 62 and 25 hundredths, 7 and 426 thousandths, 24 thousandths, 3 ten-thousandths, 4 hundredths, 2 ten-thousandths, 3 millionths.

Write the fractional part of the following numbers in the form

of decimals, viz. $6\frac{7}{10}$, $\frac{42}{100}$, $62\frac{25}{100}$, $2\frac{2}{10}$, $3\frac{5}{100}$, $262\frac{5}{1000}$, $32\frac{62}{1000}$, $2\frac{2}{1000000}$, $45\frac{5}{1000000}$, $7\frac{202}{100000000}$, $5\frac{1}{10000}$.

Write the following decimal numbers in the form of vulgar or common fractions, then reduce them to their lowest terms by ¶ XXXVII; *thus,* $2,5=2\frac{5}{10}=2\frac{1}{2}$ *in its lowest terms.*

1. 45,5	*A.* $45\frac{1}{2}$	7. 6,28	*A.* $6\frac{7}{25}$
2. 9,25	*A.* $9\frac{1}{4}$	8. 6,005	*A.* $6\frac{1}{200}$
3. 23,75	*A.* $23\frac{3}{4}$	9. 3,00025	*A.* $3\frac{1}{4000}$
4. 11,8	*A.* $11\frac{4}{5}$	10. 6,08	*A.* $6\frac{2}{25}$
5. 19,9	*A.* $19\frac{9}{10}$	11. 9,2	*A.* $9\frac{1}{5}$
6. 25,255	*A.* $25\frac{51}{200}$	12. 7,000005	*A.* $7\frac{1}{200000}$

Q. What money is adapted to decimal rules? *A.* Federal Money.

Q. What is the money unit? *A.* The dollar.

Q. How is it so adapted? *A.* As 10 dimes make a dollar, and 10 cents a dime, &c., dimes are 10ths of a dollar, cents are 100ths, and mills are 1000ths of a dollar.

Q. How are 3 dollars 2 dimes 4 cents and 5 mills written? *A.* \$3,245.

ADDITION OF DECIMALS.

¶ LXXI. *Q.* As we have seen that decimals increase from right to left in the same proportion as units, tens, hundreds, &c., how, then, may all the operations of decimals be performed? *A.* As in whole numbers.

Note. The only difficulty, which ever arises, consists in determining where the decimal point ought to be placed. This will be noticed in its proper place.

1. A merchant bought $5\frac{2}{10}$ barrels of rice at one time for $\$27\frac{825}{1000}$, at another $\frac{62}{100}$ of a barrel for \$4,255, at another $\frac{278}{1000}$ of a barrel for $\$\frac{72}{100}$, and at another $\frac{89}{100}$ of a barrel for $\$2\frac{627}{1000}$; how many barrels did he buy in all? and what did they cost him?

OPERATION.

Barrels.	Dollars.
5,2	27,825
,62	4,255
,278	0,72
,89	2,627
Ans. 6,988 *barrels for*	\$35,427

As we have seen that decimals correspond with the denominations of Federal Money, hence we may write the decimals down, placing dimes under dimes, cents under cents, &c., that is, tenths under tenths, hundredths under hundredths, &c., and add them up as in Addition of Federal Money.

From these illustrations we derive the following

RULE.

I. *How are the numbers to be written down?* *A.* Tenths under tenths, hundredths under hundredths, and so on.

II. *How do you proceed to add?* *A.* As in Simple Addition.

III. *Where do you place the separatrix?* *A.* Directly under the separating points above.

More Exercises for the Slate.

2. James bought 2,5 cwt. of sugar, 23,265 cwt. of hay, and 4,2657 cwt. of rice; how much did he buy in all? *A.* 30,0307 cwt.

3. James is $14\frac{5}{10}$ years old, Rufus $15\frac{25}{100}$, and Thomas $16\frac{75}{100}$; what is the sum of all their ages? *A.* 46,5 years.

4. William expended for a chaise $\$255\frac{2}{10}$, for a wagon $\$37\frac{25}{100}$, for a bridle $\$\frac{75}{100}$, and for a saddle $\$11\frac{255}{1000}$; what did these amount to? *A.* \$304,455.

5. A merchant bought 4 hhds. of molasses; the first contained $62\frac{4}{10}$ gallons, the second $72\frac{2657}{10000}$ gallons, the third $50\frac{2}{10}$ gallons, and the fourth $55\frac{75}{100}$ gallons; how many gallons did he buy in the whole? *A.* 240,6157 gallons.

6. James travelled to a certain place in 5 days; the first day he went $40\frac{2}{10}$ miles, the second $28\frac{37}{100}$ miles, the third $42\frac{5}{10}$ miles, the fourth $22\frac{5}{1000}$ miles, and the fifth $29\frac{42}{10000}$ miles; how far did he travel in all? *A.* 162,0792 miles.

7. A grocer, in one year, at different times, purchased the following quantity of articles, viz. 427,2623 cwt., 2789,00065 cwt., 42,000009 cwt., 1,3 cwt., 7567,126783 cwt., and 897,62 cwt.; how much did he purchase in the whole year? *A.* 11724,309742 cwt.

8. What is the amount of $\frac{6}{10}$, $245\frac{37}{100}$, $6\frac{9}{10000}$, $245\frac{75}{100000}$, $1\frac{89}{1000000}$, $\frac{6}{10000}$, $427\frac{2}{1000000}$, $4\frac{5}{10}$, $\frac{62}{10000000}$, and 1925? *A.* 2854,492472.

9. What is the amount of one, and five tenths; forty-five, and three hundred and forty-nine thousandths; and sixteen hundredths? *A.* 47,009.

SUBTRACTION OF DECIMALS.

¶ **LIV.** 1. A merchant, owing \$270,42, paid \$192,625; how much did he then owe?

OPERATION.

$270,42
$192,625
Ans. $77,795

For the reasons shown in Addition, we proceed to subtract, and point off, as in Subtraction of Federal Money.

Hence we derive the following

RULE.

I. *How do you write the numbers down?* *A.* As in Addition of Decimals.

II. *How do you subtract?* *A.* As in Simple Subtraction.

III. *How do you place the separatrix?* *A.* As in Addition of Decimals.

More Exercises for the Slate.

1. Bought a hogshead of molasses, containing 60,72 gallons; how much can I sell from it, and save 19,999 gallons for my own use? *A.* 40,721 gallons.

2. James rode from Boston to Charlestown in 4,75 minutes; Rufus rode the same distance in 6,25 minutes; what was the difference in the time? *A.* 1,5 min.

3. A merchant, having resided in Boston 6,2678 years, stated his age to be 72,625 yrs. How old was he when he emigrated to that place? *A.* 66,3572 yrs.

Note. The pupil must bear in mind, that, in order to obtain the answer, the figures in the parentheses are first to be pointed off, supplying ciphers, if necessary, then added together as in Addition of Decimals.

4. From ,65 of a barrel take ,125 of a barrel; (525) take ,2 of a barrel; (45) take ,45 of a barrel; (2) take ,6 of a barrel; (5) take ,12567 of a barrel; (52433) take ,26 of a barrel; (39) *A.* 2,13933 barrels.

5. From 420,9 pipes take 126,45 pipes; (29445) take ,625 of a pipe; (420275) take 20,12 pipes; (40078) take 1,62 pipes; (41998) take 419,89 pipes; (101) take 419,8999 pipes; (10001). *Ans.* 1536,7951 pipes.

MULTIPLICATION OF DECIMALS.

¶ LV. 1. How many yards of cloth in 3 pieces, each piece containing $20\frac{75}{100}$ yards?

OPERATION.

20,75
3
Ans. 62,25 *yds.*

In this example, since multiplication is a short way of performing addition, it is plain that we must point off as in addition, viz. directly under the separating points in the multiplicand; and, as either factor may be made the multiplicand, had there been two decimals in the multiplier also, we must have pointed off two more places for decimals, which, counting both, would make 4 Hence *we must always point off in the product as many places for decimals, as there are decimal places in both the factors.*

2. Multiply ,25 by ,5.

,25 ,5 *Ans.* ,125	In this example, there being 3 decimal places in both the factors, we point off 3 places in the product, as before directed. The reason of this will appear more evi-

dent by considering both the factors common fractions, and multiplying by ¶ XLI., thus; $,25=\frac{25}{100}$, *and* $,5=\frac{5}{10}$; *now* $\frac{25\times5}{100\times10}=\frac{125}{1000}$, *which, written decimally, is* ,125, *Ans., as before.*

3. Multiply ,15 by ,05.

OPERATION. ,15 ,05 *Ans.* ,0075	¶n this case, there not being so many figures in the product as there are decimal places in both the factors (viz. 4), we place two ciphers on the left of 75, to make as many. This will appear evident by the

following; $,15=\frac{15}{100}$ *and* $,05=\frac{5}{100}$; *then* $\frac{15\times5}{100\times100}=\frac{75}{10000}=$,0075, *Ans., the same as before.*

From these illustrations we derive the following

RULE.

I. *How do you multiply in Decimals?* *A.* As in Simple Multiplication.

II. *How many figures do you point off for decimals in the product?* *A.* As many as are in both the multiplicand and multiplier.

III. *If there be not figures enough in the product for this purpose, how would you proceed?* *A.* Prefix ciphers enough to make as many.

Q. What is the meaning of annex? *A.* To place after.

Q. What is the meaning of prefix? *A.* To place before.

More Exercises for the Slate.

4. What will 5,66 bushels of rye cost, at $1,08 a bushel? *A.* $6,1128, or $6. 11c. $2\frac{8}{10}$ m.

5. How many gallons of rum in ,65 of a barrel, each barrel containing $31\frac{5}{10}$ gallons? (20475) In ,8 of a barrel? (252) In ,42 of a barrel? (1323) In ,6 of a barrel? (189) In 1126,5

barrels? (3548475) In 1,75 barrels? (55125) In 125,626789 barrels? (39572438535). *Ans.* 39574,9238535 *gallons.*

6. What will 8,6 pounds of flour come to, at $,04 a pound? (344) At $,03 a pound? (258) At $,035 a pound? (301) At $,0455 a pound? (3913) At $,0275 a pound? (23650). *Ans.* $1,5308.

7. At $,9 a bushel, what will 6,5 bushels of rye cost? (585) What will 7,25 bushels? (6525) Will 262,555 bushels? (2362995) Will ,62 of a bushel? (558) Will 76,75 bushels? (69075) Will 1000,0005 bushels? (90000045) Will 10,00005 bushels? (9000045) *A.* $1227,307995.

DIVISION OF DECIMALS.

¶ **LVI.** In Multiplication, we point off as many decimals in the product as there are decimal places in the multiplicand and multiplier counted together; and, as division proves multiplication by making the multiplier and multiplicand the divisor and quotient, hence, *there must be as many decimal places in the divisor and quotient, counted together, as there are decimal places in the dividend.*

1. A man bought 5 yards of cloth for $8,75; how much was it a yard? $,8,75=875 *cents, or* 100*ths*; *now,* 875÷5=175 *cents, or* 100*ths,* = $1,75 *Ans.*

OR

By retaining the separatrix, and dividing as in whole numbers, thus:—

OPERATION.
5)8,75
Ans. $ 1,75

As the number of decimal places in the divisor and quotient, when counted together, must always be equal to the decimal places in the dividend, therefore, in this example, as there are no decimals in the divisor, and two in the dividend, by pointing off two decimals in the quotient, the number of decimals in the divisor and quotient will be equal to the dividend, which produces the same result as before.

2. At $2,50 a barrel, how many barrels of cider can I have for $11? $11=1100 *cents,* or 100*ths, and* $2,50=250 *cents, or* 100*ths; then, dividing* 100*ths by* 100*ths, the quotient will evidently be a whole number, thus:—*

OPERATION.

250)1100($4\frac{100}{250}$ barrels, *Ans.*
1000
——
100
250

In this example, we have for an answer 4 barrels, and $\frac{100}{250}$ of another barrel. But, instead of stopping here in the process, we may bring the remainder, 100, into 10ths, by annexing a cipher (that is, multiplying by 10), placing a decimal point at the right of 4, a whole number, to keep it separate from the 10ths, which are to follow. The separatrix may now be retained in the divisor and dividend, thus:—

OPERATION.

2,50)11,00(4,4 *Ans.*
1000
——
1000
1000
——

We have now for an answer, 4 barrels and 4 tenths of another barrel. Now, if we count the decimals in the divisor and quotient (being 3), also the decimals in the dividend, reckoning the cipher annexed as one decimal (making 3), we shall find again the decimal places in the divisor and quotient equal to the decimal places in the dividend. We learn, also, from this operation, that, *when there are more decimals in the divisor than dividend, there must be ciphers annexed to the dividend to make the decimal places equal, and then the quotient will be a whole number.*

Let us next take the 3d example in Multiplication, (¶ LV.) *and see if multiplication of decimals, as well as whole numbers, can be proved by Division.*

3. In the 3d example we were required to multiply ,15 by ,05; now we will divide the product ,0075 by ,15.

OPERATION

,15),0075(,05 *Ans.*
75

We have, in this example, (before the cipher was placed at the left of 5), four decimal places in the dividend, and two in the divisor; hence, in order to make the decimal places in the divisor and quotient equal to the dividend, we must point off two places for decimals in the quotient. But, as we have only one decimal place in the quotient, the deficiency must be supplied by prefixing a cipher.

The above operation will appear more evident by common fractions, *thus:* ,0075=$\frac{75}{10000}$, *and* ,15=$\frac{15}{100}$; *now* $\frac{75}{10000}$ *is divided by* $\frac{15}{100}$ *by inverting* $\frac{15}{100}$ (¶ XLVII.), *thus,* $\frac{100 \times 75}{15 \times 10000}$ $=\frac{7500}{150000}=\frac{5}{100}=$,05, *Ans., as before.*

From these illustrations we derive the following

RULE.

I. *How do you write the numbers down, and divide?* *A.* As in whole numbers.

II. *How many figures do you point off in the quotient for decimals?* *A.* Enough to make the number of decimal places in the divisor and quotient, counted together, equal to the number of decimal places in the dividend.

III. *Suppose that there are not figures enough in the quotient for this purpose, what is to be done?* *A.* Supply this defect by prefixing ciphers to said quotient.

IV. *What is to be done when the divisor has more decimal places than the dividend?* *A.* Annex as many ciphers to the dividend as will make the decimals in both equal.

V. *What will be the value of the quotient in such cases?* *A.* A whole number.

VI. *When the decimal places in the divisor and dividend are equal, and the divisor is not contained in the dividend, or when there is a remainder, how do you proceed?* *A.* Annex ciphers to the remainder, or dividend, and divide as before.

VII. *What places in the dividend do these ciphers take?* *A.* Decimal places.

More Exercises for the Slate.

4. At $,25 a bushel, how many bushels of oats may be bought for $300,50? *A.* 1202 bushels.

5. At $,12½, or $,125 a yard, how many yards of cotton cloth may be bought for $16? *A.* 128 yards.

6. Bought 128 yards of tape for $,64; how much was it a yard? *A.* $,005, or 5 mills.

7. If you divide 116,5 barrels of flour equally among 5 men, how many barrels will each have? *A.* 23,3 barrels.

Note. The pupil must continue to bear in mind, that before he proceeds to add together the figures in the parentheses, he must prefix ciphers, when required by the rule for pointing off.

8. At $2,255 a gallon, how many gallons of rum may be bought for $28,1875? (125) For $56,375? (25) For $112,75? (50) For $338,25? (150) *A.* 237,5 gallons.

9. If $2,25 will board one man a week, how many weeks can he be boarded for $1001,25? (445) For $500,85? (222,6) For $2006,7? (892) For $100,35? (446) For $60,75? (27) *A.* 828,4 weeks.

10. If 3,355 bushels of corn will fill one barrel, how many

barrels will 3,52275 bushels fill? (105) Will ,4026 of a bushel? (12) Will 120,780 bushels? (36) Will 63,745 bushels? (19) Will 40,260 bushels? (12) *A.* 68,17 barrels.

11. What is the quotient of 1561,275 divided by 24,3? (6425) By 48,6? (32125) By 12,15? (1285) By 6,075? (257) *Ans.* 481,875.

12. What is the quotient of ,264 divided by ,2? (132) By ,4? (66) By ,02? (132) By ,04? (66) By ,002? (132) By ,004? (66) *Ans.* 219,78.

REDUCTION OF DECIMALS.

¶ LVII. To change a Vulgar or Common Fraction to its equal Decimal.

1. A man divided 2 dollars equally among five men; what part of a dollar did he give each? and how much in 10ths, or decimals?

In common fractions, each man evidently has $\frac{2}{5}$ *of a dollar, the answer; but, to express it decimally, we proceed thus:—*

OPERATION.

```
        Numer.
Denom. 5)2,0(,4
         20
         ——
Ans. 4 tenths, = ,4
```

In this operation, we cannot divide 2 dollars, the numerator, by 5, the denominator; but, by annexing a cipher to 2, (that is, multiplying by 10,) we have 20 tenths, or dimes; then 5 in 20, 4 times; that is, 4 tenths, = ,4: Hence *the common fraction* $\frac{2}{5}$, *reduced to a decimal, is* ,4, *Ans.*

2. Reduce $\frac{3}{32}$ to its equal decimal.

OPERATION.

```
32)3,00(,09375
   288
   ———
    120
     96
    ———
    240
    224
    ———
     160
     160
     ———
```

In this example, by annexing one cipher to 3, making 30 tenths, we find that 32 is not contained in the 10ths; consequently, a cipher must be written in the 10ths' place in the quotient. These 30 tenths may be brought into 100ths by annexing another cipher, making 300 hundredths, which contain 32, 9 times; that is, 9 hundredths. By continuing to annex ciphers for 1000ths, &c., dividing as before, we obtain ,09375, *Ans. By counting the ciphers annexed to the numerator, 3, we shall find them equal to the decimal places in the quotient.*

Note. In the last answer, we have five places for decimals; but, as the 5 in the fifth place is only $\frac{5}{100000}$ of a unit, it will be found sufficiently exact for most practical purposes, to extend the decimals to only three or four places.

To know whether you have obtained an equal decimal, change the decimal into a common fraction by placing its proper denominator under it, and reduce the fraction to its lowest terms. If it produces the same common fraction again it is right; thus, taking the two foregoing examples, $,4=\frac{4}{10}=\frac{2}{5}$. Again, $,09375=\frac{9375}{100000}=\frac{3}{32}$.

From these illustrations we derive the following

RULE.

I. *How do you proceed to reduce a common fraction to its equal decimal?* *A.* Annex ciphers to the numerator, and divide by the denominator.

II. *How long do you continue to annex ciphers and divide?* *A.* Till there is no remainder, or until a decimal is obtained sufficiently exact for the purpose required.

III. *How many figures of the quotient will be decimals?* *A.* As many as there are ciphers annexed.

IV. *Suppose that there are not figures enough in the quotient for this purpose, what is to be done?* *A.* Prefix ciphers to supply the deficiency.

More Exercises for the Slate.

3. Change $\frac{1}{2}$, $\frac{3}{4}$, $\frac{1}{4}$, and $\frac{1}{25}$ to equal decimals. *A.* ,5, ,75, ,25, ,04.

4. What decimal is equal to $\frac{1}{20}$? (5) What $=\frac{4}{8}$? (5) What $=\frac{9}{12}$? (75) What $=\frac{2}{50}$? (4) *Ans.* 1,34.

5. What decimal is equal to $\frac{5}{100}$? (5) What $=\frac{2}{8}$? (25) What $=\frac{7}{14}$? (5) What $=\frac{7}{40}$? (175) What $=\frac{10}{16}$? (625) *A.* 1,6.

6. What decimal is equal to $\frac{1}{9}$? (1111) What $=\frac{4}{9}$? (4444) What $=\frac{1}{99}$? (10101) What $=\frac{1}{3}$? (3333)* *A.* ,898901. +

---

* When decimal fractions continue to repeat the same figure, like 333, &c., in this example, they are called Repetends, or Circulating Decimals. When only one figure repeats, it is called a single repetend; but, if two or more figures repeat, it is called a compound repetend: thus, ,333, &c. is a single repetend, ,010101, &c. a compound repetend.

When other decimals come before circulating decimals, as ,8 in ,8333, the decimal is called a mixed repetend.

It is the common practice, instead of writing the repeating figures several times, to place a dot over the repeating figure in a single repetend; thus, [illegible], &c.

¶ LVIII. To reduce Compound Numbers to Decimals of the highest Denomination.

Reduce 15 s. 6d. to the decimal of a pound.

OPERATION.

12)6, 0 d. 20)15, 5 s. ,775 £.	In this example, 6 d. $=\frac{6}{12}$ of a shilling, and $\frac{6}{12}$, reduced to a decimal by ¶ LVII, is equal to ,5 of a shilling, which, joined with 15 s., makes $=$15, 5 s. In the same manner, 15,5 s.$\div$20 s.$=$,775 £, *Ans.*

---

is written $\dot{1}$; also over the first and last repeating figure of a compound repetend; thus, for ,030303, &c. we write, $,\dot{0}\dot{3}$.

The value of any repetend, notwithstanding it repeats one figure or more an infinite number of times, coming nearer and nearer to a unit each time, though never reaching it, may be easily determined by common fractions; as will appear from what follows.

By reducing $\frac{1}{9}$ to a decimal, we have a quotient consisting of ,1111, &c., that is, the repetend, $,\dot{1}$; $\frac{1}{9}$, then, is the value of the repetend $\dot{1}$, the value of ,333, &c.; that is, the repetend $\dot{3}$ must be three times as much; that is, $\frac{3}{9}$ and $,\dot{4}=\frac{4}{9}$; $,\dot{5}=\frac{5}{9}$; and $,\dot{9}=\frac{9}{9}=1$ whole.

Hence, we have the following RULE for changing a single repetend to its equal common fraction,—*Make the given repetend a numerator, writing 9 underneath for a denominator, and it is done.*

What is the value of $,\dot{1}$? Of $,\dot{2}$? Of $,\dot{4}$? Of $,\dot{7}$? Of $,\dot{8}$? Of $,\dot{6}$? *A.* $\frac{1}{9}$, $\frac{2}{9}$, $\frac{4}{9}$, $\frac{7}{9}$, $\frac{8}{9}$, $\frac{6}{9}$.

By changing $\frac{1}{99}$ to a decimal, we shall have, ,010101, that is, the repetend $,\dot{0}\dot{1}$. Then, the repetend $,\dot{0}\dot{4}$, being 4 times as much, must be $\frac{4}{99}$, and $,\dot{3}\dot{6}$ must be $\frac{36}{99}$, also $,\dot{4}\dot{5}=\frac{45}{99}$.

If $\frac{1}{999}$ be reduced to a decimal, it produces $,\dot{0}0\dot{1}$. Then the decimal $,\dot{0}0\dot{4}$, being 4 times as much, is $\frac{4}{999}$, and $,\dot{0}3\dot{6}=\frac{36}{999}$. This principle will be true for any number of places.

Hence we derive the following RULE for reducing a circulating decimal to a common fraction,—*Make the given repetend a numerator, and the denominator will be as many 9s as there are figures in the repetend.*

Change $,\dot{1}\dot{8}$ to a common fraction. *A.* $\frac{18}{99}=\frac{2}{11}$.

Change $,\dot{7}\dot{2}$ to a common fraction. *A.* $\frac{72}{99}=\frac{8}{11}$.

Change $,\dot{0}0\dot{3}$ to a common fraction. *A.* $\frac{3}{999}=\frac{1}{333}$.

In the following example, viz. change $,8\dot{3}$ to a common fraction, the repeating figure is 3 that is, $\frac{3}{9}$, and ,8 is $\frac{8}{10}$; then $\frac{3}{9}$, instead of being $\frac{3}{9}$ of

Hence we derive the following

RULE.

I. *How must the several denominations be placed?* *A.* One above another, the highest at the bottom.

II. *How do you divide?* *A.* Begin at the top, and divide as in Reduction; that is, shillings by shillings, ounces by ounces, &c., annexing ciphers.

III. *How long do you continue to do so?* *A.* Till the denominations are reduced to the decimal required.

More Exercises for the Slate.

2. Reduce 7 s. 6 d. 3 qrs. to the decimal of a pound.
A. ,378125 £.
3. Reduce 5 s. to the decimal of a pound. *A.* ,25 £.
4. Reduce 3 farthings to the decimal of a pound.
A. ,003125 £.
5. Reduce 2 qrs. 3 na. to the decimal of a yard.
A. ,6875 *yd.*
6. Reduce 2 s. 3 d. to the decimal of a dollar. *A.* $,375.
7. Reduce 3 qrs. 3 na. to the decimal of a yard. *A.* ,9375 *yd.*
8. Reduce 8 oz. 17 pwts. to the decimal of a pound Troy.
A. ,7375 *lb.*
9. Reduce 8 £, 17 s. 6 d. 3 qrs. to the decimal of a pound.
A. 8,878125 £.

---

a unit, is, by being in the second place, $\frac{3}{9}$ of $\frac{1}{10}=\frac{3}{90}$; then $\frac{8}{10}$ *and* $\frac{3}{90}$ *added together, thus,* $\frac{8}{10}+\frac{3}{90}=\frac{75}{90}=\frac{25}{30}$, *Ans.*

Hence, to find the value of a mixed repetend—*First find the value of the repeating decimals, then of the other decimals, and add these results together.*

2. Change ,91$\dot{6}$ to a common fraction. *A.* $\frac{91}{100}+\frac{6}{900}=\frac{825}{900}=\frac{11}{12}$. *Proof,* $11 \div 12 =$,91$\dot{6}$.

3. Change 20$\dot{3}$ to a common fraction. *A.* $\frac{61}{300}$.

To know if the result be right, change the common fraction to a decimal again. If it produces the same, the work is right.

Repeating decimals may be easily multiplied, subtracted, &c. by first reducing them to their equal common fractions.

¶ LIX. To reduce Decimals of higher Den tions to Whole Numbers of lower Denominat

This rule is the reverse of the last,

Let us take the answer to the first example. Reduc to whole numbers of lower denominations.

OPERATION

OPERATION	
£ ,775	In this example ,775 £, reduced to sl that is, multiplied by 20, gives 15 ciphers on the right of a decimal a value;) then the decimal part ,5×1: =6 d. *Ans.* 15 s. 6 d.
20	
s. 15,500	
12	
d. 6,000	

Hence we derive the following

RULE.

I. *How do you proceed?* *A.* Multiply the given dec in Reduction; that is, pounds by what makes a pound, by what makes an ounce, &c.

II. *How many places do you point off in each prod decimals?* *A.* As many as there are decimal places given decimal.

III. *Where will you find the answer?* *A.* The sev nominations on the left hand of the decimal points will answer.

More Exercises for the Slate.

The following examples are formed by taking the ans the last rule; of course, the answers in this may be found examples of that. The examples in each are numbered to correspond.

2. Reduce ,378125 £ to whole numbers of lower d nations. (*For ans. see ex. No.* 2, ¶ LVIII.)
3. What is the value of ,25 £ of a pound?
4. What is the value of ,003125 of a pound?
5. What is the value of ,6875 of a yard?
6. What is the value of ,375 of a dollar?
7. What is the value of ,9375 of a yard?
8. What is the value of ,7375 of a pound Troy?

Application of the two foregoing Rules.

1. What will 4 yards of cloth cost, in pounds, at 7 s. yard? 7 s. 6 d., *reduced to a decimal,* =,375 £ × 4

£1,500
20
10,000 *Ans.*, 1 £ 10 s.

2. At $6 a cwt., what will 2 cwt. 2 qrs. of rice cost? *A.* $15.

3. At $20 a ton, what will 15 cwt. 2 qrs. of hay cost? *A.* $15,50

4. What cost 6 cwt. 0 qr. 7 lbs. of sugar, at $11,25 a cwt. *A.* $68,203+

5. What cost 60 gals. 1 pt. of rum, at $,78 a gallon? *A.* $46,897$\frac{5}{10}$.

6. At $1,25 a bushel, what will 36 bu. 0 pk. 4 qts. cost? *A.* $45,156$\frac{25}{100}$.

7. At $4,75 a yard, what will 26 yds. 2 qrs. of broadcloth cost? *A.* $125,87$\frac{1}{2}$.

8. At 2£, 10 s. a cwt., what will 6 cwt. 3 qrs. of rice cost? 2£, 10 s. = 2,5 £, *and* 6 *cwt.* 3 *qrs.* = 6,75 *cwt.; then,* 6,75 × 2,5 = 16,875 £ × 20 = 17,5 s. × 12 = 6 d. *Ans.* 16 £, 17 s. 6 d.

9. What will 6 gallons, 2 qts. of brandy cost, in pounds, at 15 shillings a gallon? *A.* 4 £. 17 s. 6 d.

REDUCTION OF CURRENCIES.

¶ **LX.** An apology may by some be deemed necessary for the omission, in this work, of much that is contained in other treatises, respecting what is called "the currencies of the different United States." The author, however, deems it rather necessary to apologize for introducing the subject at all. Those merely nominal currencies, originally derived from Great Britain, have long been obsolete in law, and ought to become so in practice. So long, however, as that practice continues, it may be necessary to retain a brief notice of it in elementary works.

Note. It was not intended that the following Table should be exact in every particular to a mill, but enough so, to correspond with the pecuniary calculations current among men of business; and, as such, it will be committed to memory more easily.

The design of the Table is not that it should be learned by rote, but by actual calculations from a few data; thus—as 1 far. is $\frac{1}{4}$ of a cent, then 2 farthings are $\frac{2}{4}$. Again, as 3 d. is 4 cents, and 3s. are 50 cents, then 3 s. 3 d. are 54 cents. It would be well for the teacher to direct the attention of the pupil to this object by explanations.

Repeat the

TABLE.

1 farthing	*is*	$\frac{1}{4}$ *of a penny,*	*or*	$\frac{1}{3}$ *of a cent.*
2 farthings	*are*	$\frac{2}{4}$ *of a penny,*	*or*	$\frac{2}{3}$ *of a cent.*
3 farthings	*are*	$\frac{3}{4}$ *of a penny,*	*or*	1 cent.
4 farthings	*are*	1 *penny,*	*or*	1$\frac{1}{3}$ cents.

decimals in the multiplier also, we must have pointed off two more places for decimals, which, counting both, would make 4 Hence *we must always point off in the product as many places for decimals, as there are decimal places in both the factors.*

2. Multiply ,25 by ,5.

,5 *Ans.* ,125	In this example, there being 3 decimal places in both the factors, we point off 3 places in the product, as before directed. The reason of this will appear more evi-

dent by considering both the factors common fractions, and multiplying by ¶ XLI., thus; ,25 $=\frac{25}{100}$, *and* ,5 $=\frac{5}{10}$; *now* $\frac{25\times 5}{100\times 10}=\frac{125}{1000}$, *which, written decimally, is* ,125, *Ans., as before.*

3. Multiply ,15 by ,05.

OPERATION.	
,15 ,05 *Ans.* ,0075	In this case, there not being so many figures in the product as there are decimal places in both the factors (viz. 4), we place two ciphers on the left of 75, to make as many. This will appear evident by the

following; ,15 $=\frac{15}{100}$ *and* ,05 $=\frac{5}{100}$; *then* $\frac{15\times 5}{100\times 100}=\frac{75}{10000}=$,0075, *Ans., the same as before.*

From these illustrations we derive the following

RULE.

I. *How do you multiply in Decimals?* *A.* As in Simple Multiplication.

II. *How many figures do you point off for decimals in the product?* *A.* As many as are in both the multiplicand and multiplier.

III. *If there be not figures enough in the product for this purpose, how would you proceed?* *A.* Prefix ciphers enough to make as many.

Q. What is the meaning of annex? *A.* To place after.

Q. What is the meaning of prefix? *A.* To place before.

More Exercises for the Slate.

4. What will 5,66 bushels of rye cost, at $1,08 a bushel? *A.* $6,1128, or $6. 11 c. 2$\frac{8}{10}$ m.

5. How many gallons of rum in ,65 of a barrel, each barrel containing 31$\frac{5}{10}$ gallons? (20475) In ,8 of a barrel? (252) In ,42 of a barrel? (1323) In ,6 of a barrel? (189) In 1126,5

barrels? (3548473) In 1,75 barrels? (55125) In 125,626789 barrels? (39572438535). *Ans.* 39574,9238535 *gallons.*

6. What will 8,6 pounds of flour come to, at $,04 a pound? (344) At $,03 a pound? (258) At $,035 a pound? (301) At $,0455 a pound? (3913) At $,0275 a pound? (23650). *Ans.* $1,5308.

7. At $,9 a bushel, what will 6,5 bushels of rye cost? (585) What will 7,25 bushels? (6525) Will 262,555 bushels? (2362995) Will ,62 of a bushel? (558) Will 76,75 bushels? (69075) Will 1000,0005 bushels? (90000045) Will 10,0005 bushels? (9000045) *A.* $1227,307995.

DIVISION OF DECIMALS.

¶ **LVI.** In Multiplication, we point off as many decimals in the product as there are decimal places in the multiplicand and multiplier counted together; and, as division proves multiplication by making the multiplier and multiplicand the divisor and quotient, hence, *there must be as many decimal places in the divisor and quotient, counted together, as there are decimal places in the dividend.*

1. A man bought 5 yards of cloth for $8,75; how much was it a yard? $,8,75=875 *cents, or* 100*ths*; *now,* 875÷5=175 *cents, or* 100*ths,* = $1,75 *Ans.*

OR

By retaining the separatrix, and dividing as in whole numbers, thus:—

OPERATION.	
5)8,75	As the number of decimal places in the divisor and quotient, when counted together, must always be equal to the decimal places in the dividend, therefore, in this example, as there are no decimals
Ans. $ 1,75	

in the divisor, and two in the dividend, by pointing off two decimals in the quotient, the number of decimals in the divisor and quotient will be equal to the dividend, which produces the same result as before.

2. At $2,50 a barrel, how many barrels of cider can I have for $11? $11=1100 *cents,* or 100*ths, and* $2,50=250 *cents, or* 100*ths; then, dividing* 100*ths by* 100*ths, the quotient will evidently be a whole number, thus:—*

OPERATION.
,3)3,4125

$11,375, *Ans.*

The pupil must recollect that, in division of decimals, there must be the same number of decimals in the divisor and quotient that there is in the dividend. There are 4 in the dividend, and 1 in the divisor; consequently, there must be 3 pointed off in the quotient.

2. How many pounds in $11,375?

This example being the reverse of the last, it is evident that we must multiply by ,3.

OPERATION.
$11,375
,3

A. 3£. 8 s. 3 d. 3,4125
20

8,2500
12

3,0000

It will be recollected by the pupil in pointing off, that there must be as many decimal places in the product, as there are decimal places in both multiplier and multiplicand.

From these illustrations we derive the following

RULE.

I. *How do you reduce the New England currency to federal money?* *A.* Reduce the question to the decimal of a pound, and divide by ,3.

II. *How do you reduce federal money to the same currency again?* *A.* Multiply by ,3.

More Exercises for the Slate.

3. Bought a building for 17£. 15 s. 6 d.; how many dollars will pay for it? *A.* $59,25.

4. How many pounds in $59,25? (17-15-6) In $177,75? (53-6-6) In $355,50? (106-13) In $71? (21-6) In $142? (42-12) In $568? (170-8) *Ans.* 412£. 1 s.

5. What will 15 barrels of flour cost in dollars, at 6£. 16 s. N. E. currency a barrel? (340) At 7£. 10 s. a barrel? (375) At 7£. 7 s. a barrel? (36750) At 6£. 10 s. 6 d. a barrel? (32625) At 6£. 4 s. 6 d. a barrel? (31125) *A.* $1720.

6. What will 4 acres of land cost in pounds, at $50 an acre? (60) At $49 an acre? (58-16) At $48 an acre? (57-12) At $25 an acre? (30) At $12 an acre? (14-8) At $24,50 an acre? (29-8) *Ans.* 250£. 4 s.

7. What will 4 acres of land cost in federal money, at 15£. an acre? (200) At 14£. 14 s. an acre? (196) At 14£. 8 s. an acre? (192) At 14£. 4 s. an acre? (189333) At 3£. 12 s. an acre? (48) At 7£. 4 s. an acre? (96) *Ans.* $921,333.

8. What will 15 barrels of rum cost, in pounds, at $22,75 a barrel? (102-7-6) At $23,75 a barrel? (106-17-6) At $20,75 a barrel? (93-7-6) *Ans.* 302£. 12 s. 6 d.

9. A gentleman in Virginia purchased a house for 300£. 15 s. 6 d. (1002583), 40 acres of land for 61£. 5 s. 6 d., (20425) and expended for repairs 109£. 9 s. 8 d. (364944). What did the whole amount to in federal money? *A.* $1571,777.

NEW YORK CURRENCY.—*Q. What is the currency of New York, North Carolina, and Ohio called? A.* New York currency.

Q. How many shillings make a dollar of this currency? A. 8 shillings.

Q. What part of a pound is 8 s. *A.* $\frac{8}{20}=\frac{4}{10}$; decimally $=,4$.

RULE I. *How do you proceed to reduce this currency to federal money, and federal money to the same currency again? A.* Take ,4 and proceed with it as with ,3 in the last rule.

1. Change 204£. 18 s. to dollars and cents.

204£. 18 s. = 204,9£ ÷ 4 = $512,25, *Ans.*

2. Change to federal money 409£. 16 s.; (102450) 136£. 12 s.; (34150) 413£. 16 s.; (103450) 49£. 12 s.; (124) 50£. 2 s.; (12525) 600£. (1500) *Ans.* $4149,75.

3. Change into New York currency $22,078; (8-16-7-1) $44,154; (17-13-2-3) $88,312; (35-6-5-3) $176,624; (70-12-11-3) *Ans.* 132£. 9 s. 3¼ d.

4. What will 20 yards of cloth cost, in dollars and cents, at 15 s. 6 d. a yard? (3875) At 12 s. 6 d. a yard? (3125) At 13 s. 6 d. a yard? (3375) At 17 s. 6 d. a yard? (4375) *Ans.* $147,50.

PENNSYLVANIA CURRENCY.—*Q. What is the currency of New Jersey, Pennsylvania, Delaware, and Maryland called? A.* Pennsylvania currency.

Q. How many shillings make a dollar of this currency? A 7 s. 6 d.

Q. What part of a pound is 7 s. 6 d.?

A. 7 s. 6 d. = 90 d. and 20 s. = 240 d.; then, $\frac{90}{240}=\frac{3}{8}$£.

RULE I. *How do you reduce this currency to federal money? A.* Divide by $\frac{3}{8}$; that is, multiply by 8, and divide by 3.

II. *How do you reduce federal money to the same currency again? A.* Multiply by $\frac{3}{8}$.

Exercises for the Slate.

1. Change 60£. 15 s. to federal money.

60£. 15 s. = 60,75£ × 8 = 486 ÷ 3 = $162, *Ans.*

2. Change $162 to Pennsylvania currency.

$162 × 3 = 486 ÷ 8 = 60,75£. = 60£. 15 s. *Ans.*

3. Change to dollars and cents 80£.; (213333) 250£. 16 s.

36880) 240£.; (640) 1£. 15 s. 7 d.; (4744) 50£. 7 s. 2 d.; 134288) *Ans.* $1661,165+.

4. Change to Pennsylvania currency $9,50; (3-11-3) $28,50; 10-13-9) $57; (21-7-6) $85,50; (32-1-3) $42,25; (15-16-0-2) $126,75; (47-10-7-2) $633,75; (237-13-1-2) *Ans.* 68£. 14 s. 4 d. 2 qrs.

GEORGIA CURRENCY.—*Q. What is the currency of Carolina and Georgia called?* *A.* Georgia currency.

Q. How many shillings make a dollar of this currency? *A.* s. 8 d.

Q. What part of a pound is 4 s. 8 d.? *A.* 4 s. 8 d. = 56 d., nd 20 s. = 240 d.; then, 56 d. is $\frac{56}{240}$£ = $\frac{7}{30}$£.

RULE I. *How do you reduce Georgia currency to federal money?* *A.* Divide by $\frac{7}{30}$; that is, multiply by 30, and divide y 7.

II. *How do you reduce federal money to the same currency again?* *A.* Multiply by $\frac{7}{30}$.

Exercises for the Slate.

1. Change 835£. 9 s. to federal money.

835,45£. × 30 = 2506350 ÷ 7 = $3580,50, *Ans.*

2. Change $3580,50 to Georgia currency.

$3580,50 × 7 = 25063,50 ÷ 30 = 835£. 9 s. *Ans.*

3. Change to federal money 208£. 17 s. 3 d.; (895125) 104£. 8 s. 7 d. 2 qrs.; (447562) 252£. 3 s. 1 d. 2 qrs. (1080669) *A.* $2423,356.

4. Change to pounds, shillings, &c. $447,562; (104-8-7-1) $895,125; (208-17-3) $1080,669 (252-3-1-1) *A.* 565£. 8 s. 1 d. 2 qrs.+

ENGLISH OR STERLING MONEY.—*Q. How many shillings of this money make a dollar?* *A.* 4 s. 6 d.

Q. What part of a pound is 4 s. 6 d.? *A.* 4 s. 6 d. = 54 d., and 20 s. = 240 d.; then, $1 is $\frac{54}{240}$£. = $\frac{9}{40}$£.

RULE I. *How may sterling money be reduced to federal money?* *A.* Divide by $\frac{9}{40}$; that is, multiply by 40, and divide by 9.

II. *How do you change federal money to sterling money?* *A.* Multiply by $\frac{9}{40}$.

Exercises for the Slate.

1. Change 21 £. 7 s. 6 d. to federal money.

21 £. 7 s. 6 d. = 21,375 × 40 = 855,000 ÷ 9 = $95, *Ans*

2. Change $95 to sterling money. *A.* 21 £. 7 s. 6 d.

3. Change 21 £. 7 s. 6 d. to federal money. *A.* $95.

4. Change $285 to sterling money. *A.* 64 £. 2 s. 6 d.

CANADA CURRENCY.—*Q. What is the currency of Canada and Nova Scotia called?* *A.* Canada currency.

Q. How many shillings of this currency make a dollar? *A.* 5 s.

Q. What part of a pound is 5 s.? *A.* $\frac{5}{20} = \frac{1}{4}$.

RULE I. *How do you change this currency into federal money?* *A.* Divide by $\frac{1}{4}$; that is, multiply by 4.

II. *How do you reduce federal money to Canada currency?* *A.* Multiply by $\frac{1}{4}$; that is, divide by 4.

Exercises for the Slate.

1. A gentleman, residing in Boston, contracted a debt of 200£. 17 s. in Halifax; how many dollars will pay the debt?

200£. 17 s. = 200,85£. × 4 = \$803,40, *Ans.*

2. *A*, residing in Montreal, sent 300£. Canada currency to *B*, his correspondent in New York, to purchase 120 barrels of flour. The flour cost \$12,50 per barrel; how much, in Canada currency, is the balance which is due? *A.* 75£.

3. A merchant in Quebec wrote to his correspondent in Philadelphia, to purchase a large quantity of cotton. His correspondent writes he has purchased 300 bales, each containing 275 pounds, at 10$\frac{1}{4}$ cents per pound. How many pounds Canada currency must the merchant remit to his agent to meet the purchase price? *A.* 2114£. 1 s. 3 d.

Foreign coins are estimated in the United States according to the following

TABLE.

Livre of France,	\$,18$\frac{1}{2}$.
Franc of France,	\$,18$\frac{3}{4}$.
Guilder or Florin of the U. Netherlands,	\$,40.
Mark Banco of Hamburg,	\$,33$\frac{1}{3}$.
Rix dollar of Denmark,	\$ 1,00.
Real of Plate of Spain,	\$,10.
Milrea of Portugal,	\$ 1,24.
Tale of China,	\$ 1,48.
Pagoda of India,	\$ 1,84.
Rupee of Bengal,	\$,50.

1. Reduce 500 livres of France to federal money.

1 livre = \$,18$\frac{1}{2}$; then, 500 × 18$\frac{1}{2}$ = \$92,50, *Ans.*

2. Reduce \$92,50 to livres of France. *A.* 500 livres.
3. Reduce 5000 francs to federal money. *A.* \$937,50
4. Reduce 12500 florins to federal money. *A.* \$5000.
5. Reduce \$5000 to florins. *A.* 12500 florins.

¶ LVIII. To reduce Compound Numbers to Decimals of the highest Denomination.

Reduce 15 s. 6d. to the decimal of a pound.

OPERATION.

```
12)6, 0 d.
20)15, 5 s.
     ,775 £.
```

In this example, 6 d. $=\frac{6}{12}$ of a shilling, and $\frac{6}{12}$, reduced to a decimal by ¶ LVII., is equal to ,5 of a shilling, which, joined with 15s., makes $=$15, 5 s. In the same manner, 15,5 s.$\div$20 s.$=$,775 £, *Ans.*

---

is written $\dot{1}$; also over the first and last repeating figure of a compound repetend; thus, for ,030303, &c. we write, ,$\dot{0}\dot{3}$.

The value of any repetend, notwithstanding it repeats one figure or more an infinite number of times, coming nearer and nearer to a unit each time, though never reaching it, may be easily determined by common fractions; as will appear from what follows.

By reducing $\frac{1}{9}$ to a decimal, we have a quotient consisting of ,1111, &c., that is, the repetend, ,$\dot{1}$; $\frac{1}{9}$, then, is the value of the repetend $\dot{1}$, the value of ,333, &c.; that is, the repetend $\dot{3}$ must be three times as much; that is, $\frac{3}{9}$ and ,$\dot{4}=\frac{4}{9}$; ,$\dot{5}=\frac{5}{9}$; and ,$\dot{9}=\frac{9}{9}=1$ whole.

Hence, we have the following RULE for changing a single repetend to its equal common fraction,—*Make the given repetend a numerator, writing 9 underneath for a denominator, and it is done.*

What is the value of ,$\dot{1}$? Of ,$\dot{2}$? Of ,$\dot{4}$? Of ,$\dot{7}$? Of ,$\dot{8}$? Of ,$\dot{6}$? *A.* $\frac{1}{9}$, $\frac{2}{9}$, $\frac{4}{9}$, $\frac{7}{9}$, $\frac{8}{9}$, $\frac{6}{9}$.

By changing $\frac{1}{99}$ to a decimal, we shall have, ,010101, that is, the repetend ,$\dot{0}\dot{1}$. Then, the repetend ,$\dot{0}\dot{4}$, being 4 times as much, must be $\frac{4}{99}$, and ,$\dot{3}\dot{6}$ must be $\frac{36}{99}$, also ,$\dot{4}\dot{5}=\frac{45}{99}$.

If $\frac{1}{999}$ be reduced to a decimal, it produces ,$\dot{0}0\dot{1}$. Then the decimal ,$\dot{0}0\dot{4}$, being 4 times as much, is $\frac{4}{999}$, and ,$\dot{0}3\dot{6}=\frac{36}{999}$. This principle will be true for any number of places.

Hence we derive the following RULE for reducing a circulating decimal to a common fraction,—*Make the given repetend a numerator, and the denominator will be as many 9s as there are figures in the repetend.*

Change ,$\dot{1}\dot{8}$ to a common fraction. *A.* $\frac{18}{99}=\frac{2}{11}$.

Change ,$\dot{7}\dot{2}$ to a common fraction. *A.* $\frac{72}{99}=\frac{8}{11}$.

Change ,$\dot{0}0\dot{3}$ to a common fraction. *A.* $\frac{3}{999}=\frac{1}{333}$.

In the following example, viz. change ,8$\dot{3}$ to a common fraction, the repeating figure is 3 that is, $\frac{3}{9}$, and ,8 is $\frac{8}{10}$; then $\frac{3}{9}$, instead of being $\frac{3}{9}$ of

Hence we derive the following

RULE.

I. *How must the several denominations be placed?* *A.* One above another, the highest at the bottom.

II. *How do you divide?* *A.* Begin at the top, and divide as in Reduction; that is, shillings by shillings, ounces by ounces, &c., annexing ciphers.

III. *How long do you continue to do so?* *A.* Till the denominations are reduced to the decimal required.

More Exercises for the Slate.

2. Reduce 7 s. 6 d. 3 qrs. to the decimal of a pound. *A.* ,378125 £.
3. Reduce 5 s. to the decimal of a pound. *A.* ,25 £.
4. Reduce 3 farthings to the decimal of a pound. *A.* ,003125 £.
5. Reduce 2 qrs. 3 na. to the decimal of a yard. *A.* ,6875 *yd.*
6. Reduce 2 s. 3 d. to the decimal of a dollar. *A.* $,375.
7. Reduce 3 qrs. 3 na. to the decimal of a yard. *A.* ,9375 *yd.*
8. Reduce 8 oz. 17 pwts. to the decimal of a pound Troy. *A.* ,7375 *lb.*
9. Reduce 8 £, 17 s. 6 d. 3 qrs. to the decimal of a pound. *A.* 8,878125 £.

---

a unit, is, by being in the second place, $\frac{3}{9}$ of $\frac{1}{10}=\frac{3}{90}$; then $\frac{8}{10}$ *and* $\frac{3}{90}$ *added together, thus,* $\frac{8}{10}+\frac{3}{90}=\frac{75}{90}=\frac{25}{30}$, *Ans.*

Hence, to find the value of a mixed repetend—*First find the value of the repeating decimals, then of the other decimals, and add these results together.*

2. Change $,91\dot{6}$ to a common fraction. *A.* $\frac{91}{100}+\frac{6}{900}=\frac{825}{900}=\frac{11}{12}$. *Proof*, $11\div12=,91\dot{6}$.

3. Change $20\dot{3}$ to a common fraction. *A.* $\frac{61}{300}$.

To know if the result be right, change the common fraction to a decimal again. If it produces the same, the work is right.

Repeating decimals may be easily multiplied, subtracted, &c. by first reducing them to their equal common fractions.

From these illustrations we derive the following

RULE.

I. *How are the numbers to be written down?* *A.* Tenths under tenths, hundredths under hundredths, and so on.

II. *How do you proceed to add?* *A.* As in Simple Addition.

III. *Where do you place the separatrix?* *A.* Directly under the separating points above.

More Exercises for the Slate.

2. James bought 2,5 cwt. of sugar, 23,265 cwt. of hay, and 4,2657 cwt. of rice; how much did he buy in all? *A.* 30,0307 cwt.

3. James is $14\frac{5}{10}$ years old, Rufus $15\frac{25}{100}$, and Thomas $16\frac{75}{100}$; what is the sum of all their ages? *A.* 46,5 years.

4. William expended for a chaise $\$255\frac{2}{10}$, for a wagon $\$37\frac{25}{100}$, for a bridle $\$\frac{75}{100}$, and for a saddle $\$11\frac{255}{1000}$; what did these amount to? *A.* $304,455.

5. A merchant bought 4 hhds. of molasses; the first contained $62\frac{4}{10}$ gallons, the second $72\frac{2657}{10000}$ gallons, the third $50\frac{2}{10}$ gallons, and the fourth $55\frac{75}{100}$ gallons; how many gallons did he buy in the whole? *A.* 240,6157 gallons.

6. James travelled to a certain place in 5 days; the first day he went $40\frac{2}{10}$ miles, the second $28\frac{37}{100}$ miles, the third $42\frac{5}{10}$ miles, the fourth $22\frac{5}{1000}$ miles, and the fifth $29\frac{42}{10000}$ miles; how far did he travel in all? *A.* 162,0792 miles.

7. A grocer, in one year, at different times, purchased the following quantity of articles, viz. 427,2623 cwt., 2789,00065 cwt., 42,000009 cwt., 1,3 cwt., 7567,126783 cwt., and 897,62 cwt.; how much did he purchase in the whole year? *A.* 11724,309742 cwt.

8. What is the amount of $\frac{6}{10}$, $245\frac{37}{1000}$, $6\frac{8}{10000}$, $245\frac{75}{100000}$, $1\frac{89}{1000000}$, $\frac{6}{10000}$, $427\frac{2}{1000000}$, $4\frac{5}{10}$, $\frac{62}{10000000}$, and 1925? *A.* 2854,492472.

9. What is the amount of one, and five tenths; forty-five, and three hundred and forty-nine thousandths; and sixteen hundredths? *A.* 47,009.

SUBTRACTION OF DECIMALS.

¶ **LIV.** 1. A merchant, owing $270,42, paid $192,625; how much did he then owe?

OPERATION.	
$270,42 $192,625 *Ans.* $77,795	*For the reasons shown in Addition, we proceed to subtract, and point off, as in Subtraction of Federal Money.*

Hence we derive the following

RULE.

I. *How do you write the numbers down?* *A.* As in Addition of Decimals.

II. *How do you subtract?* *A.* As in Simple Subtraction.

III. *How do you place the separatrix?* *A.* As in Addition of Decimals.

More Exercises for the Slate.

1. Bought a hogshead of molasses, containing 60,72 gallons; how much can I sell from it, and save 19,999 gallons for my own use? *A.* 40,721 gallons.

2. James rode from Boston to Charlestown in 4,75 minutes; Rufus rode the same distance in 6,25 minutes; what was the difference in the time? *A.* 1,5 min.

3. A merchant, having resided in Boston 6,2678 years, stated his age to be 72,625 yrs. How old was he when he emigrated to that place? *A.* 66,3572 yrs.

Note. The pupil must bear in mind, that, in order to obtain the answer, the figures in the parentheses are first to be pointed off, supplying ciphers, if necessary, then added together as in Addition of Decimals.

4. From ,65 of a barrel take ,125 of a barrel; (525) take ,2 of a barrel; (45) take ,45 of a barrel; (2) take ,6 of a barrel; (5) take ,12567 of a barrel; (52433) take ,26 of a barrel; (39) *A.* 2,13933 barrels.

5. From 420,9 pipes take 126,45 pipes; (29445) take ,625 of a pipe; (420275) take 20,12 pipes; (40078) take 1,62 pipes; (41998) take 419,89 pipes; (101) take 419,8999 pipes; (10001). *Ans.* 1536,7951 pipes.

MULTIPLICATION OF DECIMALS.

¶ **LV.** 1. How many yards of cloth in 3 pieces, each piece containing $20\frac{75}{100}$ yards?

OPERATION.	
20,75 3 *Ans.* 62,25 *yds.*	In this example, since multiplication is a short way of performing addition, it is plain that we must point off as in addition, viz. directly under the separating points in the multiplicand; and, as either factor may be made the multiplicand, had there been two

Hence, *To find the interest of any sum of dollars, pounds, shillings, or eagles, for one year or more, we have the following*

RULE.

I. *How do you proceed?* *A.* Multiply by ½ the number of months, and cut off two figures at the right, (for dividing by 100.)

Mental Exercises.

1. What is the interest of $8 for 4 months? *A.* 16 cents.
2. What is the interest of $4 for 4 months? *A.* 8 cents.
3. What is the interest of $2 for 6 months? *A.* 6 cents.
4. What is the interest of $20 for 2 months? *A.* 20 cents.
5. What is the interest of $80 for 10 months? *A.* $4,00.
6. What is the interest of $40 for 1 yr., or 12 mo.? *A.* $2,40.
7. What is the interest of $8 for 1 yr. 4 mo.? *A.* 64 cents.
8. What is the interest of $5 for 1 yr. 6 mo.? *A.* 45 cents.
9. What is the interest of $1 for 4 years? *A.* 24 cents.
10. What is the interest of $8 for 2 mo.? *A.* 8 cts. What is the amount? *A.* $8,08.
11. What is the interest of $6 for 1 yr.? *A.* 36 cts. What is the amount? *A.* $6,36.
12. What is the interest of $1 for 4 yrs.? *A.* 24 cts. What is the amount? *A.* $1,24.
13. What is the interest of 100 £. for 2 months? *A.* 1£. What is the amount? *A.* 101 £.
14. What is the interest of 10 £. for 1 yr. 8 mo.? *A.* 1 £ What is the amount? *A.* 11 £.
15. What is the interest of $2,50 for 2 mo.? *A.* 2 cts. 5 m What is the amount? *A.* $2,525.
16. What is the interest of $6,50 for 2 mo.? *A.* 6½ cts. What is the amount? *A.* $6,56½.

Exercises for the Slate.

1. What is the interest of $240,30 for 3 yrs. 4 mo.?

3 *yrs.* 4 *mo.* = 40 *mo.* ÷ 2 = 20, *half the number of months.*

OPERATION.	
240,30 20 $48,0600, *Ans.*	In this example, as there are two places for cents in the multiplicand, there will be two also in the product; then, cutting off two more figures, (for dividing by 100,) we have $48,06, *Ans.*

2. What is the interest of $400 for 2 yrs. 6 mo.? *A.* $60.
3. What is the amount of $500 for 4 yrs. 1 mo.? *A.* $622,50.
4. What is the interest of $75 for 2 yrs. 6 mo.? (1125) Of $250? (3750) Of $800? (120) Of $95? (1425). Of $650? (9750) *A.* $280,50.

5. What is the interest of $1500 for 4 mo.? (30) For 6 mo.? (45) For 10 mo.? (75) For 1 yr. 2 mo.? (105) For 1 yr. 8 mo.? (150) For 4 yrs. 2 mo.? (375) For 6 yrs. 6 mo.? (585) *A.* $1365.

6. What is the amount of $75 for 2 yrs. 6 mo.? (8625) Of $250? (28750) Of $800? (920) Of $95? (10925) Of $650? (74750) *A.* $2150,50.

7. What is the amount of $615,75 for 5 yrs.? (800475) For 11 yrs. 1 mo.? (1025223) For 7 yrs. 2 mo.? (880522) *A.* $2706,22 +.

8. What is the interest of $7650 for 3 yrs. 3 mo.? (149175) For 3 yrs. 4 mo.? (1530) For 6 yrs. 6 mo.? (298350) For 2 yrs. 2 mo.? (99450) *A.* $6999,75.

9. What is the amount of $7,50 for 10 yrs. 1 mo.? (12037) For 2 yrs. 3 mo.? (8512) For 1 mo.? (7537) For 11 mo.? (7912) For 1 yr. 7 mo.? (8212) *A.* $44,21 +.

¶ **LXII.** Since days are always either 30ths of a month, or some greater part, as halves, 3ds, 4ths, 5ths, &c.; thus, 1 day $= \frac{1}{30}$; 2 days $= \frac{2}{30}$, which, being reduced to its lowest terms, is $\frac{1}{15}$; 3 days $= \frac{3}{30}$ or $\frac{1}{10}$; 5 days $= \frac{5}{30}$, or $\frac{1}{6}$; 20 days $= \frac{20}{30}$, or $\frac{2}{3}$; it follows, that, if these parts be diminished in the same proportion as the months, that is, if half the fractional part be taken for a multiplier, the product will express the interest for the days in cents, or 100ths, which, divided by 100, as before, will be the interest required.

To halve any thing, we divide by 2.

Note. It will be recollected, that, to divide a fraction by 2, we can

Multiply the denominator, or divide the numerator.

1. What is the interest of $60 for 15 days? $60

15 days $= \frac{15}{30}$ *or* $\frac{1}{2}$ *mo.* $\div 2 = \frac{1}{4}$, *multiplier.* $\frac{1}{4}$

Ans. $,15

2. What is the interest of $24 for 10 days? $24

10 days $= \frac{10}{30}$ *mo.* $= \frac{1}{3} \div 2 = \frac{1}{6}$, *multiplier* $\frac{1}{6}$

Ans. $,04

3. What is the interest of $120,60 for 20 days? $120,60

20 days $= \frac{20}{30}$ *mo.* $= \frac{2}{3} \div 2 = \frac{1}{3}$, *multiplier.* $\frac{1}{3}$

Ans. $,4020

4. What is the interest of $360,60 for 19 days? $360,60

19 days $= \frac{19}{30}$ *mo.* $\div 2 = \frac{19}{60}$, *multiplier.* $\frac{19}{60}$

Ans. $1,14,1$\frac{9}{10}$ m.

Hence, *to find the interest of any sum for days, we have the following*

RULE.

I. *How do you proceed first?* *A.* I find what fractional part of a month the days are, and reduce the fraction to its lowest terms.

II. *What do you make the multiplier?* *A.* Half of this fraction.

III. *How do you halve the fraction?* *A.* Halve the numerator, or double the denominator.

IV. *After you have multiplied by the fraction, what is to be done with the product, to get the interest?* *A.* Cut off two figures, as before.

Mental Exercises.

1. What is the interest of $120 for 15 days? *A.* 30 cts.
2. Interest of $60 for 15 days? *A.* 15 cts. For 10 da.? *A.* 10 cts.
3. Interest of $18 for 20 da.? *A.* 6 cts. For 10 da.? *A.* 3 cts.
4. Interest of $120 for 1 da.? *A.* 2 cts. For 2 da.? *A.* 4 cts.
5. Interest of $60 for 3 da.? *A.* 3 cts. For 6 da.? *A.* 6 cts.

Exercises for the Slate.

1. What is the interest of $1200 for 2 da.? $1200

2 days $= \frac{2}{30} \div 2 = \frac{1}{30}$, *multiplier.* $\frac{1}{30}$

Ans. 40 cts.

2. What is the interest of $600 for 20 da.? $600

20 days $= \frac{20}{30} = \frac{2}{3} \div 2 = \frac{1}{3}$, *multiplier.* $\frac{1}{3}$

Ans. $2,00

3. What is the interest of $2400 for 15 da.? *A.* $6.
4. What is the interest of $3600 for 10 da.? *A.* $6.
5. What is the interest of $726 for 20 da.? *A.* $2,42.
6. What is the interest of $1200 for 1 da.? (20) For 3 da.? (60) For 4 da.? (80) For 5 da.? (1) For 10 da.? (2) For 15 da.? (3) For 20 da.? (4) For 25 da.? (5). *A.* $16,60.

7. What is the interest of $120 for 8 yrs. 4 mo. 15 da.? $120

8 yrs. 4 mo. $= 100 \div 2 = 50$; *and 15 da* $= \frac{1}{2}$ *mo.* $\div 2 = \frac{1}{4}$; *then, the multiplier for the days and months is* $50\frac{1}{4}$. $50\frac{1}{4}$

6000
30

Ans. $60,30

8. What is the interest of $1200,60 for 1 yr. 10 mo. 24 da.?

$1200,60

1 yr. 10 mo. = 22 ÷ 2 = 11 mo.; and 24 da. $\frac{24}{30} = \frac{4}{5} \div 2 = \frac{2}{5}$; the multiplier, then, is $11\frac{2}{5}$.

$11\frac{2}{5}$

1320660
48024

Ans. $136,868$\frac{4}{10}$ *m.*

Note. When the days are an even number, it will oftentimes be found convenient to find what fractional part of a month the days will be, without halving the fraction afterwards; thus, for 20 days, take 10 days = $\frac{10}{30} = \frac{1}{3}$, the multiplier.

9. What is the interest of $180 for 29 days? *A.* 87 cts.

10. What is the amount of $180,60 for 2 yrs. 4 mo. 20 da.? *A.* $206,486.

11. What is the amount of $36,60 for 2 yrs. 1 mo. 5 da.? *A.* $41,205$\frac{5}{12}$.

12. What is the interest of $300 for 1 yr. 6 mo. 15 da.? (2775) For 2 yrs. 6 mo. 15 da.? (4575) For 3 yrs. 4 mo. 10 da.? (6050) For 4 yrs. 4 mo. 5 da.? (7825) *A.* $212,25.

13. What is the interest of $600,50 for 2 yrs.? (7206) For 1 yr. 8 mo.? (6005) For 2 yrs. 8 mo. 1 da.? (9618) For 5 yrs. 7 mo. 12 da.? (202368) For 8 yrs. 4 mo. 4 da.? (30065) *A.* $731,308.

14. What is the interest of $700 for 1 yr? (42) For $6\frac{1}{2}$ mo.? (2275) For 4 mo.? (14) For 20 da.? (2333). *A.* $81,083.

15. What is the amount of $60000 for 3 da.? (60030) For 8 da.? (60080) For 9 da.? (60090). *A.* $180200.

16. What is the interest of $60 for 2 mo. 1 da.? (61) For 2 mo. 2 da.? (62) For 2 mo. 3 da.? (63). *A.* $1,86.

17. What is the interest of $60 for 2 mo. 7 da.? (67) For 2 mo. 8 da.? (68) For 2 mo. 12 da.? (72). *A.* $2,07.

18. What is the interest of $1200 for 12 yrs. 11 mo. 29 da.? *A.* $935,80.

The foregoing example, although it is as difficult a one as usually occurs, is solved by one third of the usual number of figures required by other methods.

¶ **LXIII.** It is evident, that, when the rate is either more or less than 6 per cent., the interest for the given rate will be a certain part of 6 per cent.; thus, 5 per cent. will be $\frac{5}{6}$ as much as 6 per cent., 4 per cent. $\frac{4}{6}$ as much, 7 per cent. $\frac{7}{6}$ as much, &c.

To get $\frac{4}{6}$, $\frac{7}{6}$ of any number, we multiply by the numerator, and divide by the denominator; and, as the denominator will always be 6, and the numerator the given rate, hence,

To find the interest of any sum, when the rate is not 6 per cent., we have the following

RULE.

I. *How do you proceed?* *A.* Find the interest for 6 per cent. as before.

II. *How do you proceed next?* *A.* Multiply the interest of 6 per cent. by the given rate, and always divide by 6.

1. What is the interest of $600 for 1 yr. 2 mo. and 15 days, at 5 per cent.?

```
              600
               7½ mo.
             ----
             4200
              150
             ----
             4350 int. at 6 per cent
                5
          -------
          6)21750
          -------
Ans. $36,25 int. at 5 per cent
```

2. What is the interest of $240 for 2 yrs. 6 mo. at 1 per cent.? (6) At 2 per cent.? (12) At 4 per cent.? (24) At 6 per cent.? (36) At 10 per cent.? (60) At 5½ per cent.? (33) *A.* $171.

3. What is the interest of $480 for 3 yrs. 2 mo., at 15 per cent.? (228) At 20 per cent.? (304) At 10¾ per cent.? (16340) At 15½ per cent.? (23560) At 7 per cent.? (10646) *A.* $1037,40.

4. What is the interest of $600 for 15 mo., at 2¾ per cent.? (20625) At 3¾ per cent.? (28125) *A.* $48,75.

5. What is the interest of $600 from January 1st to March 1st? (6) From January 15th to May 15th? (12) From January 15th to September 15th? (24) *A.* $42.

6. What is the amount of $500 from March 10th, 1824, to March 10th, 1827? (590) From March 29th, 1820, to March 29th, 1826? (680) From March 16th, 1820, to March 16th, 1824? (620) *A.* $1890.

7. What is the interest of $60 from June 1st, 1826, to November 1st, 1827? (510) From April 1st, 1825, to August 16th, 1826? (495) From July 4th, 1825, to August 19th, 1828? (1125) *A.* $21,30

8. What is the interest of $300 from September 5th, 1826, to September 25th, 1826? (1) From August 9th, 1826, to December 24th, 1827? (2475) *A.* $25,75.

9. What is the amount of $180 from October 1st, 1826, to December 1st, 1830? *A.* $225.

¶ **LXIV.** A concise and practical Rule for the State of New York, in which the interest is established by law at 7 per cent. It has been remarked, that 7 per cent. is $\frac{7}{6}$ of 6 per cent.; that is, $\frac{1}{6}$ more than 6 per cent: Hence,

Q. To find the interest at 7 per cent., what is the

RULE?

A. Add $\frac{1}{6}$ of the interest, at 6 per cent. (found as before), to itself; the sum will be the interest at 7 per cent.

Note. The interest for any rate per cent. may be found in the same manner by subtracting, when the given rate is under 6 per cent., and adding, when it is more.

1. What is the interest of $360 for 20 days, at 7 per cent.?

New Method.	*Old Method.*
$360	$360
$\frac{1}{3}$	7
$\frac{1}{6}$) $1,20 *at 6 per cent.*	1=$\frac{1}{12}$)2520
20	210
Ans. $1,40 *at 7 per cent.*	15=$\frac{1}{2}$)105
	5=$\frac{1}{3}$) 35
	Ans. $1,40

2. What is the interest of $60 for 2 yrs. 4 mo., at 7 per cent.? *A.* $9,80.

3. What is the amount of $120,60 for 1 yr., 6 mo., 10 da., at 7 per cent.? *A.* $133,497$\frac{5}{10}$ m.

4. What is the amount of $241,20 for 6 mo. 20 da.? (25058) For 1 mo. 1 da.? (242653) For 1 yr. 4 mo. 5 da.? (263946) For 2 yrs. 6 mo. 25 da.? (284582) *A.* $1041,761+.

¶ **LXV.** Since 6 per cent. is $6 on $100, that is, $\frac{6}{100}$ of the principal, and 5 per cent. $\frac{5}{100}$, &c., hence,

To calculate the interest at any rate per cent., when the time is 1 year, we proceed as follows:

RULE: *Multiply by the given rate, and cut off two figures, as before.*

1. What is the interest of $220,40 for 1 yr., at 9 per cent?

$$
\begin{array}{r}
\$\,220,40 \\
9 \\
\hline
\$\,19,8360 \ \textit{Ans.}
\end{array}
$$

2. What is the interest of $1200,30 for 1 yr., at 12½ per cent.? *A.* $150,03,7.

3. What is the amount of $80,10, for 1 yr., at 2½ per cent.? (82102) At 5 per cent.? (84105) At 10 per cent.? (8811) At 4½ per cent.? (83704) At 19¾ per cent.? (95919) *A.* $433,94+.

¶ LXVI. COMMISSION. Q. *When an allowance of so much per cent. is made to a person called either a correspondent, factor, or broker, for buying, or assisting in buying and selling goods for his employer, what is it called?* *A.* Commission.

RULE. *Since commission, insurance, buying and selling stocks, and loss and gain, are rated at so much per cent., without regard to time, how may all these be calculated?* *A.* Multiply by the rate per cent., and cut off two figures, as in the last rule.

1. What would you demand for selling $400 worth of cotton, for 2½ per cent. commission?

$400 × 2½, *and cutting off two figures,* = $10, *commission, Ans.*

2. My correspondent informs me that he has purchased goods to the amount of $5000; what will his commission amount to, at 2½ per cent.? *A.* $125.

3. What must I be allowed for selling 300 pounds of indigo, at $2,50 per lb., for 2 per cent. commission? (15) For 2¾ per cent.? (20625) For 5 per cent.? (3750) For 6½ per cent.? (4875) For 7 per cent.? (5250). *A.* $174,37½.

INSURANCE. Q. *What is the allowance of so much per cent. made to persons, to make good the losses sustained by fire, storms, &c. called?* *A.* Insurance.

Q. *By what name is the instrument that binds the contracting parties called?* *A.* Policy.

Q. *What is the sum paid for insurance called?* *A.* Premium.

1. What will be the premium for insuring an East India ship, valued at $25000, at 15½ per cent.? *A.* $3875.

2. What is the premium for insuring $2600, at 20 per cent.? (520) At 30 per cent.? (780) At 18½ per cent.? (481) 26½ per cent.? (689) *A.* $2470.

STOCK. Q. *What is the general name for all moneys invested trading companies, or the funds of government, called?* *Stock.*

Q. When $100 of stock sells for $100, how is the stock said to be? A. At par.

Q. When is it said to be above par, and when below par? A. When $100 stock sells for more than $100, it is said to be above par; when for less than 100, below par.

Q. When it is above par, what is it said to be? A. So much advance.

1. What is the value of $2500 of stock, at 106 per cent.; that is, 6 per cent. advance? *Ans.* $2500 × 106 = $2650.

2. What is the value of $1000 of insurance stock, at 95 per cent.; that is, 5 per cent. below par? *A.* $950.

3. What is the value of $1200 of bank stock, at 3 per cent. below par; that is, 97 per cent.? (1164) At 112 per cent., or 12 per cent. advance? (1344) At 87½ per cent.? (1050) At 12½ per cent. advance; that is, 112½ per cent.? (1350) *A.* $4908.

LOSS AND GAIN. 1. Bought a piece of broadcloth for $80; how much must I sell it for, to gain 10 per cent.; that is, 10 per cent. advance, which is 110 per cent. on the cost? $80 × 110 = $88, *Ans.*

2. Bought a hogshead of molasses for $50, and 5 gallons having leaked out, I sold the remainder at 10 per cent. loss; that is, 10 per cent. below par, being 90 per cent. on the cost; what did I get for it? *A.* $45.

3. If I pay $50 for a piece of broadcloth, how must I sell the same so as to gain 20 per cent.; that is, 20 per cent. advance, or 120 per cent on the cost? *A.* $60.

4. Bought rum at $1,25 per gallon; and, by accident, so much leaked out, that I am content to lose 20 per cent.; how must I sell it per gallon? *A.* $1.

5. A merchant bought 400 barrels of flour for $3500; how must he sell it per barrel, to gain 25 per cent.? *A.* $10,93¾.

6. Bought sugar at 15 cents per lb.; at what rate must I sell it a lb. so as to gain 20 per cent.? (18) So as to gain 25 per cent.? (1875) 30 per cent.? (195) 40 per cent.? (21) 45 per cent.? (2175) 50 per cent.? (225) 65 per cent.? (2475) 75 per cent.? (2625) 90 per cent.? (285) 100 per cent., or to double my money? (30) *A.* $2,31.

7. Bought 100 tierces of rice, each tierce weighing 300 lbs. net, at 6¼ cents per lb.; (1875) 30 pipes of wine for $1,12½ per gallon; (425250) 3 hhds. of rum for 90 cents per gallon; (17010) 40 barrels of flour for $7½ per barrel; (300) and 40 bushels of salt for 7 s. 6 d. or $1,25 per bushel; (50) how much must all the said articles be sold for, to gain 50 per cent., being 150 per cent. on the first cost? *A.* $9971,40.

¶ LXVII. *Time, Rate per cent., and Amount given, to find the Principal.*

1. What sum of ready money, put at interest for 1 yr. 8 mo. at 6 per cent., will amount to $220?

The amount of $1 for 1 year and 8 mo. is $1,10; then $220 ÷ $1,10 = $200, *Ans.*

RULE. *How, then, would you proceed to find the principal?*
A. Divide the given amount by the amount of $1, at the given rate and time.

2. What principal, at 6 per cent., in 5 years, will amount to $650?

In this example, in dividing $650 by $1,30, we annex two ciphers to 500, to make the decimal places in the divisor and dividend equal. (See ¶ LVI.)

A. $500.

3. What principal, at 6 per cent., in 1 year 2 mo., will amount to $642. *A.* $600.

4. What principal will amount to $691,50 in 2 yrs., 6 mo., 15 da., at 6 per cent.? *A.* $600.

5. A correspondent has in his hands $210, to be laid out in goods; after deducting his own commission of 5 per cent., how much will remain to be laid out.

It is evident, that the commission which he received, added to the money laid out, must make $210; hence, $210 may be considered the amount, and the money laid out the principal; consequently, the question does not differ materially from the foregoing. In such questions as these, in which time is not regarded, the amount of $1 is the rate per cent. added to $1.

It will be recollected that 6 per cent. is 6 cents on 100 cents, or $1; 5 per cent., 5 cents; the amount, then, of $1, at 5 per cent., is 5 cents added to $1, making $1,05; then $210 ÷ $1,05 = $200, *Ans.*

6. A factor receives $1040 to be laid out in goods, after deducting his own commission of 4 per cent.; how much does his commission amount to?

The sum laid out, found as before, is $1000; then, 1040 — 1000 = $40, commission, *the Answer.*

7. A factor receives $2100, from which he wishes to deduct his commission of 5 per cent.; what will his commission amount to? *A.* $100.

DISCOUNT. 1. William owes Rufus $1272 to be paid in 1 year, without interest; but Rufus, wanting his money immediately, says to William, I am willing to allow you 6 per cent., the lawful interest, if you will pay me now; what sum ought William to pay Rufus?

It is evident that he ought to pay just such a sum, as, put at interest, would in 1 year amount to $1272; or, in other words, such a principal as would amount to $1272. This question, therefore, is solved in the same manner as the preceding.

$1272 ÷ $1,06 = 1200, the Ans.

Q. What is an allowance made for the payment of a sum of money before it becomes due called? A. Discount.

Q. What is the sum called, which, put at interest, would, in the given time and rate, amount to the given sum or debt? A. The present worth.

Q. In calculating interest, what would the present worth be called? *A.* The principal.

Q. What would the given sum, or debt, be called? *A.* The amount.

Q. What is the discount of any sum equal to? *A.* The interest of its present worth for the same time.

Q. As operations in discount are substantially the same as in the preceding paragraph, what is the rule, which was there given, that is applicable to discount?

RULE. Divide the given sum, or debt, by the amount of $1, at the given rate and time; the quotient will be the present worth.

Q. How is the discount found? *A.* By subtracting the present worth from the given sum or debt.

Note. It will be recollected that, when no per cent. is mentioned, 6 per cent. is understood.

2. What is the present worth of $133,20, due 1 yr. 10 mo. hence? *A.* $120.

PROOF. 3. What is the amount of $120 for 1 yr. 10 mo.? (*Perform this example by the rule for calculating interest.*) *A.* $133,20.

4. What is the discount of $660, due 1 yr. 8 mo. hence? *A.* $60.

PROOF. 5. What is the interest of $600 for 1 yr. 8 mo.? *A.* $60.

6. What is the discount of $460, due 2 yrs. 6 mo. hence? *A.* $60.

7. What is the present worth of $1350, due 5 yrs. 10 mo. hence? *A.* $1000.

8. Bought goods to the amount of $520 on 8 mo. credit; how much ready money ought I to pay as an equivalent? *A.* $500.

9. Bought goods in Boston, amounting to $1854, for which I gave my note for 8 mo.; but, being desirous of taking it up, at the expiration of 2 months, what sum does justice require me to pay? *A.* $1800.

10. What is the discount of $615, due 5 mo. hence? *A.* $15.

11. What is the present worth of $1260, due 10 mo. hence? *A.* $1200.

12. What is the present worth of $1272, due 2 yrs. hence, discounting at 3 per cent.? *A.* $1200.

13. What is the present worth of $51,50, due 6 mo. hence?

(50) Of $204, due 4 mo. hence? (200) Of $13000, due 5 yrs. hence? (10000) Of $9440, due 3 yrs. hence? (8000)
A. $18250.

14. What is the present worth of $515, due 6 mo. hence? (500) Due 1 yr. hence? (485849) Due 15 mo. hence? (479069) Due 20 mo. hence? (468181) Due 4 yrs. hence? (415322)
A. $2348,421 +.

¶ LXVIII. *Time, Rate per cent., and Interest, being given, to find the Principal.*

1. What sum of money, put at interest 1 yr. 8 mo. at the rate of 6 per cent., will gain $20,60 interest?

The interest of $1 for 1 yr. 8 mo. = 10 cts.; then, $20,60 ÷ $,10 = $206, Ans.

Rule. *How, then, would you proceed?* *A.* Divide the given gain or interest by the interest of $1 at the given rate and time; the quotient will be the principal required.

2. A certain rich man has paid to him, every year, $48000 interest money; how much money must he have at interest? or what principal will gain $48,000 in 1 year, at 6 per cent.?
A. $800000.

3. If a man's salary be $12000 a year, what principal at interest for 1 yr. at 6 per cent. would gain the same?
A. $200000.

4. Paid $45, the lawful interest on a note, for 2 yrs. 6 mo.; what was the face or principal of the note? *A.* $300.

¶ LXIX. *The Principal, Interest, and Time, being given, to find the Rate per cent.*

1. If I have $2000 at interest, and at the end of the year I should receive $120 interest, what rate per cent. would that be?

The interest of $2000 at 1 per cent. for 1 year is $20; therefore, $120 ÷ 20 = $6, that is, 6 per cent., the rate required.

Rule. *How, then, do you proceed to find the rate per cent.?* *A.* Divide the given interest by the interest of the given sum, at 1 per cent. for the given time; the quotient will be the required rate.

2. If I receive $60 for the use of $600, 1 yr. and 8 mo., what is the rate per cent.? *A.* 6 per cent.

3. If I pay $200 for the use of $2000 for 2 yrs. 6 mo., what is the rate per cent.? *A.* 4 per cent.

When the prices of goods are given, to find what is the Rate per cent. of Gain or Loss.

1. A merchant bought cloth for $1,20 a yard, and sold it for $1,50; what was the gain per cent.?

In this example, we are required to find the rate per cent. The process, then, of finding it, is substantially the same as in the foregoing examples.

It has been remarked, that 6 per cent. is 6 cents on 100 cents; that is, the interest is $\frac{6}{100}$ of the principal; which, written decimally, is ,06; 5 per cent is $\frac{5}{100}$ = ,05; 25 per cent. is $\frac{25}{100}$ = ,25; that is, the rate may always be considered a decimal carried to two places, or 100ths. In the last example, by subtracting $1,20 from $1,50, we have 30 cents gain on a yard, which is $\frac{30}{120}$ of the first cost; $\frac{30}{120}$ = ,25 = 25 per cent., *the Answer.*

Rule. *How, then, do you proceed to find the rate per cent. of gain or loss?* *A.* Make a common fraction by writing the gain or loss for a numerator, and the cost of the article the denominator; then change it to a decimal.

2. A merchant bought molasses for 24 cents a gallon, which he sold for 30 cents; what was his gain per cent.?

A. ,25 = 25 per cent.

3. A grocer bought a hhd. of rum for $75, from which several gallons having leaked out, he sold the remainder for $60; what did he lose per cent.?

In this example the decimal is ,2; which, carried to two places, is ,20 = 20 per cent., *the Answer.*

4. A man bought a piece of cloth for $20, and sold it for $25; what did he gain per cent.? *A.* 25 per cent.

5. A grocer bought a barrel of flour for $8, and sold it for $9; what was the gain per cent.?

As two decimal places only are assigned to the rate per cent., ,125 is 12$\frac{5}{10}$ = 12½ per cent., that is, the third place is so many tenths of 1 per cent.; thus, 1 per cent. is ,01, and ½ per cent. is ,005 = $\frac{5}{10}$, or ½ of 1 per cent.

A. 12½ per cent.

6. A merchant bought a quantity of goods for $318,50, and sold them again for $299,39; what was his loss per cent.?

A. 6 per cent.

7. What is the gain per cent. in buying rum at 40 cents a gallon, and selling it at 42 cents a gallon? (5) At 44 cents? (10) At 46 cents? (15) At 50 cents? (25) At 54 cents? (35) At 60 cents? (50) *A.* 140 per cent.

8. Bought a hhd. of molasses, containing 112 gallons, at 26 cents a gallon, and sold it for $,286 a gallon; what was the whole gain, and what was the gain per cent.?

A. $2,912, and the gain ,1 = 10 per cent.

9. Bought flour at $9 a barrel, and sold it for $10,80 a barrel; what was the gain per cent.? *A.* 20 per cent.

10. If I buy a horse for $150, and a chaise for $250, and sell the chaise for $350, and the horse for $100, what is my gain per cent.? *A.* $,125 = 12½ per cent.

11. If I buy cotton at 15 cents a pound, and sell it for 16½ cents, what should I gain in laying out $100? *A.* $10.

12. Bought 20 barrels of rice for $20 a barrel, and paid for freight 50 cents a barrel; what will be my gain per cent. in selling it for $25,62$\frac{5}{10}$ a barrel? *A.* 25 per cent.

¶ LXX. *The Principal, Rate per cent., and Interest, being given, to find the Time.*

1. William received $18 for the interest of $200 at 6 per cent.; how long must it have been at interest?

The interest on $200 for 1 yr. at 6 per cent. is $12; hence $18 ÷ 12 = 1,5 = 1½ years, the required time, Ans.

Q. *What, then, is the* RULE? *A.* Divide the given interest by the interest of the principal for 1 year at the given rate, the quotient will be the time required, in years and decimal parts of a year.

2. Paid $36 interest on a note of $600, the rate being 6 per cent.; what was the time? *A.* 1 year.

3. Paid $200 interest on a note of $1000; what was the time, the rate being 5 per cent.? *A.* 4 years.

4. On a note of $60, there was paid $9,18 interest, at 6 per cent; how long was the note on interest?

A. 2,55 yrs. = 2 yrs. 6 mo. 18 da.

COMPOUND INTEREST.

¶ LXXI. 1. Rufus borrows of Thomas $500, which he agrees to pay again at the end of 1 year, together with the interest, at 6 per cent.; but, being prevented, he wishes to keep the $500 another year, and pay interest the same as before. How much interest ought he to pay Thomas at the end of the two years?

In this example, if Rufus had paid Thomas at the end of the first year, the interest would have been $500 × 6 = $30, which, added to the principal, $500, thus, 500 + 30, = 530, the sum or amount justly due Thomas at the end of the first year; but, as it was not paid then, it is evident, that, for the next year, (2d year,) Thomas ought to receive interest on $530, (being the amount of the first year). The interest of $530 for 1 year is 530 × 6 = $31,80, which, added to $530, = 561,80, the amount for 2 years; hence, $561,80 — $500 = $61,80, *compound interest, the Answer.*

This mode of computing interest, although strictly just, is not authorized by law

Q. When the interest is added to the principal, at the end of 1 year, and on this amount the interest calculated for another year, and so on, what is it called? *A.* Compound Interest.

Q. How, then, may it be defined? *A.* It is interest on both principal and interest.

Q. What is Simple Interest? *A.* It is the interest on the principal only.

Hence we derive the following

RULE.

I. *How do you proceed?* *A.* Find the amount of the principal for the 1st year, by multiplying as in simple interest; then of this amount for the 2d, and so on.

II. *How many times do you multiply and add?* *A.* As many times as there are years: the last result will be the amount.

III. *How is the compound interest found?* *A.* By subtracting the given sum or first principal from the product.

More Exercises for the Slate.

2. What is the compound interest of $156 for 3 yrs.?

$156 = *given sum, or first principal.*
6

9,36 = *interest, and*
156 = *principal, added together.*

165,36 = *amount, or principal for 2d year.*
6

9,9216 = *compound interest 2d year, and*
165,36 = *principal 2d year, added together.*

175,2816 = *amount, or principal for 3d year*
6

10,516896 = *compound interest 3d year, and*
175,2816 = *principal 3d year, added together*

185,798496 = *amount.*
156 = *first principal subtracted.*

Ans. $29,798, *rejecting the three last figures, as of trifling value*

3. What will be the amount of $500 for 4 years, at compound interest? *A.* $631,238$\frac{4}{10}$+.

4. What is the amount of $500 for 4 years, at simple interest? *A.* $620.

5. What will be the amount of $700 for 5 years, compound interest? *A.* 936,757$\frac{9}{10}$+.

6. What will be the amount of $700 for 5 years, at simple interest? *A.* $910.

7. What will be the amount of $1000 for 3 years, at compound interest? (1191016) $1500 for 6 years? (21277786) $2000 for 2 years? (224720) $400 for 7 years? (601452) *A.* $6167,446$\frac{6}{10}$+.

8. What is the compound interest of $150 for 2 years? (1854) $1600 for 4 years? (4199631) $1000 for 3 years? (191016) $5680 for 4 years? (1490869) $500 for 3 years? (95508) *A.* $2215,896$\frac{1}{10}$+.

9. What is the compound interest of $600,50 for 2 years, at 2 per cent? (242602) At 3 per cent? (365704) At 4 per cent.? (490008) At 5 per cent.? (615512) At 7 per cent? (870124) At 10 per cent.? (126105) *A.* $384,50.

10. What is the difference between the simple interest of $200 for 3 years, and the compound interest for the same time? *A.* $2,203$\frac{2}{10}$.

11. What is the compound interest of $600 for 2 years 6 months?

In calculating the compound interest for months and days, first find the amount for the years, and on that amount calculate the interest for the months and days; this interest, added to the amount for the years, will be the interest required.

A. $94,38,4+.

12. What is the compound interest of $500 for 3 yrs. 4 mo.? *A.* 107,418$\frac{1}{10}$+.

13. What is the difference between the simple and compound interest of $200 for 3 yrs.? (22032) For 4 yrs. 6 mo.? (60702) For 2 yrs. 8 mo. 15 da.? (17706) *A.* $10,044.

As the amount of $2 is twice as much as $1, $4, 4 times as much, &c., hence, we may make a table containing the amount of the 1£, or $1, for several years, by which the amount of any sum may be easily found for the same time.

TABLE,

Showing the amount of 1£, or $1, for 20 years, at 5 and 6 per cent., at compound interest.

Years.	5 per cent.	6 per cent	Years.	5 per cent.	6 per cent.
1	1,05000	1,06000	11	1,71033	1,89829
2	1,10250	1,12360	12	1,79585	2,01219
3	1,15762	1,19101	13	1,88564	2,13292
4	1,21550	1,26247	14	1,97993	2,26090
5	1,27628	1,33822	15	2,07892	2,39655
6	1,34009	1,41851	16	2,18287	2,54035
7	1,40710	1,50363	17	2,29201	2,69277
8	1,47745	1,59384	18	2,40661	2,85433
9	1,55132	1,68947	19	2,52695	3,02559
10	1,62889	1,79084	20	2,65329	3,20713

14. What is the compound interest of $20,15 for 4 years, at 6 per cent.?

By the Table, $1, at 6 per cent. for 4 years, is $1,26247, × $20,15 = $25,438, amount, from which $20,15 being subtracted, leaves $5,28 $8\frac{7}{10}$ +.

15. What is the amount of $10,50, at 5 per cent. for 2 years? (115762) For 6 years? (140709) For 8 years? (155132) For 15 years? (218286) For 17 years? (240661) For 20 years? (278595). $114,914 $\frac{5}{10}$ +, *A.*

Any sum, at simple interest, will double itself in 16 years 8 months; but at compound, in a little more than half that time; that is, in 11 years, 8 months and 22 days. Hence, we see that there is considerable difference in a few years, and when compound interest is permitted to accumulate for ages, it amounts to a sum almost incredible. If 1 cent had been put at compound interest at the commencement of the Christian era, it would have amounted, at the end of the year 1827, to a sum greater than could be contained in six millions of globes, each equal to our earth in magnitude, and all of solid gold, while the simple interest for the same time would have amounted to only about one dollar. The following question is inserted, more for the sake of exemplifying the preceding statement, than for the purpose of its solution. The amount, however, at compound interest, may be found, without much perplexity, by ascertaining the amount of 1 cent for 20 years, found by the Table, then making this amount the principal for 20 years more, and so on for the whole number of years.

16. Suppose 1 cent had been put at interest at the commencement of the Christian era, what would it have amounted to at simple, and what at compound interest, at the end of the year 1827? *A.* Simple, $1,106 $\frac{2}{10}$; compound, $1726164740475525294707009149747119599766203545 $\frac{56}{100}$ nearly.

EQUATION OF PAYMENTS.

¶ LXXII. *Q. What is the meaning of* equation? *A.* The art of making equal.

Q. What is equation of payments? A. It is the method of finding an equal or mean time for the payment of debts, due at different times.

1. In how many months will $1 gain as much as $2 will gain in 6 months? *A.* 6 × 2 = 12 months.

2. How long will it take $1 to gain as much as $5 will gain in 12 months? *A.* 60 months.

3. How many months will it take $1 to be worth as much as the use of $10, 20 months? *A.* 200 months.

4. A merchant owes 2 notes, payable as follows: one of $8, to be paid in 4 months; the other of $6, to be paid in 10 months; but he wishes to pay both at once: in what time ought he to pay them?

4 × 8 = 32; *therefore,* $8 *for* 4 *mo.* = $1 *for* 32 *mo., and*
10 × 6 = 60; *therefore,* $6 *for* 10 *mo.* = $1 *for* 60 *mo.*
14 92

Therefore, he might have $1, 92 months, and he may keep $14, $\frac{1}{14}$ part as long; that is, $\frac{1}{14}$ of 92 months, which is 92 ÷ 14, = 6 mo. $13\frac{1}{14}$ da., *Ans.*

Q. Hence, *to find the mean time of payment, what is the* Rule? *A.* Multiply each payment by the time, and the sum of these several products, divided by the sum of the payments, will be the answer.

Note This rule proceeds on the supposition, that what is gained by keeping the money after it is due, is equal to what is lost by paying it before it is due But this is not exactly true, for the gain is equal to the interest, while the loss is equal only to the discount, which is always less than the interest However, the error is so trifling, in most cases which occur in business, as not to make any material difference in the result.

5. A owes B $200 to be paid in 6 months, $300 in 12 months, $500 in 3 months; what is the equated time for the payment of the whole? *A.* $6\frac{3}{10}$.

6. What is the equated time for paying $2000, of which $500 is due in 3 months, $360 in 5 months, $600 in 8 months, and the balance in 9 months? *A.* $6\frac{960}{2000} = 6\frac{12}{25}$ months.

7. A merchant owes $600, payable as follows: $100 at 2 months, $200 at 5 months, and the rest at 8 months; but he wishes to pay the whole debt at one time: what is the just time for said payment? *A.* 6 months.

8. I owe as follows, viz. to A $1200, payable in 4 months; to B $700, payable in 10 months; to C $650, payable in 2 years; to D $1000, payable in $3\frac{1}{2}$ years; to E $1270, payable in 20

months; and to F $500, payable in 4 years; now, what would be the equated time for paying the whole? *A.* $22\frac{1760}{5320}$ months.

Questions on the foregoing.

1. A man bought a barrel of flour for 2£ 15 s. 6 d., a hhd. of molasses for 6£ 15 s., and a barrel of brandy for 8£ 15 s.; what did the whole cost? *A.* 18£ 5 s. 6 d.

2. What will 9600 yards of cloth cost, at $,50 a yard? (4800) At $,33⅓? (3200) At $,25? (2400) At $,20? (1920) At $,16⅔? (1600) At $,12½? (1200) At $,6¼? (600) At $,5? (480). *A.* $16200.

3. What is the product of 2 s. 6 d. multiplied by 2? (5) By 4? (10) By 7? (17-6) By 10? (1-5) By 12? (1-10) *A.* 4£ 7 s. 6 d.

4. Divide 21£ 19 s. 9 d. equally among 6 men? *A.* 3£ 13 s. 3½ d.

5. Reduce $\frac{7}{8}$, $\frac{3}{4}$, and $\frac{1}{5}$, to the least common denominator. *A.* $\frac{35}{40}$, $\frac{30}{40}$, $\frac{8}{40}$.

6. Change 2000 francs to federal money. *A.* $375.

7. $\frac{25}{1000}+,3+\frac{50}{1000}+673+\frac{5}{100000}$. *A.* 673,37505.

8. Change 4500£, English or sterling money, to dollars of 4 s. 6 d. each. *A.* $20000.

9. What is the interest of $21,20 for 6 months? (636) For 3 months 15 days? (371) For 1½ month? (159) For 10 days? (35) For 5 days? (17) For 4 days? (14) *A.* $1,23 2+.

10. What is the amount of $300, at 7 per cent., for 1 year? (321) At 3 per cent.? (309) At 5½ per cent.? (31650) At 9¾ per cent.? (32925) At 12½ per cent.? (33750) *A.* $1613,25.

11. What is the discount of $315 for 10 months, at 6 per cent.? (15) Of $550 for 1 yr. 8 mo.? (50) Of $2660 for 5 yrs. 6 mo.? (660) Of $121,402 for 8 yrs. 4 mo. 15 da.? (40602) *A.* $765,602.

12. What is the compound interest of $560 for 4 yrs.? (146987) For 2 yrs. 6 mo.? (88092) For 3 yrs.? (106968) *A.* 342,047.

13. A merchant bought goods amounting to $368,925 ready money, and sold them again for $488,75, payable in 2 yrs. 6 mo.; how much did he gain, discounting at 6 per cent.?

(*Find the present worth of $488,75 first, then subtract to find the gain.*)

A. $56,07 5.

14. Bought corn for $,60, and sold it for $,72; what was the gain per cent.? *A.* 20 per cent.

15. Bought 40 gallons of molasses, at 27 cents a gallon; but, by accident, 4 gallons leaked out; at what rate must I sell the remainder per gallon to lose nothing? and how much to gain on the whole cost 20 per cent.?

A. 30 *cts.*; *and, to gain 20 per cent.,* 36 *cts.*

RULE OF THREE,

OR

SIMPLE PROPORTION.

¶ **LXXIII.** 1. What will 3 yds. of cloth come to at 20 cents a yard? What will 5 yards? Will 7? 8? 12?

2. If 2 gallons of molasses cost 50 cents, what will 3 gallons cost? (*Find what 1 gallon will cost first. It is 25 cents. Then, 3 gallons are 3 times 25, = 75 cents. Proceed in the same manner with other sums of like nature.*) What will 5 gallons cost? What will 8?

3. If 4 lbs. of sugar cost 48 cents, what will 2 lbs. cost? (*Find what 1 lb. will cost first.*) What will 6 lbs. cost? What will 8? What will 12? What will 20?

4. If 2 bushels of corn cost a dollar, how much is it a bushel? What will 3 bushels cost? What will 4? What will 6? What will 8?

5. If 20 yards of cloth cost 60 cents, how much is it a yard? What will 6 cents buy? Will 18 cents buy? Will 30 cents? Will 90? Will 300?

6. How many pounds of cheese will 12 cents buy, if 4 lbs. cost 48 cents? How many will 24 cents buy? How many will 60 cents? How many will 108 cents?

7. If 4 dollars buy 2 barrels of cider, how many barrels will 6 dollars buy? (*Find the value of 1 first.*) How many will 8 dollars buy? How many will 12? How many will 24? How many will 36? How many will 48? How many will 60? How many will 100? How many will 150? How many will 300? How many will 400? How many will 500? How many will 800? How many will 2000? How many will 3000? How many will 40000?

8. If you pay 16 cents for 4 oranges, how many cents will buy 6? How many 36? How many 48? How many 60?

9. If 100 oranges cost 400 cents, how many cents will 4 cost? What will 8 cost? What will 25 cost? What will 30 cost? What will 50 cost?

10. If 4 tons of hay will keep 2 cattle over the winter, how many tons will keep 6 cattle the same time? How many 8? How many 10? How many 20? How many 40? How many 60? How many 80?

11. If 500 cattle eat 1000 tons of hay in one winter, what will 2 cattle eat? What will 3? What will 5? What will 20? What will 50? What will 200?

12. If 2 penknives cost 25 cents, what will 3 cost? What will 4? What will 8? What will 12? What will 16?

13. If you pay 26 cents for 2 inkstands, how many cents will buy 3? Will buy 4? 5? 6? 7? 8? 9? 10? 20? 30? 40?

14. If $\frac{2}{8}$ of a yard of broadcloth cost 4 dollars, what will $\frac{4}{8}$ cost? (*If $\frac{2}{8}$ cost 4 dollars, $\frac{1}{8}$ will cost 2 dollars.*) What will $\frac{5}{8}$? What $\frac{6}{8}$? What $\frac{7}{8}$? What $\frac{8}{8}$? What $1\frac{1}{8}$? What $1\frac{2}{8}$? What $1\frac{3}{8}$? What $1\frac{4}{8}$? What $1\frac{5}{8}$? What $1\frac{6}{8}$? What 2 yards? What $\frac{9}{8}$? (*If $\frac{8}{8}$ make 1 yard, then 1 yard and $\frac{1}{8}$ are $\frac{9}{8}$.*) What will $\frac{10}{8}$ of a yard cost? What $\frac{11}{8}$? What $\frac{12}{8}$? What $\frac{16}{8}$? What $\frac{24}{8}$? What 3 yards?

15. The interest of 100 dollars for 1 year is 6 dollars at 6 per cent.; what is it for 2 years? For 3? For 5? For 7? For 9? For 12? For 20?

16. If 6 men can do a piece of work in 12 days, how long will it take 1 man to do the same? (*1 man will be 6 times as long as 6.*) How long will it take 2 men? (*2 men will do it quicker than 1 man.*) 3 men? 12 men?

17 If 4 men build a wall in 20 days, how many men would it require to build the same in 40 days? (*$\frac{1}{2}$ as many men.*) In 80 days?

After the same manner perform the following

Exercises for the Slate.

1. If 20 yards of cloth cost \$40, what will 8 yards cost?

1 yard is $\frac{1}{20}$ of \$40; that is, $40 \div 20 = \$2$ a yard; then, 8 yards are $8 \times 2 = \$16$, Ans.

2. What can you buy 15 tons of hay for, if 3 tons cost \$36? (*Find what 1 ton will cost first.*) *A.* \$180.

3. If 2 bushels of oats cost 40 cents, what will 24 bushels cost? *A.* \$4,80.

4. What will 25 lbs. of sugar cost, at 17 cents a pound?

$17 \times 25 = \$4{,}25$, *Ans.*

5. If \$4,25 buy 25 lbs. of sugar, how much is it a pound?

A. 17 cents.

6. If 3 pair of shoes cost \$4,50, what will 12 pair cost? (18) What will 8? (12) What will 15? (2250) What will 16? (24)

A. \$76,50.

7. If 2 pair of stockings cost 50 cents, what will 3 pair cost? (75) What will 15? (375) What will 25? (625) What will 80? (20) What will 96? (24) What will 267? (6675)

A. \$121,50.

8. What will 600 bushels of rye cost, at 84 cents a bushel? (504) What will 10? (840) What will 40? (3360). What

will 800? (672) What will 1000? (840) What will 2? (168). *A.* $2059,68.

9. If 60 cents buy 4 lbs. of tobacco, how much will 30 cents buy? (2) How much will 90 cents? (6) How much will 120 cents? (8) How much will $2,10? (14) How much will $2,40? (16) *A.* 46 lbs.

10. If 1 pair of gloves cost 75 cents, what will 1 dozen pair cost? (9) What will 1½ doz.? (1350) What will 2 doz.? (18) What will 2½ doz.? (2250) What will 3 doz.? (27)
A. $90.

11. If 3 doz. pair of shoes cost 27 dollars, what will 1 pair cost? (75) What will 2½ doz.? (2250) What will 2 doz.? (18) What will 1½ doz.? (1350) What will 1 doz.? (9)
A. $63,75.

12. If 5 tons of hay will keep 25 sheep over the winter, how many sheep can be kept on 7 tons, at the same rate? (35) On 8 tons? (40) On 15 tons? (75) On 60 tons? (300) On 80? (400). *A.* 850.

13. Boarding at $2,25 a week, how long will $9 last me? (4) How long will $13,50? (6) How long will $18? (8) How long will $20,25? (9) How long will $49,50? (22) *A.* 49 weeks.

14. If a man receive $50 for 2 months' wages, what is that a year? (300) What will 8 months' come to? (200) 16 months' come to? (400) 1½ years' come to? (450) 2 years' come to? (600) 10 years' come to? (3000) *A.* $4950.

15. What will 6 pieces of cloth, each piece containing 20 yards, come to, at $1,50 a yard? (180) What will 1 piece come to? (30) What will 3 pieces? (90) What will 5 pieces? (150) What will 10 pieces? (300) *A.* $750.

16. Bought 5 hhds. of rum, each containing 60 gallons, for $2 a gallon; what do they come to? (600) What will 4 hhds. come to? (480) What will 20 hhds.? (2400) *A.* $3480.

17. William's income is $1500 a year, and his daily expenses are $2,50; how much will he have saved at the year's end?
A. $587,50.

18. If William's income had been $2000, how much would he have saved? (108750) If $2500? (158750) If $3600? (268750) If $4000? (308750) *A.* $8450.

19. If a hhd. of molasses cost $20,16, how much is it a gallon? (*Divide by the number of gallons in a hhd.*) (32) How much is it a quart? (*Divide by the number of quarts in a hhd.*) (8) How much is it a pint? (4) How much is it a gill? (1)
A. 45 *cents.*

The foregoing questions have been solved by a method termed *analysis* This method is thought to accord with the natural operations of the human mind. Men in business scarcely recognise any other. The formality of statements is rarely if ever made by them; and, when it is made, they do it more *for the sake of testing* the correctness of the other method, than for any practical purpose. They may have adopted a statement in the commencement of their business, from the circumstance of having been taught it at school; but the longer they continue in business, the less occasion they have for it. There is another method, which consists in ascertaining the ratio or relation which one number has to another. This is used more or less by all, but more extensively by scientific men, and those well versed in mathematical principles.

20. If 8 pair of shoes cost 63 cents, what will 24 pair cost?

$\frac{1}{8}$ of 63 = $7\frac{7}{8}$ cents, the price of 1 pair, which we multiply by 24 to get the price of 24 pair; thus, 24 × $7\frac{7}{8}$ = \$1,89. But, since $7\frac{7}{8}$ is a fraction, it would be more convenient to multiply by 24 first, and divide by 8 afterwards, as this cannot make any difference; and, that we may make no mistake in the operation, we will make a statement, by writing the 63 cents on the right, as a third term (see operation); on the left of which we write the multiplier, 24, as a 2d term, and, as a first term, the divisor, 8: then, 63 × 24 = 1512 ÷ 8 = \$1,89, the *Ans.*, as before.

OPERATION.

pair.	*pair*	*cents.*
8 ,	24 ,	63
		24
		252
		126
		8)1512
	Ans.,	\$ 1,89

21. If 3 yards of cloth cost 24 cents, what will 6 yards cost?

OPERATION.

yds.	*yds.*	*cts.*
3 ,	6 ,	24
		6
		3)144
	Ans.,	\$,48

24 × 6 = 144 ÷ 3 = 48, *the Ans.*

Or, as we know that 6 yards cost 2 times as much as 3 yds., that is, $\frac{6}{3}$ = 2, by simply multiplying 24 by 2, it makes 48, the answer, the same as before. This is a much shorter process; and, could we discover the principle, it would oftentimes render operations very simple and short. In searching for this, we shall naturally be led to the consideration of ratio, or relation; that is, the relation which necessarily exists between two or more numbers.

Ratio. *Q. What part of* 6 *is* 5? *A.* $\frac{5}{6}$.

Q. What part of 5 *is* 6? *A.* $\frac{6}{5}$.

Q. What part of 3 *is* 4? *A.* $\frac{4}{3}$.

Q. What part of 4 *is* 3? *A.* $\frac{3}{4}$.

Q. What is the finding what part one number is of another called? A. Finding the ratio, or relation of one number to another. *What is ratio, then? A.* The number of times one number, or quantity, is contained in another.

Q. What part of 10 *is* 9? *or, what is the ratio of* 10 *to* 9? *A.* $\frac{9}{10}$.

Q. What is the ratio of 17 *to* 18? *A.* $\frac{18}{17}$.

Q. What is the ratio of 18 *to* 17? *A.* $\frac{17}{18}$.

Q. What part of 3 *oz. is* 12 *oz.? or, what is the ratio of* 3 *oz. to* 12 *oz.? A.* $\frac{12}{3}=4$, *ratio.*

Q. What part of four yards is 9 *yds.? or, what is the ratio of* 4 *to* 9? *A.* $\frac{9}{4}=2\frac{1}{4}$.

Q. Hence, to find the ratio of one number to another, how do you proceed? A. Make the number which is mentioned *last* (*whether it be the larger or smaller*), the numerator of a fraction, and the other number the denominator; that is, always divide the second by the first.

1. What part of $1 is 50 cents? or, what is the ratio of $1 to 50 cents?

A. $1 = 100 cents; *then,* $\frac{50}{100}=\frac{1}{2}$, *the ratio, Ans.*

2. What part of 5 s. is 2 s. 6 d.? or, what is the ratio of 5 s. to 2 s. 6 d.?

2 s. 6 d. = 30 d., *and* 5 s. = 60 d.; *therefore,* $\frac{30}{60}=\frac{1}{2}$, *the ratio, Ans.*

3. What is the ratio of £1 to 15 s.? *A.* $\frac{15}{20}=\frac{3}{4}$, *the ratio.*

4. What is the ratio of 2 to 3? *A.* $\frac{3}{2}$. Of 4 to 20? *A.* 5. Of 20 to 4? *A.* $\frac{1}{5}$. Of 8 to 63? *A.* $7\frac{7}{8}$. Of 200 to 900? *A.* $4\frac{1}{2}$. Of 800 to 900? *A.* $1\frac{1}{8}$. Of 2 quarts to 1 gallon? *A.* 2.

Let us now apply the principle of ratio, which we were in pursuit of, to practical questions.

PROPORTION. **22.** If 2 melons cost 8 cents, what will 10 cost?

It is evident, that 10 *melons will cost* 5 *times as much as* 2; *that is, the ratio of* 2 *to* 10 *is* $\frac{10}{2}=5$; *then,* $5\times8=40$, *Ans.* But, by stating the question as before, we have the following proportions:—

OPERATION.

melons.	*melons.*	*cents.*
2 ,	10 ,	8
		10
		2)80
		$,40

In this example, we make a new discovery, viz. that the ratio of 8 *to* 40 (*which is* $\frac{40}{8}=5$), *is the same as* 2 *to* 10, *which is also* 5; *that is,* 2 *is the same part of* 10 *that* 8 *is of* 40.

Q. When, then, numbers bear such relations to each other, what are the numbers said to form? Ans. A proportion.

Q. *How may proportion be defined, then?* A. It is an equality of ratios.

Q. *How many numbers must there be to form a ratio?* A. Two.

Q. *How many to form a proportion?* A. At least, three.

To show that there is a proportion between three or more numbers, we write them thus:—

melons. melons. cents. cents.

2 : 10 : : 8 : 40,

which is read, 2 is to 10 as 8 is to 40; or, 2 is the same part of 10 that 8 is of 40; or, the ratio of 2 to 10 is the same as that of 8 to 40.

Q. *What is the meaning of* antecedent? A. Going before.

Q. *What is the meaning of* consequent? A. Following.

Q. *What is the meaning of* couplet? A. Two, or a pair.

Q. *What may both terms of a ratio be called?* A. A couplet.

Q. *What may each term of a couplet be called, as* 3 *to* 4? A. The 3, being first, may be called the *antecedent;* and the 4, being after the 3, the *consequent.*

Q. *In the following proportion, viz.* 2 : 10 : : 8 : 40, *which are the antecedents, and which are the consequents?* A. 2 and 8 are the antecedents, and 10 and 40 the consequents.

Q. *What are the ratios in* 2 : 10 : : 8 : 40?

In the last proportion, 2 and 40, being the first and last terms, are called *extremes;* and 10 and 8, being in the middle, are called the *means.* Also, in the same proportion, we *know* that the extremes 2 and 40, multiplied together, are equal to the product of the means, 10 and 8, multiplied together, thus; $2 \times 40 = 80$, and $10 \times 8 = 80$. Let us try to explain the *reason* of this. In the foregoing proposition, the first ratio, $\frac{10}{2}$, ($= 5$), being equal to the second ratio, $\frac{40}{8}$, ($= 5$), that is, the fractional ratios being equal, it follows, that reducing these fractions to a common denominator will make their numerators alike; thus, $\frac{10}{2}$ and $\frac{40}{8}$ become $\frac{80}{16}$ and $\frac{80}{16}$; in doing which, we multiply the numerator 40 (one extreme) by the denominator 2 (the other extreme), also the numerator 10 (one mean) by the denominator, 8, (the other mean); hence the reason of this equality. Q. *When, then, any four numbers are proportional, what may we learn respecting the product of the extremes and means?* A. That the product of the extremes will always be equal to the product of the means. Hence, with any three terms of a proportion being given, the fourth or absent term may easily be found. Let us take the last example:

melons. melons. cts. cts.

..... 10 : : 8 : 40

Multiplying together 8 and 10, the two means, makes 80; then $80 \div 40$, the known extreme, gives 2, the other extreme required, or first term. *Ans.* 2.

Again, let us suppose the 10 absent, the remaining terms are

melons.	*melons.*	*cts.*	*cts.*
2 :	..... : :	8	40

By multiplying together 40 and 2, the extremes, we have 80; which, divided by 8, the known mean, gives 10, the 2d term, or mean, required. Let us exemplify this principle more fully by a practical example.

23. If 10 horses consume 30 bushels of oats in a week, how many bushels will serve 40 horses the same time?

In this example, knowing that the number of bushels eaten are in proportion to the number of horses, we write the proportion thus:

OPERATION.

horses.	*horses.*	*bushels.*
10 :	40 : :	30
		40
	1|0)	120|0
		120 *bushels, Ans.*

By multiplying together 40 and 30, the two means, we have 1200, which, divided by the known extreme, 10, gives 120; that is, 120 bushels, for the other extreme, or 4th term, that was required. Let us apply the principle of ratio in finding the 4th term in this example. The ratio of 10 to 40 is $\frac{40}{10} = 4$, that is, 40 horses will consume 4 times as many bushels as 10; then 4×30 bu. = 120 bushels, the 4th term, or extreme, as before.

Q. When any three terms of a proportion are given, what is the process of finding the fourth term called? A. The Rule of Three.

Q. How, then, may it be defined? A. It is the process of finding, by the help of three given terms, a fourth term, between which and the third term there is the same ratio or proportion as between the second and first terms.

It will sometimes be necessary to change the order of the terms; but this may be determined very easily by the nature of the question, as will appear by the following example:—

24. If 8 yards of cloth cost $4, what will 2 yards cost?

OPERATION.

yds.	*yds.*	$
8 :	2 : :	4
		2
	8)	8
		$1

In this example, since 2 yards will cost a less sum than 8 yards, we write 2 yards for one mean, which thus becomes the multiplier, and 8 yards, the known extreme, for the divisor; for the less the multiplier, and the greater the divisor, the less will be the quotient; then, $2 \times 4 = 8 \div 8 = \$1$, *Ans.* But multiplying by the ratio will be much easier, thus; the ratio of 8 to 2 is $\frac{2}{8} = \frac{1}{4}$; then, $4 \times \frac{1}{4} = \$1$, *Ans., as before.*

From these illustrations we derive the following

RULE.

I. *Which of the three given terms do you write for a third term?* A. That which is of the same kind with the answer.

II. *How do you write the other two numbers, when the answer ought to be greater than the third term?* A. I write the greater for a second term, and the less for a first term.

III. *How do you write them, when the answer ought to be less than the third term?* A. The less for a second term, and the greater for a first term.

IV. *What do you do, when the first and second terms are not of the same denomination?* A. Bring them to the same by Reduction Ascending or Descending.

V. *What is to be done, when the third term consists of more than one denomination?* A. Reduce it to the lowest denomination mentioned, by Reduction.

VI. *How do you proceed in the operation?* A. Multiply the second and third terms together, and divide their product by the first term; the quotient will be the fourth term, or answer, in the same denomination with the third term.

VII. *How may this process of multiplying and dividing be, in most cases, materially shortened?* A. By multiplying the third term by the ratio of the first and second, expressed either as a fraction in its lowest terms, or as a whole number.

VIII. *If the result, or fourth term, be not in the denomination required, what is to be done?* A. It may be brought to it by Reduction.

IX. *If there be a remainder in dividing by the first term, or multiplying by the ratio, what is to be done with it?* A. Reduce it to the next lower denomination, and divide again, and so on, till it can be reduced no more.*

---

* As this rule is commonly divided into *direct* and *inverse*, it may not be amiss, for the benefit of some teachers, to explain how they may be distinguished; also, to give the rule for each.

The Rule of Three Direct is when *more requires more, or less requires less.* It may be known thus; more requires more, when the third term is *more* than the first, and requires the fourth term, or answer, to be *more* than the second; and less requires less, when the third term is *less* than the first, and requires the fourth term, or answer, to be *less* than the second.

RULE 1. *State the question, that is, place the numbers so that the first and third terms may be of the same name, and the second term of the same name with the answer, or thing sought.*

2. *Bring the first and third terms to the same denomination, and reduce the second term to the lowest denomination mentioned in it.*

3. *Divide the product of the second and third terms by the first term; the quotient will be the answer to the question, in the same denomination with the second term which may be brought into any other denomination required.*

More Exercises for the Slate.

25. If 600 bushels of wheat cost $1200, what will 3600 bushels cost? and what is the ratio of the 1st and 2d terms?

Perform the foregoing example, and the following, first, without finding the ratio; then, by finding the ratio, and multiplying by it.

A. $7200. *The ratio,* $\frac{3600}{600} = 6 \times 1200 =$ $7200, *the same.*

26. How many bushels of wheat may be bought for $7200, if 600 bushels cost $1200? *A.* 3600 bushels. *Ratio,* 6; *then,* $6 \times 600 = 3600$ *bushels.*

27. If $7200 buy 3600 bushels of wheat, what will 600 bushels cost? *A.* $1200. *Ratio,* $\frac{1}{6}$.

28. If board for 1 year, or 52 weeks, amount to $182, what will 39 weeks come to? *A.* $136,50. *Ratio,* $\frac{3}{4} \times 182 =$ $136,50, *the same.*

29. If 30 bushels of rye may be bought for 120 bushels of potatoes, how many bushels of rye may be bought for 600 bushels of potatoes? *A.* 150 bushels rye. *Ratio,* 5.

30. If 4 cwt. 1 qr. of sugar cost $45,20, what will 21 cwt. 1 qr. cost? (*Bring 4 cwt. 1 qr., and 21 cwt. 1 qr. into quarters first.*) *A.* $226. *Ratio,* 5.

31. If I buy 60 yards of cloth for $120, what is the cost per yard? (2) What is the cost per ell Flemish. (150) What per ell English? (250) What per ell French? (3) *A.* $9.

32. Bought 4 tuns of wine for $322,56; what did 1 pipe cost?

The Rule of Three Inverse is, *when more requires less, or less requires more,* and may be known thus; more requires less, when the third term is *more* than the first, and requires the fourth term, or answer, to be *less* than the second; and less requires more, when the third term is *less* than the first, and requires the fourth term to be *more* than the second.

RULE. *State and reduce the terms as in the Rule of Three Direct; then multiply the first and second terms together, and divide their product by the third term; the quotient will be the answer, in the same denomination with the middle term.*

Note. Although the distinction of *direct* and *inverse* is frequently made, still it is totally useless. Besides, this mode of arranging the proportional numbers is very erroneous, and evidently calculated to conceal from the view of the pupil the true principles of ratio, and, consequently, of proportion, on which the solution proceeds. The following example will render the absurdity more apparent.

A certain rich farmer gave 20 sheep for a sideboard; how many sideboards may be bought for 100 sheep?

20 sheep : 1 sideboard : : 100 sheep : 5 sideboards, *the 4th term, or Ans.*

It must appear evident, to every rational mind, that there can be no analogy between 20 sheep and 1 sideboard, or 100 sheep and 5 sideboards. With the same propriety it may be asked, what ratio or analogy there is between such heterogeneous quantities as 2 monkeys and 5 merino shawls; or between 7 lobsters and 4 bars of music; the one is equally as correct as the other.

(4032) What did 1 hhd.? (2016) What did 1 tierce? (1344) What did 1 bbl.? (1008) *Ans.* $84. What did 2 quarts cost? (16) What did 3 pints cost? (12) What did 4 gills cost? (4)

A. $,32.

33. Bought 6 tuns of wine for $500,50; what did 1 pipe cost? *A.* $41,708.+.

34. When a merchant compounds with his creditors for 40 cents on the dollar, how much is A's part, to whom he owes $2500? how much is B's part, to whom he owes $1600?

A. A's, $1000; B's, $640.

35. A, failing in trade, owes the following sums, viz. to B $1600,60, to C $500, to D $750,20, to E $1000, to F $230; and his property, which is worth no more than $1020,20, he gives up to his creditors; how much does he pay on the dollar? and what is the amount of loss sustained by all?

A. 25 *cents on* $1; *and the amount of loss is* $3060,60.

36. Bought 4 tierces of rice, each weighing 7 cwt. 2 qrs. 16 lbs.; what do they come to at $9,35 per cwt.? (285842) At $2,50 per qr.? (305714) At 10 cents per pound? (34240)

A. $933,956 +.

37. Bought, by estimation, 300 yards of cloth, for $450,60; but, by actual measurement, there were no more than 275 yds. 2 qrs.; for how much must I sell the measured yards per yd., so as to neither make nor lose? (1635) How much must I sell 20 yds. for, so as to lose nothing? (32711) How much 25 yds. 2 qrs.? (41707) How much 30 yds. 1 qr. 3 na.? (49782)

A. $125,835 +.

38. If a staff 6 feet long cast a shade on level ground 9 feet, what is the height of that steeple whose shade measures at the same time 198 feet? *A.* 132 *feet in height. Ratio*, 22.

39. If 3 cwt. 2 qrs. 16 lbs. of sugar cost $51, what will 10 cwt. 3 qrs. 20 lbs. cost? (*Ratio*, $3 \times 51 = 153$) What will 21 cwt. 3 qrs. 12 lbs.? (6, *ratio*, $\times\ 51 = 306$) What will 43 cwt. 2 qrs. 24 lbs.? (*Ratio*, 12; then, $12 \times 51 = 612$) *A.* $1071.

Note. The following examples may be performed either by analysis, or by finding the ratio, or by the common rule. Perhaps it would be well to let the pupil take his choice. The one by ratio is recommended.

40. If $100 gain $6 in a year, how much will $20 gain in the same time? (120. *Ratio*, $\frac{1}{5}$) How much will $10 gain? (60. *Ratio*, $\frac{1}{10}$) How much will $50 gain? (3. *Ratio*, $\frac{1}{2}$) How much will $75 gain? (450. *Ratio*, $\frac{3}{4}$) What will $200 gain? (12. *Ratio*, 2) How much will $300 gain? (18. *Ratio*, 3) How much will $500 gain? (30. *Ratio*, 5) How much will $800 gain? (48. *Ratio*, 8) How much will $1000 gain? (60. *Ratio*, 10) How much will $1250 gain? (75. Ratio, $12\frac{1}{2}$) How much will $2000 gain? (120. Ratio, 20) *Ans.* $372,30.

41. If 12 men can build a wall in 20 days, how many can do the same in 5 days? *A.* 48 men. *Ratio*, 4.

42. If 60 men can build a wall in 4 days, how many men can do the same in 20 days? *A.* 12 men. *Ratio*, $\frac{1}{5}$.

43. If 4 men can build a wall in 120 days, in how many days will 12 men do the same? (40) Will 16 men? (30) Will 20 men? (24) Will 24 men? (20) *A.* 114 days.

44. If a man perform a journey in 30 days, by travelling 6 hours each day, in how many days will he perform it by travelling 10 hours each day? (10 *hours will require a less number of days than* 6 *hours; that is, the multiplying or* 2d *term must be the smallest.*) *A.* 18 days. *Ratio*, $\frac{3}{5} \times 30 = 18$, *the same.*

45. If a field will keep 2 cows 20 days, how long will it keep 5 cows? (8) Will it keep 8 cows? (5) Will it keep 10 cows? (4) Will it keep 20 cows? (2) *A.* 19 days.

46. If 60 bushels of grain, at $1 per bushel, will pay a debt, how many bushels, at $1,50 a bushel, will pay the same? (40) How many bushels at $1,20? (50) How many at 80 cents? (75) At 50 cents (120) At 40 cents? (150) At 30 cents? (200) *A.* 635 bushels.

47. How much in length that is 6 inches in breadth will make a square foot? (12 *inches in length and* 12 *in breadth make* 1 *square foot; then,* 6 *inches in breadth will require more in length; that is,* 6 : 12 : : 12) (24) How many 4 inches in breadth? (36) How many 8 inches in breadth? (18) How many 16 inches in breadth? (9) *A.* 87 inches.

48. If a man's income be $1750½ a year, how much may he spend each day to lay up $400 a year? *A.* $3,70.

49. If 6 shillings make $1, New England currency, how much will 4 s. 6 d. make, in federal money? (,75) Will 2 s 6 d.? (,41⅔) Will 1 s. 6 d.? (,25) Will 3 s. 9 d.? (,62½)

A. $2,04⅙.

50. A merchant bought 26 pipes of wine on 6 months' credit but, by paying ready money, he got it 3 cents a gallon cheaper; how much did he save by paying ready money? *A.* $98,28.

51. Bought 400 yards 2 qrs. of plaid for $406,80, but could sell it for no more than $300; what was my loss per ell French? *A.* $,40.

52. If 120 gallons of water, in 1 hour, fall into a cistern containing 600 gallons, from which, by 1 pipe, 20 gallons run out in 1 hour, and by another 50 gallons, in what time will the cistern be filled? *A.* 12 hours.

53. A merchant bought 40 pieces of broadcloth, each piece containing 45 yards, at the rate of $6 for 9 yards, and sold it again at the rate of $15 for 18 yards; how much did he make in trading? *A.* $300.

54. A borrowed of B \$600 for 3 years; how long ought A to lend B \$800 to requite the favour? (2-3) How long ought he to lend him \$900? (2) How long \$500? (3-7-6) How long \$1200? (1-6) *A.* 9 years, 4 mo. 6 days.

55. A gentleman bought 3 yards of broadcloth $1\frac{1}{2}$ yards wide; how many yards of flannel, which is only $\frac{3}{4}$ yd. wide, will line the same?

It is evident it will take more cloth which is only $\frac{3}{4}$ yd. wide, than if it were $1\frac{1}{2}$ yd. wide; hence $1\frac{1}{2}$ must be the middle term.

A. 6 yds. *Ratio,* 2.

56. A regiment of soldiers, consisting of 800 men, are to be clothed, each suit containing $4\frac{5}{8}$ yds. of cloth, which is $1\frac{3}{4}$ yd. wide, and lined with flannel $\frac{7}{8}$ yd. wide; how many yards of flannel will be sufficient to line all the suits?

A. 8633 yds. 1 qr. $1\frac{1}{3}$ na.

FRACTIONS. 57. If $\frac{1}{8}$ of a barrel of flour cost $\frac{5}{16}$ of a dollar, what will $\frac{3}{4}$ of a barrel cost?

By analysis. It is plain, that if we knew the price of 1 barrel, $\frac{3}{4}$ of a barrel would cost $\frac{3}{4}$ as much. If $\frac{1}{8}$ of a barrel cost $\frac{5}{16}$ of a dollar, $\frac{8}{8}$, or 1 barrel, will cost 8 times as much, that is,

$$\frac{8\times 5}{16}=\frac{40}{16}\times\frac{3}{4}=\$1\tfrac{7}{8},\ \textit{Ans.}$$

OR, *as $\frac{3}{4}$ is more than $\frac{1}{8}$, we may make $\frac{3}{4}$ the 2d, or multiplying term, as in the foregoing examples, thus:—*

bbls. bbls. \$

$\frac{1}{8}:\frac{3}{4}::\frac{5}{16}$. *Then,* $\frac{5}{16}\times\frac{3}{4}=\frac{15}{64}\div\frac{1}{8}$ (*Inverting* $\frac{1}{8}$ *by* ¶ XLVII. *then,* $\frac{15\times 8}{64\times 1}$) $=\frac{120}{64}=\$1\frac{7}{8}$, *Ans.*

OR, *multiplying by the ratio, thus; the ratio of $\frac{1}{8}$ to $\frac{3}{4}$ is $\frac{3}{4}\div\frac{1}{8}=\frac{24}{4}=6$, ratio; then,* $6\times\frac{5}{16}=\frac{30}{16}=\$1\frac{7}{8}$, *Ans., as before.*

OR, *which is obviously the same, having inverted the 1st, or dividing term, multiply all the fractions together; that is, proceed as in Division of Fractions,* (¶ XLVII.) *thus,* $\frac{8\times 3\times 5}{1\times 4\times 16}=\frac{120}{64}=\$1\frac{7}{8}$, *Ans., as before.*

The pupil may perform the following examples by either of the preceding methods, but the one by analysis is recommended, it being the best exercise for the mind.

58. If 3 lbs. of butter cost $\frac{3}{8}$ of a dollar, what cost $\frac{1}{2}$ lb.?

A. \$$\frac{1}{16}$

59. If $\frac{1}{3}$ of a bushel of wheat cost $\frac{3}{16}$ of a dollar, what will bushel cost? *A.* $\frac{9}{16}$.

60. If $1\frac{1}{4}$ yds. of cloth cost $\$\frac{7}{12}$, what will 1 yd. cost? *A.* $\$\frac{1}{3}$.

61. At $\$\frac{1}{16}$ a pound, what will 40 pounds cost? *A.* $\$2\frac{1}{2}$.

62. If $\frac{3}{4}$ yd. cost $\$2\frac{3}{25}$, what will 1 yd. cost? *A.* $\$2,82\frac{2}{3}$.

63. If $\frac{1}{2}$ of $\frac{3}{4}$ yd. cost $\$2$, what is it a yard? *A.* $\$5\frac{1}{3}$.

64. If $\frac{2}{3}$ of $\frac{7}{8}$ of $\frac{5}{16}$ of $\$1$ buy 20 apples, how many apples ll $\$5$ buy? *A.* $487\frac{13}{21}$ apples.

65. If $\frac{5}{6}$ oz. of gold be worth $\$1,50$, what is the cost of 1 oz.? $\$1,80$.

66 If $16\frac{7}{8}$ yds. will make 8 coats, how many yards will it take 1 coat? *A.* $2\frac{7}{64}$ yds.

67. If $\frac{1}{2}$ of $\frac{3}{4}$ of a gallon cost $\$\frac{5}{8}$, what will $5\frac{1}{2}$ gallons cost? $\$9\frac{1}{6}$.

68. If 6 yds. cost $\$5\frac{3}{5}$, what will $14\frac{2}{3}$ yds. cost? *A.* $\$13\frac{31}{45}$.

69. If $\frac{1}{2}$ of $\frac{7}{8}$ cwt. of sugar cost $\$\frac{9}{10}$, what will 40 cwt. cost? $\$82\frac{2}{7}$.

70. If $\frac{3}{4}$ yd. of silk cost $\frac{3}{4}$ of $\$\frac{5}{8}$, what is the price of 50 yds.? $\$31\frac{1}{4}$.

71. If 1 cwt. of flour cost $\$\frac{15}{16}$, what will $\frac{5}{112}$ cwt. cost? $\$\frac{75}{1792}$.

72. If 3 yds. of cloth, that is $2\frac{1}{2}$ yds. wide, will make a cloak, w much cloth, that is only $\frac{3}{4}$ yd. wide, will make the same rment?

The narrower the cloth, the more yards it will take; hence we make the eater the second term, thus; $\frac{3}{4}$ *yd.* : $2\frac{1}{2}$ *yds.* : : 3 *yds.* : 10 *yds., Ans.*

73. If I lend my friend $\$960$ for $\frac{8}{9}$ of a year, how much ought to lend me $\frac{9}{10}$ of a year to requite the favour?

He ought not to lend me so much as I lend him, because I am to keep the ney longer than he; therefore, make $\frac{8}{9}$ *the middle term.* *A.* $\$853\frac{1}{3}$.

74. If 12 men do a piece of work in $12\frac{1}{4}$ days, how many men ll do the same in $6\frac{1}{8}$ days? *A.* 24 men. *Ratio*, 2.

75. A merchant, owning $\frac{2}{3}$ of a vessel, sells $\frac{3}{5}$ of his share for 500; what was the whole vessel worth?

$\frac{3}{5}$ *of* $\frac{2}{3} = \frac{6}{15} = \frac{2}{5}$; *then, as* $\frac{2}{5}$ *of the vessel is* $\$500$, $\frac{1}{5}$ is $\$250$, *nd* $\frac{5}{5}$, *or the whole vessel, is* $5 \times 250 = \$1250$.

Or thus; $\frac{3}{5}$ *of* $\frac{2}{3}$: 1 : : 500 : $\$1250$, *Ans., as before.*

76. *If* $1\frac{1}{2}$ lb. indigo cost $\$3,84$, what will 49,2 lbs. cost at the *ne rate?* *A.* $\$125,952$.

77. If $29¾ buy 59½ yds. of cloth, what will $60 buy? *A.* 120 yds.

78. How many yds. of cloth can I buy for $75½, if 267¾ yds. cost $37¾? *A.* 535½ yds. *Ratio*, 2.

COMPOUND PROPORTION.

¶ **LXXIV.** 1. If 40 men, in 10 days, can reap 200 acres of grain, how many acres can 14 men reap in 24 days?

By analysis. If 40 *men, in* 10 *days, reap* 200 *acres*, 1 *man, in the same time, will reap* $\frac{1}{40}$ *of* 200 *acres, that is*, 5 *acres in* 10 *days; and in* 1 *day, he will reap* $\frac{1}{10}$ *of* 5 *acres* $= \frac{5}{10} = \frac{1}{2}$ *an acre a day; then* 14 *men in* 1 *day will reap* 14 *times as much; which is*, $14 \times \frac{1}{2} = 7$ *acres; and in* 24 *days*, 24 *times* 7 *acres*, $=$ 168 *acres, Ans.*

Perform the following sums in the same manner.

2. If 4 men mow 96 acres in 12 days, how many acres can 8 men mow in 16 days?

First find how many acres 1 *man will mow in* 12 *days; then, in* 1 *day.* *A.* 256 acres.

3. If a family of 8 persons, in 24 months, spend $480, how much would 16 persons spend in 8 months? *A.* $320.

4. If a man travel 60 miles in 5 days, travelling 3 hours each day, how far will he travel in 10 days, travelling 9 hours each day?

$\frac{1}{5}$ *of* $60 = 12$, *and* $\frac{1}{3}$ *of* $12 = 4$ *miles, the distance which he travels in* 1 *hour; then*, 4 *miles* $\times$ 9 *hours* $= 36 \times 10$ *days* $=$ 360 *miles, the Ans.*

It will oftentimes be found convenient to make a statement, as in Simple Proportion. Take the last example.—In solving this question, we found the answer, which is miles, depended on two circumstances; the number of days which the man travels, and the number of hours he travels each day.

Let us, in the first place, find how far he would go in 5 days, supposing he travelled the same number of hours each day. The question will then be:

If a man travel 60 miles in 5 days, how many miles will he travel in 10 days? This will give the following proportion, to which, and the next following proportion, the answers, or fourth terms, are to be found by the Rule of Three; thus,

5 days : 10 days : : 60 miles : miles;

which gives for the fourth term, or answer, 120 miles. In the next place, we will consider the difference in hours; then the question will be,

If a man, by travelling 3 hours a day for a certain number of days, travel

miles, how many miles will he travel, in the same number of days, if he
rel 9 hours a day; which will give the following proportion:—

3 hours : 9 hours : : 120 miles : miles;

ch gives for the fourth term, or answer, 360 miles.

n performing the foregoing examples, we, in the first operation, multiplied
y 10, and divided the product by 5, making 120. In the next operation, we
tiplied 120 by 9, and divided the product by 3, making 360, the answer.
, which is precisely the same thing, we may multiply the 60 by the product
he multipliers, and divide this result by the product of the divisors; by
ch process the two statements may be reduced to one; thus,

5 days : 10 days }
3 hours : 9 hours } : : 60 miles : miles;

n this example, the product of the multipliers, or second terms, is $9 \times 10 =$
and the product of the divisors, or first terms, is $3 \times 5 = 15$; *then,* 60×90
$5400 \div 15 = 360$ *miles, the Ans., as before.*

Note. It will be recollected, that the ratio of any two terms is the second
ded by the first, expressed either as a fraction, or by its equal whole num-

Or, by comparing the different terms, we see that 60 miles
the same proportion to the fourth term, or answer, that 5 days
to 10 days, and that 3 hours has to 9 hours; hence we may
reviate the process, as in Simple Proportion, by multiplying
third terms by the ratio of the other terms, thus:

The ratio of 5 *to* 10 *is* $\frac{10}{5} = 2$, *and of* 3 *to* 9 *is* $\frac{9}{3} = 3$. *But*
ltiplying 60 *miles by the product of the ratios* 2 *and* 3, *that*
6, *is the same as multiplying* 60 *by them separately; then,*
$\times\ 60 = 360$ *miles, Ans., as before.*

Note. This method, in most cases, will shorten the process very mate-
ly, and in no case will it be any longer; for, when the ratios are fractions,
tiplying the third term by them (according to the rule for the multiplica-
of fractions) will, in fact, be the same process as by the other method.

Q. *From the preceding remarks, what does Compound Pro-*
tion, or Double Rule of Three, appear to be? A. It is find-
the answer to such questions as would require two or more
tements in Simple Proportion; or, in other words, it is when
relation of the quantity required, to the given quantity of
same kind, depends on several circumstances combined.

Q. *The last question was solved by multiplying the third term*
the product of the ratios of the other terms; what, then, may
product be called, which results from multiplying two or
re ratios together? A. Compound Ratio.

om the preceding remarks we derive the following

RULE.

. *What number do you make the third term?* A. That which
f the same kind or denomination with the answer.

I. *How do you arrange all the remaining terms?* A. Take
two which are of the same kind, and, if the answer ought

to be greater than the third term, make the greater the second term, and the smaller the first; but, if not, make the less the second term, and the greater the first; then take any other two terms of the same kind, and arrange them in like manner, and so on till all the terms are used; that is, proceed according to the directions for stating in Simple Proportion.

III. *How do you proceed next?* *A.* Multiply the third term by the continued product of the second terms, and divide the result by the continued product of the first terms; the quotient will be the fourth term, or answer.

IV. *How may the operation, in most cases, be materially shortened?* *A.* By multiplying the third term by the continued product of the ratios of the other terms.

More Exercises for the Slate.

1. If 25 men, by working 10 hours a day, can dig a trench 36 feet long, 12 feet broad, and 6 feet deep, in 9 days, how many hours a day must 15 men work, in order to dig a trench 48 feet long, 8 feet broad, and 5 feet deep, in 12 days?

$$\left.\begin{array}{lcl} 15\ \textit{men} & : & 25\ \textit{men} \\ 36\ \textit{length} & : & 48\ \textit{length} \\ 12\ \textit{breadth} & : & 8\ \textit{breadth} \\ 6\ \textit{depth} & : & 5\ \textit{depth} \\ 12\ \textit{days} & : & 9\ \textit{days} \end{array}\right\} :: 10\ \textit{hours} : \quad \textit{hours.}$$

In this example, the second terms, $25 \times 48 \times 8 \times 5 \times 9, = 432000$, *and the first terms,* $15 \times 36 \times 12 \times 6 \times 12, = 466560$. *Then, the third term,* $10 \times 432000, = 4320000 \div 466560 = 9\ h.\ 15\frac{5}{9}\ m.$, *the fourth term, or Answer*

Or, *by multiplying the third term by the ratios, thus: the ratio of* 15 *to* 25 *is* $\frac{25}{15} = \frac{5}{3}$, *of* 36 *to* 48 *is* $\frac{4}{3}$, *of* 12 *to* 8 *is* $\frac{2}{3}$, *of* 6 *to* 5 *is* $\frac{5}{6}$, *of* 12 *to* 9 *is* $\frac{3}{4}$, *whose products, multiplied by the third term, are* $\frac{5 \times 4 \times 2 \times 5 \times 3 \times 10\ h.}{3 \times 3 \times 3 \times 6 \times 4} = \frac{6000}{648}\ h. = 9\ h.\ 15\frac{5}{9}\ m.$, *Ans., as before.*

This method, it will be perceived, is much shorter than the former. But, had we selected terms whose ratios would be whole numbers, the process would have been shorter still, as is the case in the next question. The present example, however, may be rendered more simple by rejecting equal terms, as in ¶ XLI.; thus, *the ratios* $\frac{5 \times 4 \times 2 \times 5 \times 3 \times 10\ h.}{3 \times 3 \times 3 \times 6 \times 4} = \frac{5 \times 2 \times 5 \times 10\ h.}{3 \times 3 \times 6} = \frac{500}{54} = 9\ h.\ 15\frac{5}{9}\ m.$, *Ans., as before.*

Let the pupil perform the following examples by the common rule of proportion first, then by multiplying by the ratio, and lastly by analysis.

2. If 5 men can build 10 rods of wall in 6 days, how many rods can 20 men build in 18 days?

Men 5 : 20 } : : 10 *rods* : *rods.*
Days 6 : 18 }

In this example, the ratios of 5 *to* 20, *and of* 6 *to* 18, *are* 3 *and* 4; *then*, $3 \times 4 \times 10$ *rods* = 120 *rods, Ans.*

The same by analysis. 1 *man will build* $\frac{1}{5}$ *of* 10 *rods* = $\frac{10}{5}$ *in* 6 *days, and in* 1 *day* $\frac{1}{6}$ *of* $\frac{10}{5} = \frac{10}{30} = \frac{1}{3}$; *that is,* 1 *man will build* $\frac{1}{3}$ *of a rod a day; then,* 20 *men* $\times$ 18 *days* $\times \frac{1}{3} = 120$ *rods, Ans., as before.*

3. If 4 men receive $24 for 6 days' work, how much will 8 men receive for 12 days work? *A.* $96.

4. If 4 men receive $24 for 6 days' work, how many men may be hired 12 days for $96? *A.* 8 men.

5. If 8 men, in 12 days, receive $96, how much will 4 men receive for 6 days' work? *A.* $24.

6. If 8 men receive $96 for 12 days' work, how long may 4 men be hired for $24? *A.* 6 days.

7. If 9 persons in a family spend $1512 in 1 year (or 12 mo.), how much will 3 of the same persons spend in 4 months? *A.* $168.

8. If $2000 will support a garrison of 150 men 3 months, how long will $6000 support 4 times as many men? (The *ratios* are 3 and $\frac{1}{4}$; then, $3 \times \frac{1}{4} \times 3$ mo. $= \frac{9}{4}$.) *A.* $2\frac{1}{4}$ mo.

9. If $100 gain $6 in 1 year, what will $900 gain in 8 months? *A.* $36.

10. If $100 gain $6 in 1 year, in what time will $900 gain $36? *A.* 8 months.

11. If the transportation of 12 cwt. 3 qrs., for 400 miles, cost $57,12, what will the transportation of 10 tons, for 75 miles, amount to? *A.* $168.

12. An usurer put out $150 at interest, and when it had been on interest 8 months, he received, for principal and interest, $160; at what rate per cent. per annum did he receive interest?

By cancelling the ratios $\frac{3}{2}$ *and* $\frac{2}{3}$, *the third term will be the answer. A.* 10 *per cent.*

Questions on the foregoing.

1. What will 2 yds. of cloth cost, at 50 cents (or $$\frac{1}{2}$) a yard? What will 10 yds.? What will 100 yds.? What will 5 yds.? What will 9 yds.?

2. At 25 cents (or $$\frac{1}{4}$) a yard, what will 4 yds. cost? What will 12 yds.? What will 40 yds.? What will 300 yds.?

3. At $,33$\frac{1}{3}$ (or $$\frac{1}{3}$), what will 6 yds. cost? What will 9 *yds.*? Will 24 yds.? Will 300 yds.? Will 7 yds.? Will 25 *yds.*?

4. At \$,16$\frac{2}{3}$ (or \$$\frac{1}{6}$), what will 12 yds. cost? Will 13 yds cost? Will 14 yds. cost? Will 25 yds. cost? Will 120 yds.? Will 300 yds.?

5. At \$,12$\frac{1}{2}$ (or \$$\frac{1}{8}$) a yard, what will 16 yds. cost? Will 96 yds.? Will 97 yds.? Will 100 yds.?

6. At \$,06$\frac{1}{4}$ (or \$$\frac{1}{16}$), what will 33 yds. cost? Will 66 yds.?

7. What will 4 qts. of molasses cost at 2 cents a gill?

8. How many shillings in 4£? In 8£? In 3£ 5 s.

9. How many minutes in 8 hours? In 12 hours?

10. How many cents is $\frac{1}{8}$ of a dollar? Is \$$\frac{1}{16}$? Is \$$\frac{1}{8}$? Is \$$\frac{1}{4}$? Is \$$\frac{1}{5}$? Is \$$\frac{1}{8}$? Is \$$\frac{3}{16}$? Is \$$\frac{3}{8}$? Is \$$\frac{7}{16}$? Is \$$\frac{9}{16}$? Is \$$\frac{5}{8}$? Is \$$\frac{11}{16}$? Is \$$\frac{3}{4}$? Is \$$\frac{13}{16}$? Is \$$\frac{7}{8}$? Is \$$\frac{15}{16}$?

11. How many pence in 3 s. 9 d.? In 8 s. 6 d.?

12. How many cents in 2 s.? In 3 s.? In 3 s. 9 d.?

13. How many cents will buy 3 slates, at 1 s. 9 d. each?

14. How many cents are 18d.? Are 2 s. 6 d.? Are 5 s. 3d.?

15. What kind of a fraction is $\frac{3}{4}$? Is 14$\frac{1}{2}$? Is $\frac{1}{2}$ of $\frac{2}{3}$? Is $\frac{2\frac{5}{3}}{}$? Is $\frac{6}{7}$?

16. What mixed number is equal to $\frac{12}{5}$? $\frac{47}{8}$?

17. How many 5ths in 8$\frac{2}{5}$? In 4$\frac{1}{5}$?

18. How many pounds in $\frac{40}{20}$£? In $\frac{80}{20}$£?

19. How many pounds and shillings in 10$\frac{8}{20}$£? ($\frac{1}{20}$£ *is* 1 s.) In 8$\frac{5}{20}$£?

20. How many shillings in $\frac{36}{3}$ d.? In $\frac{144}{12}$ d.?

21. What fraction is equal to $\frac{4}{8}$? $\frac{8}{12}$? $\frac{4}{20}$?

22. What kind of a fraction is ,5? Is 14,3?

23. What will be the cost of 3 books, at \$,5 apiece? At \$,25?

24. What decimal fraction is equal to $\frac{1}{2}$? $\frac{3}{4}$? $\frac{1}{5}$?

25. Bought ,4 of a bushel of rye at one time, at another ,25, and at another ,35; how much did I buy in all?

26. How much is $\frac{3}{4}$ of a shilling? $\frac{1}{6}$ s.? ($\frac{1}{6}$ s. *is* $\frac{1}{6}$ *of* 12 d. = 2 d.) $\frac{1}{2}$ shilling?

27. How much is $\frac{1}{2}$ of $\frac{1}{2}$? Is $\frac{1}{2}$ of $\frac{1}{3}$? Is $\frac{1}{3}$ of $\frac{1}{12}$?

28. How much is 3 times $\frac{1}{3}$? 3 times $\frac{2}{3}$? 3 times $\frac{1}{5}$?

29. A man bought $\frac{1}{4}$ of a barrel of flour for 3 dollars; how much was it a barrel?

30. 3 is $\frac{1}{4}$ of what number?

31. 5 is $\frac{1}{6}$ of what number?

32. 10 is $\frac{1}{3}$ of what number?

33. 7 is $\frac{1}{8}$ of what number?

34. $\frac{1}{3}$ of 6 is $\frac{1}{4}$ of what number?

35 $\frac{1}{3}$ of 12 is $\frac{1}{5}$ of what number?

36. $\frac{2}{3}$ of 12 is $\frac{1}{5}$ of what number?

37. $\frac{3}{4}$ of 12 is $\frac{1}{3}$ of what number?

38. If a man save $\frac{1}{8}$ of a dollar a day, how many dollars will he save in 16 days? How many in 17 days? In 25 days?

39. In $\frac{16}{8}$ how many times 1?

40. In $\frac{17}{8}$ how many times 1?

41. In $\frac{25}{8}$ how many times 1?

42. In $\frac{144}{12}$, how many times 1?

43. William gave away 1 apple and $\frac{1}{2}$, which was $\frac{1}{3}$ of all he had; how many had he? *$\frac{3}{3}$ is 3 times as much as $\frac{1}{3}$; hence, if $1\frac{1}{2}$ is $\frac{1}{3}$ of a certain number, the number must be 3 times $1\frac{1}{2}$; thus, $1\frac{1}{2}=\frac{3}{2}$; then $\frac{3}{2}\times 3=\frac{9}{2}=4\frac{1}{2}$, Ans.*

44. $2\frac{1}{4}$ is $\frac{1}{5}$ of what number? (*If $2\frac{1}{4}$ is $\frac{1}{5}$, $\frac{5}{5}$, or the number, is 5 times as much; that is, 5 times $2\frac{1}{4}=$*) $11\frac{1}{4}$, *Ans.*

45. $1\frac{3}{4}$ is $\frac{1}{3}$ of what number?

46. $1\frac{2}{3}$ is $\frac{1}{7}$ of what number?

47. William gave $,25 for 2,4 yds. of ribbon; how much was it a yard?

48. Divide 3,5 by ,7. *A.* 5.

59. Divide 3,5 by ,07. *A.* 50.

50. Multiply 5 by ,7. *A.* 3,5.

51. Multiply 50 by ,07. *A.* 3,5.

52. How much will $\frac{1}{2}$ of a peck of salt cost, at $1 a bushel?

53. How many drams is $\frac{1}{3}$ of an ounce?

54. How much is $\frac{1}{8}$ of a yard?

55. How much is $\frac{3}{4}$ of a day?

56. How much is $\frac{3}{12}$ of a minute?

57. If you pay 6 cents a quart for ale, how much is that a gallon?

58. If a man earn $3 a week, how much is that for each working day in a week?

59. If a man travel 6 miles in 2 hours, how far will he travel in 16 hours?

60. What is the ratio of 2 to 6? Of 6 to 2? Of 9 to 6? Of 6 to 9? Of 10 to 100?

61. If 20 bushels of apples cost $10, what will 5 bushels cost?

62. What is the ratio of 20 to 5?

63. If 6 gallons of water fall into a cistern, containing 12 gallons, in 1 hour, and 3 gallons leak out in an hour, how long will it take the cistern to be filled?

. *If 4 men,* in one day, consume 3 loaves of bread, how *aves* will 12 men consume in 4 days?

65. If 6 men, in 12 days, reap 18 acres of grain, how many acres will 12 men reap in 4 days?

66. What part of 1 month is 15 days? Is 1 day? Is 2 days? Is 3 days? Is 5 days? Is 10 days? Is 20 days?

67. What is the interest of $20 for 4 months? For 10 months? For 3 months? For 1 year 4 months?

68. What is the amount of $40 for 2 mo.? For 5 mo.? For 10 mo.? For 1 yr. 8 mo.?

69. What is the interest of $60 for 15 days? For 10 days? For 1 day? For 5 days? For 2 mo. 15 days? For 4 mo. 10 days? For 1 yr. 8 mo. 15 days?

70. What must you pay a broker, who gives you $20 in Boston bills, in exchange for Providence bills, at ½ per cent commission? At ¾ per cent.? At 1 per cent.?

71. What is my demand for selling $600 worth of cotton, and guarantying the payment, at 5 per cent. commission? At 7 per cent. commission? At 4 per cent.?

72. What is the amount of $1 for 1 yr. 8 mo.?

73. What is the amount of $200 for 1 yr. 8 mo.?

74. What, then, is the present worth of $220, due 1 yr. 8 mo hence?

75. What is the present worth of $530, due 1 yr. hence? What is the discount?

76. A merchant, having bought cloth for 50 cents a yard, wishes to mark it so as to gain 10 per cent.; what price must he put on it? (,50 × 110 = $,55, *Ans.*) What price must he put on it to gain 4 per cent.? 8 per cent.? 20 per cent?

77. William bought a sled for $3, and sold it so as to gain 10 per cent.; what did he get for it?

Exercises for the Slate.

1. Bought 1 gallon of molasses for 62½ cents, 1 quarter of flour for $1,12½, 3 lbs. of tea for $1,05, 3 yards of flannel for $,87½, and 1 skein of silk for 6¼ cents; what was the amount of the whole? *A.* $3,73¾.

2. Bought 144 lbs. of raisins for $16; what was that a pound? *A.* $,111+.

3. What is the cost of 3600 yards of flannel, at $,12½ a yard? (450) At $33⅓? (1200) At $,20? (720) *A.* $2370.

4. In 20 lbs. of silver, how many spoons, each weighing 5 oz. 10 pwt.? *A.* 43 spoons, and 3 oz. 10 pwt. rem.

5. A bought 4 hogsheads of molasses for $84, sold 1 hogshead for $25, and the remainder at the rate of 4 cents a pint; how much did he make on the whole? *A.* $1,48.

6. Bought a hogshead of sugar, weighing 12 cwt. 1 qr. 15 lbs., and sold at one time 2 cwt. 2 qrs. 8 lbs., at another, 5 cwt. 15 lbs.,

and at another, 1 cwt. 3 qrs.; how much remained unsold? *A.* 2 cwt. 3 qrs. 20 lbs.

7. Multiply $\frac{3}{4}$ by $\frac{7}{8}$. *A.* $\frac{21}{32}$.

8. Divide $\frac{3}{4}$ by $\frac{7}{8}$. *A.* $\frac{6}{7}$.

9. Multiply $\frac{7}{8}$ by $\frac{3}{4}$. *A.* $\frac{21}{32}$.

10. Divide $\frac{7}{8}$ by $\frac{3}{4}$. *A.* $\frac{28}{24}=1\frac{1}{6}$.

11. There are 4 pieces of cloth, one containing $8\frac{1}{2}$ yds., another $16\frac{7}{8}$ yds., another $12\frac{1}{4}$ yds., and another $7\frac{1}{8}$ yds.; how many yards in the 4 pieces? *A.* 44 yds. 3 qrs. $3\frac{1}{2}$ na.

12. What is the difference between $\frac{1}{2}$ and $\frac{1}{3}$? *A.* $\frac{1}{6}$. Between $\frac{3}{4}$ and $\frac{1}{3}$? *A.* $\frac{5}{12}$. Between $\frac{7}{8}$ and $\frac{5}{16}$? *A.* $\frac{9}{16}$.

13. Reduce $\frac{4}{7}$ of a guinea to the fraction of a pound. *A.* $\frac{3}{5}$.

14. What is the value of ,003125£? *A.* 3 qrs.

15. How much does ,025 make multiplied by ,325? (8125) By $\frac{24}{1000}$? (6) By $3\frac{7}{100}$? (7675) By 276? (69) By $25\frac{5}{10000}$? (6250125) *A.* 7,6104875.

16. A farmer sent a load of hay to market, which, with the cart, weighed 29 cwt. 3 qrs. 16 lbs.; the weight of the cart was $10\frac{3}{4}$ cwt.; what did the hay come to, at $15 a ton?

A. $14,357 +.

17. A merchant bought sugar in a hogshead, both of which weighed 8 cwt. 15 lbs.; the hogshead alone weighed 1 cwt. 1 qr.; what was the cost of the sugar, at $11\frac{1}{4}$ cents a pound?

A. $86,73$\frac{3}{4}$.

The two preceding questions are proper examples in a rule usually termed *Tare* and *Trett*.

18. Bought 50 yards of broadcloth for $160,50, but, not proving so good as I expected, I am willing to lose $42 on the sale of it; what must I demand per ell French? *A.* $3,555.

19. What is my demand for selling 600 bales of cotton, at 40\frac{3}{4}$ a bale, for $2\frac{1}{4}$ per cent. commission? (550125) For $4\frac{1}{2}$ per cent.? (110025) For 7 per cent.? (171150) For $5\frac{3}{8}$ per cent.? (13141875) *A.* $4676,06$\frac{1}{4}$.

20. What is the interest of $200,50 for 2 yrs. 6 mo.? (30075) For 5 yrs. 3 mo. 15 days? (63658) For $6\frac{1}{2}$ mo.? (6516) For 1 yr. 3 mo. 19 days? (15672) *A.* $115,921.

21. What is the difference between the compound and simple interest of $200 for 1 yr. 6 mo.? (36) For 3 yrs. 4 mo.? (2967) For 2 yrs. 6 mo. 15 days? (1523) *A.* $4,85.

22. What is the amount of $60 for 10 yrs. 3 mo. 19 days? (9709) For 8 yrs. 9 mo. 9 days? (9159) *A.* $188,68.

23. Bought calico for 25 cts. a yard; how must I mark it so as to gain 10 per cent. on each yard? (275) 12 per cent.? (28) 15 per cent.? (287) 20 per cent.? (30) *A.* $1,142.

24. What is the difference between the discount of $227,66

for 2 yrs. 3 mo. 20 days, and the interest of the same sum for the same time? *A.* $3,832.

25. Which is the most, the compound interest of $520 for 5 yrs., or the discount of the same sum for the same time?

A. Compound interest, by $55,877.

26. If 300 men, in 6 months, perform a piece of work when the days are 12 hours long, how many men will do the same in 4 months, when the days are only 8 hours long? *A.* 675 men.

27. What is the difference of time between April 1st, 1826, and June 15th, 1829? (3,2,14) Between March 19th, 1829, and August 20, 1826? (2,6,29) Between July 5, 1800, and February 16, 1826? (25,7,11) *A.* 31 yrs., 4 mo., 24 days.

28. What is the interest of $120,60, from June 1, 1828, to June 16, 1829? (7537) From May 15, 1824, to August 29, 1830? (45506) From October 10, 1825, to November 1, 1828? (2213) *A.* $75,173+.

¶ LXXV. *In computing Interest on Notes,*

When a settlement is made within a short time from the date or commencement of interest, it is generally the custom to proceed according to the following

RULE.

Find the amount of the principal, from the time the interest commenced to the time of settlement, and likewise the amount of each payment, from the time it was paid to the time of settlement; then deduct the amount of the several payments from the amount of the principal.

Exercises for the Slate.

1. For value received, I promise to pay Rufus Stanly, or order, Three Hundred Dollars, with interest. April 1, 1825.

$300. PETER MOSELY.

On this note were the following endorsements:—

		Time.
Oct. 1, 1825, received	$100	
April 16, 1826, . .	$ 50	3, 2, 6, 1, 11, 15, 4.
Dec. 1, 1827, ...	$120	

What was due April 1, 1828? *Ans.*, $60,73.

CALCULATION.

The first principal on interest from April 1, 1825, ...		$300,00
Interest to April 1, 1828, (36 *mo.*), ...		54,00
Amount of principal		$354,00
First payment, Oct. 1, 1825, ...	$100,00	
Interest to April 1, 1828, (30 *mo.*)	15,00	
Second payment, April 16, 1826, ...	50,00	
Interest to April 1 1828, (23½ *mo.*)	5,87	
Third payment, Dec. 1, 1827, ...	120,00	
Interest to April 1, 1828, (4 *mo.*), ...	2,40	
Amount of payments deducted, ...		$293,27
Remains due, April 1, 1828, ...		$60,73

2. For value received, I promise to pay Peter Trusty, or order, Five Hundred Dollars, with interest. July 1, 1825.

$500. JAMES CARELESS.

ENDORSEMENTS.

		Time.
July 16, 1826, received	$200	
Jan. 1, 1827, ...	$ 40	3, 15, 2, 1,
March 16, 1827, ...	$230	6, 15, 1, 4.

What remained due July 16, 1828? *Ans.*, $75,15.

3. For value received, I promise to pay William Stimpson, or order, One Thousand Dollars, with interest. Jan. 16, 1820.

$1000. PETER CARRFUL.

ENDORSEMENTS.

		Time
March 16, 1821, received	$600	
May 1, 1822, ...	$120	3, 1, 10,
July 16, 1822, ...	$180	8, 15, 6.

What remained due Jan. 16, 1823? *Ans.*, $203,50.

Many have discarded the preceding rule, under a false impression that it does not proceed on the ground of simple interest; in other words, that the "interest" (allowed the debtor) "will, in a course of years, completely expunge, or, as it may be said, eat up the debt." This objection, as it respects simple interest, it is believed, may be completely obviated. We will cite an example, which is generally considered as illustrating this objection in a forcible manner:—

A lends B 100 dollars, at 6 per cent. interest, and takes his note of hand. B does no more than pay A at every year's end 6 dollars (which is then justly due to B for the use of his money), and has it endorsed on his note. At the end of 10 years, B takes up his note, and the sum he has to pay is reckoned thus:—the principal, 100 dollars, on interest 10 years, amounts to 160 dollars; there are nine endorsements, of 6 dollars each, upon which the debtor claims interest; one for 9 years, the second for 8 years, the third for 7 years, and so down to the time of settlement; the whole amount of the several endorsements and their interest (as any one may see by casting it), is $70, 20 cts.; this, subtracted from 160 dollars, the amount of the debt, leaves, in favour of the creditor, $89, 40 cts., or $10, 20 cts. less than the original principal, of which he has not received a cent, but only its annual interest.

Extend it to 28 years, and A, the creditor, would fall in debt to B, without receiving a cent of the 100 dollars which he lent him.

In this example, in the first place, there is not actually due A $6 at each year's end, to be by him put out at interest; for this would be more than simple interest. In the next place, the objection proceeds on the supposition that the annual endorsements of $6 were closely locked up in A's coffers, or lying dormant; for, had the $6 been put out at interest, as it might have been, the amount of it would of course be the same as the amount of the endorsements: hence the principal of the note would remain in reality unimpaired. In this way B gets interest for the money advanced or endorsed, and A loses nothing by it, for the payments are at his disposal. The correctness of this rule may be further illustrated by the same example. If the $6 interest be reserved, and the payment of $6 be deducted yearly from the principal, the interest being cast on the remainder at every year's end, and reserved as before, then, at the time of settlement, A could not reasonably claim any more, on the ground of simple interest, than what remained of the principal, and the sum of the interest of the principal from time to time, added together. For it certainly is not reasonable that he should have interest on any more principal than he actually had during the whole time. This process produces the same result as the former. Hence, we infer, that, if the preceding rule be correct on the ground of simple interest (which must be allowed from the preceding facts), it follows, that any other mode, producing a different result from this, must be incorrect.

The Connecticut and Massachusetts rules, which are by law established, *produce results* widely different from the above, and that, too, in favour of the *lender.* They must, then, give the lender more than simple interest, that is, compound; which is not recognised by statute of either state, and which, it is said, the law specially guards against—a strange discrepancy in legislative enactments. We can illustrate this best by a practical example :—

Messrs. Brown and Ives, Providence, loan to government $200000 at 6 per cent. interest: the avails of the post office in said town, being about $1000, are to be annually applied to the payment of the debt, and to operate as endorsements. The interest of $200000 for one year is $12000, and the amount, $212000. Then, by the Connecticut mode, the payment of $1000 is to be deducted from $212000, which leaves $211000 on interest for the 2d year; that is, $11000 more than the original debt immediately accumulating interest, besides the $1000, which is at their disposal. Here, then, it may be asked, what is this but interest on interest, that is, compound interest?

The Massachusetts mode is equally as incorrect as the Connecticut, if not more so; for, by that, the principal is permitted to accumulate interest, till the payment, or sum of the payments, is equal to the interest then due; in which case the payments are to lie dormant, be they ever so many, and ever so large; and to remain so till the time of settlement, be it at ever so remote a period. But when the sum of the payments is equal to the interest then due, it is first to cancel an equal sum of interest, thereby placing the payment at the disposal of the lender, which he can, if he chooses, put at interest, while the borrower gets nothing at all for the use of his money, provided it does not at any time exceed the interest. In this way, the lender most surely gets interest on interest, that is, compound interest.

Under such circumstances, it would be politic in the borrower not to make any payments at all; for, as sure as he does, if it is not more than one cent, and has it endorsed on the note, by the Connecticut mode, the interest of one year is immediately incorporated with the principal, both drawing interest, compounding the interest yearly, provided payments are made yearly. By the Massachusetts mode, the lender would have the interest on any sum of endorsements less than the interest till the time of settlement, while the interest is running on against the borrower; and when it is equal, it is to be placed in the hands of the lender, which he may put at interest, if he chooses. From the preceding remarks it is fairly to be inferred, that the borrower would find it for his advantage to deposit his payments in the Savings' Bank at 5 per cent. interest, instead of making endorsements, and thus continue to do till he is able to pay the note with the interest. It would also be for the advantage of the lender even to hire the borrower to make endorsements as frequent as possible. By the Connecticut mode, the trifling sum of $1, endorsed on a note, may, in many cases, be equivalent to the annual rent of 20 brick buildings, or more, in Broadway, in the city of New York.

These rules, however, although not on the ground of simple interest, must have a place, as they are established by law.

MASSACHUSETTS RULE.

"Compute the interest on the principal sum to the first time when a payment was made, which, either alone, or together with the preceding payments (if any), exceeds the interest then due; add that interest to the principal, and from the sum subtract the payment, or the sum of the payments, made at that time, and the remainder will be a new principal, with which proceed as with the first principal, and so on, to the time of settlement."

1. For value received, I promise to pay Jason Park, or order, Six Hundred Dollars, with interest. March 1, 1822.

$600. STEPHEN STIMPSON.

ENDORSEMENTS.

		Time.
May 1, 1823, received	$200	
June 16, 1824, . . .	$ 80	1, 2, 1, 1, 15, 1,
Sept. 17, 1825, . . .	$ 12	8, 15, 1, 7,
Dec. 19, 1825, . . .	$ 15	15, 10, 15.
March 1, 1826, . . .	$100	
Oct. 16, 1827, . . .	$150	

What was there due August 31, 1828? *Ans.* $194,41.

The principal, $600, *on interest from March* 1, 1822, ...		$600,00
Interest to May 1, 1823, (14 *mo.*)		42,00
Amount,		$642,00
Payment, May 1, 1823, *a sum greater than the interest*, . . .		200,00
Due May 1, 1823, *forming a new principal*, ...		$442,00
Interest on $442, *from May* 1, 1823, *to June* 16, 1824, (13½ *mo.*) .		29,83
Amount,		$471,83
Payment, June 16, 1824, *a sum greater than the interest then due*,		80,00
Due June 16, 1824, *forming a new principal*, ...		$391,83
Interest on $391,83, *from June* 16, 1824, *to March* 1, 1826, (20½ *mo.*)		40,16
		$431,99
Payment, a sum less than the interest then due, ...	$ 12	
Payment, a sum less than the interest then due, ...	$ 15	
Payment, a sum greater than the interest then due, ...	$100	
		$127,00
Due March 1, 1826, *forming a new principal*, ...		$304,99
Interest on $304,99, *from March* 1, 1826, *to Oct.* 16, 1827, (19½ *mo.*)		29,73
Amount,		$334,72
Payment, Oct. 16, 1827, *a sum greater than the interest then due*, .		150,00
Due Oct. 16, 1827, *forming a new principal*, ...		$184,72
Interest on $184,72, *from Oct.* 16, 1827, *to August* 31, 1828, *being the time of settlement*, (10½*mo.*)		9,69
Balance due Aug. 31, 1828, ...		$194,41

2. For value received, I promise to pay Asher L. Smith, or order, Nine Hundred Dollars, with interest. June 16, 1820.

$900. WILLIAM MORRIS.

ENDORSEMENTS.

		Time.
July 1, 1821, received	$150	1, 15, 1,
Sept. 16, 1822, . . .	$ 90	2, 15, 2,
Dec. 10, 1824, . . .	$ 10	11, 1, 6,
June 1, 1825, . . .	$ 20	15, 1, 6.
Aug. 16, 1825, . . .	$200	
March 1, 1827, . . .	$300	

What remained due Sept. 1, 1828? *Ans.* $483,07

CONNECTICUT RULE.

Established by the Supreme Court of the State of Connecticut in 1804.

"*Compute the interest to the time of the first payment; if that be one year or more from the time the interest commenced, add it to the principal, and deduct the payment from the sum total.*

*If there be after payments made, compute the interest on the balance due to the next payment, and then deduct the payment as above; and, in like manner, from one payment to another, till all the payments are absorbed; provided the time between one payment and another be one year or more. But if any payments be made before one year's interest hath accrued, then compute the interest on the principal sum due on the obligation, for one year, add it to the principal, and compute the interest on the sum paid, from the time it was paid up to the end of the year; add it to the sum paid, and deduct that sum from the principal and interest, added as above.**

"*If any payments be made of a less sum than the interest arisen at the time of such payment, no interest is to be computed, but only on the principal sum for any period.*"

1. For value received, I promise to pay Peter Trusty, or order, One Thousand Dollars, with interest. June 16, 1824.
$1000. JAMES PAYWELL.

ENDORSEMENTS.

			Time.
July 1, 1825,	received	$250	1, 15, 1, 1, 15,
Aug. 16, 1826,	. . .	$157	1, 8, 15, 1, 2,
Dec. 1, 1826,	. . .	$ 87	15, 6, 10.
Feb. 16, 1828,	. . .	$218	

What was due August 26, 1828? *Ans.* $507,86.

$1000,00 *principal of the note.*
62,50 *interest to July* 1, 1825, (12½ *months.*)

$1062,50 *amount.*
250,00 1*st payment deducted.*

$812,50 *due July* 1, 1825.
54,84 *interest to August* 16, 1826, (13½ *months.*)

$867,34 *amount.*
157,00 2*d payment deducted.*

$710,34 *due August* 16, 1826.
42,62 *interest for* 1 *year.*

$752,96
90,69 { *amount of* 3*d payment to August* 16, 1827 (*the end of the year*), *being* 8½ *months.*

$662,27 *due December* 1, 1826.
48,01 *interest to February* 16, 1828, (1 *year* 2½ *months.*)

$710,28 *amount.*
218,00 4*th payment.*

$492,28 *due February* 16, 1828.
15,58 *interest to August* 26, 1828, (6⅓ *months.*)

$507,86 *due August* 26, 1828, *the time of settlement.*

* *If a year does not extend beyond the time of payment; but if it does, then find the amount of the principal, remaining unpaid, up to the time of settle-*

2. For value received, I promise to pay John P. Smith, or order, Eight Hundred and Seventy-five Dollars, with interest. January 10, 1821.

$875. HARRY THOMSON.

ENDORSEMENTS

			Time.
Aug. 10, 1824,	received	$260	3, 7, 1, 4, 6, 1, 9½,
Dec. 16, 1825,	. . .	$300	
March 1, 1826,	. . .	$ 50	1, 4, 1, 2.
July 1, 1827,	. . .	$150	

What was due September 1, 1828? *Ans.* $474,95.

¶ LXXVI. Practice in Compound Numbers.

Operations in compound numbers, as pounds, shillings, for instance, may be shortened by taking aliquot parts, as in *Practice of Federal Money*, ¶ XXVIII.

1. What is the cost of 28 bushels of salt, at 10 s. a bushel?

In this example, 10 s. = ½ £ ; *then,* ½ *of* 28 *bushels is the cost in pounds ; thus,* 28 ÷ 2 = 14 £, *Ans.*

2. What will 40 bushels of wheat cost, at 5 s. = ¼ £ a bushel? At 10 s. = ½ £ ? At 4 s. = ⅕ £ ? At 1 s. = $\frac{1}{20}$ £ ?

I. Hence, *when the price is an aliquot, or even part of a pound, we divide the number of gallons, yards, &c. by this aliquot part, as in* ¶ XXVIII.

Exercises for the Slate.

1. What will 8640 yds. of cloth cost, at 10 s. = ½ £ a yd.? (4320) At 6 s. 8 d. = ⅓ £ ? (2880) At 5 s. = ¼ £ ? (2160) At 4 s. = ⅕ £ ? (1728) At 3 s. 4 d. = ⅙ £ ? (1440) At 2 s. 6 d. = ⅛ £ ? (1080) At 1 s. 8 d. = $\frac{1}{12}$ £ ? (720) At 1 s. 4 d. = $\frac{1}{15}$ £ ? (576) At 1 s. 3 d. = $\frac{1}{16}$ £ ? (540) At 1 s. = $\frac{1}{20}$ £ ? (432) At 10 d. = $\frac{10}{240}$ = $\frac{1}{24}$ £ ? (360) At 8 d. = $\frac{1}{30}$ £ ? (288) At 5 d. = $\frac{1}{48}$ £ ? (180) At 2½ d. = $\frac{1}{96}$ £ ? (90) *A.* 16794 £.

Note. The aliquot parts of a pound, in the following examples, may be found in the former examples.

2. What cost 20 gallons of brandy, at 6 s. 8 d. per gallon? (6,13,4).
3. What cost 8 yds. of broadcloth, at 10 s. per yard? (4)
4. What cost 25 bushels of rye, at 5 s. per bushel? (6,5)
5. What cost 30 bushels of oats, at 2 s. 6 d. per bushel? (3,15)

Ans. 20 £ 13 s. 4 d.

---

ment, likewise the amount of the payment or payments from the time they were paid to the time of settlement, and deduct the sum of these several amounts from the amount of the principal.

6. What will 51 bbls. of cider cost, at 7 s. 6 d. per bbl.?

$$\frac{1}{2})\frac{1}{4}|51$$

12£ 15 s.	= *cost, at* 5 s. *per bbl.*
6£ 7 s. 6 d.	= *cost, at* 2 s. 6 d. *per bbl*
Ans. 19£ 2 s. 6 d.	= *cost, at* 7 s. 6 d. *per bbl.*

Or,

5 s. = ¼£)51

2 s. 6 d. = ½ of 5 s.; then, ½)12£ 15 s.	= *cost, at* 5 s. *per bbl.*
6£ 7 s. 6 d.	= *cost, at* 2 s. 6 d. *per bbl.*
Ans. 19£ 2 s. 6 d.	= *cost, at* 7 s. 6 d. *per bbl.*

II. Hence, when the price is not an aliquot part of 1£, *we may first find what is the greatest even part, and then take parts of this part, and so on, for several times.*

7. What will 20 yds. of cloth cost, at 12 s. 6 d. per yard? (12,10)

8. What will 40 yards of cloth cost, at 15 s. per yard? (30)

9. What will 36 bushels of corn cost, at 7 s. 6 d. per bushel? (13,10)

10. What cost 12 bbls. of ale, at 17 s. 6 d. per bbl.? (10,10)

A. 66£ 10 s.

11. What will 5 cwt. 3 qrs. 21 lbs. of sugar cost, at $9,60 per cwt.?

2 qrs. = ½ cwt.	$9,60
	5
	48,00 = *cost of* 5 cwt.
1 qr. = ½ of 2 qrs.; then, ½	4,80 = *cost of* 2 qrs.
14 lbs. = ½ of 1 qr.; then, ½	2,40 = *cost of* 1 qr.
7 lbs. = ½ of 14 lbs.; then, ½	1,20 = *cost of* 14 lbs.
	,60 = *cost of* 7 lbs.
Ans.	$57,00 = *cost of* 5 cwt. 3 qrs. 21 lbs.

12. At $2,50 per yard, what will 5 yds. 2 qrs. of broadcloth cost? (1375) What will 4 yds. 1 [illegible]. cost? (10625) Will 6 yds. 3 qrs.? (16875) *Ans.* $41,25.

13. 5 cwt. 3 qrs. 16 lbs., at $4,20 per cwt.? (2475)

14. 3 cwt. 1 qr. 7 lbs., at $3,60 per cwt.? (11925)

15. 4 yds 2 qrs., at $2,10 per yd.? (945)

16. 4 gals. 2 qts., at $3,40 per gal.? (1530) *A.* $61,425.

FELLOWSHIP.

¶ LXXVII. 1. Two boys, William and Thomas, trading with marbles, in company, gained 80 cents; William owned ¾ of the marbles, and Thomas ¼; what was each one's part of *the gain?*

2. James and Rufus, owning a sled, in company, sold it for \$3 more than it cost, that is, \$3 gain; Rufus owned $\frac{2}{3}$ of it, and James $\frac{1}{3}$: what was each one's share of the gain?

Q. *What is the* Rule *of* Fellowship? *A.* When men are trading in company, it ascertains the gain or loss to be shared by each.

Q. *What is called* Stock, *or* Capital? *A.* The money advanced or put in trade.

Q. *What is called* Dividend? *A.* The gain or loss to be shared by each.

1. Three men, A, B, and C, traded in company; A put in \$200, B \$400, and C \$600; they gained \$300: what was each man's share of the gain?

In this example, it is evident, that B ought to have twice as much of the gain as A, for his stock is twice as much, and C 3 times as much as A; that is, each man's gain or loss ought to have the same relation to the whole gain or loss, as the money he put in has to the money they all put in. The same principle will apply in all cases in which a number is to be divided into parts, which shall have a given relation, or ratio, to each other, as the dividing a bankrupt's estate among his creditors, apportioning taxes, &c. Hence, from the foregoing example, we derive the following proportions:

		\$		\$		\$		\$	
A's *stock,*	\$200	1200	:	200	: :	300	:	50,	A's *gain.*
B's *stock,*	\$400	1200	:	400	: :	300	:	100,	B's *gain.*
C's *stock,*	\$600	1200	:	600	: :	300	:	150,	C's *gain.*
Whole stock,	\$1200								

By ratios. These are $\frac{200}{1200}, \frac{400}{1200}, \frac{600}{1200} = \frac{1}{6}, \frac{1}{3}, \frac{1}{2}$; *then,* $300 \times \frac{1}{6}$, that is, $\frac{1}{6}$ of \$300 = \$50, A's; $\frac{1}{3}$ *of* 300 = \$100, B's, *and* $\frac{1}{2}$ of 300 = \$150, C's *gain.*

Or, by analysis. If \$1200 *stock gain* \$300, *then* \$1 *stock will gain* $\frac{1}{1200}$ *of* \$300, $= \frac{300}{1200} =$ \$$\frac{1}{4}$. *Now, if* \$1 *gain* \$$\frac{1}{4}$, *then* $\frac{1}{4}$ *of* \$200, *A's stock,* = \$50, *A's gain;* $\frac{1}{4}$ *of* \$400, *B's stock,* = \$100, *B's gain; and* $\frac{1}{4}$ *of* \$600, *C's stock,* = \$150, *C's gain.*

This last method will generally be found the shortest, and best adapted to business; especially when there are several statements, in which all the first terms are alike, and all the third terms are alike.

Proof. It is plain, that, if the work be right, the amount of the shares of the gain or loss must be equal to the whole gain or loss; thus, in the last example, A's is \$50 + B's \$100 + C's \$150 = \$300, *the whole gain.*

What, then, is the Rule? As the whole stock : to each man's stock : : the whole gain : to each man's gain.

2. Three merchants, A, B, and C, gained, by trading in com-

pany, $200; A's stock was $150, B's $250, and C's $400: what was the gain on $1, and what was each man's *share* of the gain. *A. The gain on* $1 *is* $¼; *then*, ¼ *of* $150 = $37,50, A's; ¼ *of* $250 = $62,50, B's; *and* ¼ *of* $400 = $100, C's.

3. A, B, and C, freight a ship with 270 tons; A shipped on board 96 tons, B 72, and C 102; in a storm, the seamen were obliged to throw 90 tons overboard: what was the loss on 1 ton? and how many tons did each lose? *A. The loss on* 1 *ton is* ⅓ *of a ton;* A's, 32; B's, 24; C's, 34.

4. A and B trade in company with a joint capital of $600; A put in $350,50, and B $249,50, and, by trading, they gained $120: what is the gain on $1, and what is each person's share of the gain? *Ans.* $⅕; A's, $70,10; B's, $49,90.

5. A ship valued at $25200, was lost at sea, of which ⅓ belonged to A, ½ to B, and the remainder to C; what is the loss on $1, and how much will each man sustain, supposing $18000 of her to be ensured? *Ans.* 2/7; A's, $2400; B's, $3600, and C's, $1200.

Perform the following examples, in the same manner, by finding how much it is for $1, or unity.

6. A detachment, consisting of 5 companies, was sent into a garrison, in which the duty required 228 men a day; the first company consisted of 162 men; the second, 153; the third, 144; the fourth, 117; and the fifth, 108; how many men must each company furnish, in proportion to the whole number of men? *A. The proportion for* 1 *man is* ⅓; *then*, ⅓ *of* 162 = 54, *first company; the second*, 51; *the third*, 48; *the fourth*, 39; *and the fifth*, 36 *men.*

7. Two men, A and B, traded in company, with a joint capital of $1000; they gained $400, of which A took $300, and B the remainder; what was each person's stock? *Ans.* $1 *gain requires* 2½ *stock;* A's, $750; B's, $250.

8. Divide $1000 between 4 persons, so that their shares may be to each other as 1, 2, 3, 4. *A.* $100, $200, $300, $400.

9. A bankrupt is indebted to A $350, to B $1000, to C $1200, to D $420, to E $85, to F $40, and to G $20; his whole estate is worth no more than $1557,50: what will be each creditor's part of the property?

In adjusting claims of this nature, it is the general practice to find how much the debtor pays on $1, which is, in this case, $½. *Ans.* A, $175; B, $500; C, $600; D, $210; E, $42,50; F, $20; G, $10.

10. A wealthy merchant, at his death, left an estate of $30000, to be divided among his 5 children, in such a manner that their shares shall be to each other as their ages, which are 7, 10, 12, 15, 16 years; what was the share of each?

Ans. $3500, $5000, $6000, $7500, $8000.

11. A and B invest equal sums in trade, and clear $220, of which A is to have 8 shares, because he spent all his time in managing the concerns, and B, only 3 shares: what is each man's gain? and how much is A allowed for his trouble? *Ans.* $160; *A's share,* $100 *for his trouble;* $60, *B's share.*

12. If a town raise a tax of $1920, and all the property in town be valued at $64000, what will that be on $1? and what will be A's tax, whose property is valued at $1200? *Ans.* $,03 *on a dollar;* *A's tax,* $36.

In assessing taxes, we must first make an inventory of all the property, both real and personal, of the whole town, and also of each individual who is to be taxed; and, as the whole number of polls are rated at so much each, the tax on all the polls must first be taken out from the whole tax, and the remainder is to be assessed on the property. Then, to find how much any individual must be taxed for his property, we need only find how much the remainder of the whole tax is on $1, and multiply his inventory by it.

Note. In some states taxes are assessed only on the real and personal estate of the inhabitants, no poll taxes being allowed.

13. A certain town is taxed $2140; the whole property of the town is valued at $500000; there are 200 polls, which are taxed $,70 each; A's property is valued at $1400, and he pays for 2 polls:

	$	polls		$ cts.	polls
C's, *at*	1200, *pays for*	2	H's, *at*	825,50, *pays for*	3
D's, ...	1265, ...	1	I's, ...	800,40, ...	2
E's, ...	2125, ...	3	J's, ...	375,25, ...	1
F's, ...	3621, ...	2	K's, ...	265,30, ...	2

What will be the tax on $1? and what will be A's tax?

200 *polls* × $,70 = $140, *amount of the poll taxes, and* $2140 — $140 = $2000, *which is to be assessed on the property.* $500000 : $2000 : : $1 : $,04, *tax on* $1. *Then, to find* A's *tax, his inventory being* $1400, *we proceed thus:—*

$1400 × $,04 = $56
2 *polls at* $,70 = $ 1,40

$57,40 A's *whole tax, Ans.*

What will be C's tax? (4940) What D's? (5130) What E's? (8710) What F's? (14624) What H's? (3512) What I's? (334160) What J's? (1571) What K's? (12012)

Ans. $430,298

COMPOUND FELLOWSHIP.

¶ **LXXVIII.** 1. Two men hired a pasture for $9; A put in 2 oxen for 6 months, and B 3 oxen for 5 months; what ought each to pay for the pasture?

2 oxen for 6 months is the same as ($2 \times 6 =$) *12 oxen for 1 month; and 3 oxen for 5 months is the same as* ($3 \times 5 =$) *15 oxen for 1 month.*

The shares of A and B are the same as if A had put in 12 oxen, and B 15, for 1 month each; hence the relation of 12 to 15 is the same as in Simple Fellowship, thus,

$$\left.\begin{array}{r} 2 \times 6 = 12 \\ 3 \times 5 = 15 \\ \overline{27} \end{array}\right\} \begin{array}{l} 27 : 12 :: 9 : \$4, \text{ A's.} \\ 27 : 15 :: 9 : \$5, \text{ B's.} \end{array}$$

Q. *How, then, does Compound differ from Simple Fellowship?* A. Compound regards time, Simple does not.

Q. *From the preceding example, what appears to be the*

RULE?

I. Multiply each man's stock by the time it is continued in trade.

II. Then—As the sum of the products : each man's product :: the whole gain or loss : each man's gain or loss.

More Exercises for the Slate.

2. Three merchants, A, B, and C, enter into partnership: A puts in \$60 for 4 mo., B \$50 for 10 mo., and C \$80 for 12 mo.; but by misfortune they lose \$50: how much loss must each man sustain?

Ans. A's, \$7,058+; B's, \$14,705+; C's, \$28,235+

3. Three butchers hire a pasture for \$48: A puts in 80 sheep for 4 mo., B 60 sheep for 2 mo., and C 72 sheep for 5 mo.; what share of the rent must each man pay?

Ans. A's, \$19,20, B's, \$ 7,20, C's, \$21,60.

4. Two merchants entered into partnership for 16 mo.: A at first put in stock to the amount of \$600, and, at the end of 9 months, put in \$100 more; B put in at first \$750, and, at the expiration of 6 months, took out \$250; with this stock they gained \$386; what was each man's part?

Ans. A's, \$200,797, B's, \$185,202.

5. On the first of January, A began trade with \$760, and, on the first of February following, he took in B with \$540; on the first of June following, he took in C with \$800; at the end of the year they found they had gained \$872; what was each man's share of the gain?

Ans. A's *share*, \$384,929; B's, \$250,71; C's, \$236,36

MENSURATION.

¶ LXXXIX. SQUARE MEASURE.

Q. *What are your ideas of a square?* A. It is any thing which is as long as it is wide.

Q. *What kind of a figure does this on the right appear to be?* A. A square figure.

Q. *Why?* A. Because the side AB is as long as the side BC.

Q. *How many sides has this figure, and what is their length?*

Q. *How many equal corners has it?* A. Four.

Q. *What are these corners generally called?* A. Angles.

A B D C

Q. *How, then, would you describe a square figure?* A. It has four equal sides, and four equal angles.

Q. *In the above figure, if each side be 1 foot in length, what ought it to be called?* A. 1 square foot.

A 3 feet = 1 yard. B
3 feet = 1 yard.
D C

Q. *If the sides of a square be each 1 yard in length, as in the figure on the right, what ought it to be called?* A. 1 square yard.

Q. *In this square I perceive there are several smaller squares, contained in the larger. If you count all the smaller squares, allowing each one to be 1 foot, how many square feet or square yards will they make?*

Q. *Why?* A. Because there are 9 small squares, each containing 1 square foot, which make 9 square feet, that is, 1 square yard.

Q. *How many square feet, then, make 1 square yard?* A. 9.

Q. *If we multiply 3 feet (the length of 1 side) by the width, 3 feet, making 9, the same result is produced as before. What, then, will multiplying the length of any square by the breadth, or the length into itself, give?* A. The square feet, square inches, &c., contained in the figure.

Q. *How many square inches in a figure 2 inches long and 2 inches wide?* $2 \times 2 = 4$, *Ans. How many in a figure 4 inches long and 4 inches wide? 12 inches square, that is, 12 inches long, and 12 inches wide? 8 inches square? 6 inches square? 20 inches square? 30 inches square?*

Q. *How many square feet in a figure 1 foot, or 12 inches, square?* A. 1 square foot.

Q. *How many square inches in 1 square foot? and why?* A. 144 sq. in.; because 12 in. $\times$ 12 in. = 144.

Q. *How many square feet in 1 square yard? and why?* A. 9 sq. ft.; because 3 ft. $\times$ 3 ft. = 9.

Q. *How many square yards in 1 square rod? and why?* A. $30\frac{1}{4}$ sq. yds.; because $5\frac{1}{2}$ yds. $\times$ $5\frac{1}{2}$ yds. = $30\frac{1}{4}$

Q. *How many square feet in 1 square rod? and why?* *A.* 272¼ sq. ft., because 16½ ft. (the number of feet in 1 rod in length) × 16½ ft. = 272¼.

Q. *This figure on the right is called a parallelogram: what, then, are your ideas of a parallelogram?* *A.* That it is a figure, which is longer than it is wide.

Q. *We see by this figure, that there are two kinds of parallelograms, viz.* ABCD *and* ABEF. *By inspecting these they will be found to be equal: how, then, may a parallelogram be defined?* *A.* It is a figure which has its opposite sides of equal length, and its opposite angles equal.

Q. *If this figure had been square, and each side 2 feet in length, it is plain that it would have contained 4 square feet; but, allowing the longest side to be 2 feet, and the shortest side only 1 foot, it will, of course, contain but ½ as many square feet.—How many, then, does it contain?* *A.* 2 ft. (length) × 1 ft. (breadth) = 2 sq. ft.

Q. *If a figure 1 inch in breadth and 1 inch in length contains 1 square inch, how many square inches will a figure 1 inch wide and 2 inches long contain? 3 inches long? 4 inches long? 8 inches long? 12 inches long? 20 inches long?*

Q. *If a figure 1 foot wide and 1 foot long contains 1 square foot, how many square feet will a figure 1 foot wide and 2 feet long contain? 3 feet long? 4 feet long? 8 feet long? 10 feet long?*

Q. *How, then, do you proceed to find the square feet, inches, &c. of a square or parallelogram?* *A.* Multiply the length by the breadth.

1. How many square feet in a room 10 feet long and 2 feet wide? (10 × 2 = 20 sq. ft., *Ans.*) In a room 8 feet wide and 12 feet long? 20 feet long?

2. How many square rods in a piece of land 4 rods wide and 8 rods long? 10 rods long? 11 rods long? 12 rods long? 10 rods long and 4 rods wide?

Q. *When a piece of land, in any shape, contains 40 square rods, what is it called?* *A.* 1 rood.

3. How many square rods in a piece of land 40 rods long and 2 rods wide? 4 rods wide?

Q. *When a piece of land, in any shape, contains 160 square rods, what is it called?* *A.* 1 acre.

4. How many square rods in a piece of land 20 rods long, and 2 rods wide? How many such pieces will make an acre, or 160 square rods?

5. How wide must a piece of land be, which is 80 rods long, to make an acre? 40 rods long? 20 rods long?

6. How many square feet of boards are contained in the floor of a room 10 feet square? 20 feet square? 10 feet wide and 20 feet long? 20 feet wide and 30 feet long?

7. How many square yards in a figure 3 feet long and 3 feet wide? 6 feet square? 10 feet long and 9 feet wide? 6 feet long and 2 feet wide? ($2 \times 6 = 12 \div 9 = 1\frac{1}{3}$ yds., *Ans.*) In a figure 10 feet long and 4 wide? *A.* $4\frac{4}{9}$ yds.

8. How many square yards in 9 square feet? In 108? In 72? In 99? In 27? In 80? In 37?

Q. *How, then, must square feet, square inches, &c. be divided?* A. Square inches by square inches, square feet by square feet, &c.

Q. We are now prepared to answer that interesting question which occurs in geography, viz. the difference between miles square and square miles. The figures on the right are introduced for the purpose of its illustration. Examine them attentively, then tell me, for instance, What is the difference between 5 square miles and 5 miles square? *A.* 5 square miles means 5 miles in length and only 1 in breadth; but 5 miles square means 5 miles in length and 5 miles in breadth, making 5 times as many miles as only 1 in breadth; that is, 25 square miles.

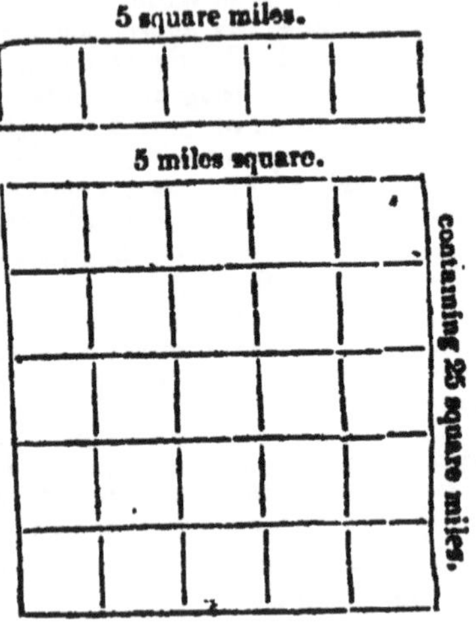

From these illustrations we derive the following general

RULE.

I. *How do you proceed to find the contents of a square or parallelogram?* *A.* Multiply the length by the breadth.

Exercises for the Slate.

1. In a room 16 feet long and 11 feet wide, how many square feet? *A.* 176.

2. How many acres in a piece of land 560 rods long and 32 rods wide? $560 \times 32 = 112$ square acres, *Ans.*

The pupil must recollect that square inches must be divided by square inches, square yards by square yards, &c.

3. How many acres in a piece of land 370 rods wide and 426 rods long? *A.* 985 acres, 20 rods.

4. How many rods long must a piece of land be, which is 80 rods wide, to make 1 acre? (2) How many rods wide to make 4 acres? (8) How many rods wide to make 200 acres? (400) *A.* 410 rods.

5. How many square feet of boards are contained in the floor of a room 40 ft. 6 in. long and 10 ft. 3 in. wide? (*Reduce the inches to the decimal of a foot.*) *A.* 415,125 ft. = 415⅛ feet.

6. How many acres are contained in the road from Boston to Providence, allowing the distance to be 40 miles, and the average width of the road 4 rods? *A.* 320 acres.

7. How many square feet are contained in a board 12 inches long and 12 inches wide? (1) 12 inches wide and 24 inches long? (2) 3 feet long? (3) 20 feet long? (20) *A.* 26 feet.

8. How many square feet in a board 1 ft. 6 in. wide and 18 ft. 9 in. long? *A.* 28,125 ft. = 28⅛ feet.

9. How many yards of carpeting, that is 1¼ yd. wide, will cover a floor 21 ft. 3 in. long and 13 ft. 6 in. wide?

A. 25½ yards.

10. How many feet of boards will it take to cover the walls of a house 30 ft. 6 in. wide, 40 ft. 9 in. long, and 20 ft. high? and what will they come to at $10 per 1000 feet? *A.* 2850 feet; *cost* $28½.

11. How many shingles will it take to cover the roof of a barn 40 feet long, allowing the length of the rafters to be 16 ft. 6 in., and 6 shingles to cover 1 square foot? what will they cost, at $1,25 per 1000? *A.* 7920 shingles; *cost* $9,90.

12. What will a lot of land, 300 rods wide and 600 rods long come to, at $15 an acre? *A.* $16875.

13. What will a lot of land, 1 mile square, come to, at $20,75 per acre? *A.* $13280.

¶ LXXX. SOLID, OR CUBIC MEASURE.

Q. *When a block is 1 inch long, 1 inch thick, and 1 inch wide, how many solid inches is it said to contain?* *A.* 1 solid or cubic inch.

Q. *How many solid feet does a block, that is 1 foot long, 1 foot thick, and 1 foot wide, contain?* *A.* 1 solid or cubic foot.

Q. *If a block 1 foot thick, 1 foot wide, and 1 foot long, contains 1 solid foot, how many solid feet does such a block that is 2 feet long contain? 3 feet long? 5 feet long? 10 feet long? 20 feet long? 30 feet long?*

Q. *How many solid feet does a block 2 feet long, 2 feet thick, and 1 foot wide, contain? 2 feet wide? 3 feet wide?*

Q. *How many solid inches does a block 3 inches long, 2 inches wide, and 1 inch thick, contain? 2 inches thick? 4 inches thick? 10 inches thick?*

Q. *How, then, would you proceed to find how many solid feet, inches, &c. are contained in a solid body?* *A.* Multiply the length, breadth and depth together.

1. How many solid feet in a block 4 feet thick, 2 feet wide, and 5 feet long? *Ans.* $4 \times 2 \times 5 = 40$ solid feet.

2. How many solid or cubic feet in a block 12 inches long, 12 inches wide, and 12 inches thick? *A.* 1 solid foot.

Q. *When a load of wood contains 128 solid feet, what is it called?* A. 1 cord.

3. How many solid feet in a pile of wood 8 feet long, 4 feet wide, and 4 feet high? A. $128 = 1$ cord. How many cords of wood in a pile 8 feet long, 4 feet wide, and 8 feet high?

A. 256 solid feet $= 2$ cords.

Q. *In common language, we say of a load of wood brought to market, if it is 8 feet long, 4 feet high, and 4 feet wide, that it is a cord, or it contains 8 feet of wood. But this would make 128 solid feet; what, then, is to be understood by saying of such a load of wood, that it contains 8 feet of wood? or, in common language, "there is 8 feet of it."*

A. As 16 solid feet, in any form, are $\frac{1}{8}$ of 128 feet, that is, $\frac{1}{8}$ of a cord, it was found convenient, in reckoning, to call every 16 solid feet 1 cord foot; then, 8 such cord feet will make 128 solid feet, or 1 cord, for 8 times 16 are 128.

Q. *How, then, would you bring solid feet into cord feet?* A. Divide by 16.

4. How many cord feet in a pile of wood 8 feet long, 2 feet high, and 1 foot wide? How many in a load 8 feet long, 2 feet high, and 2 feet wide? 8 feet long, 4 feet wide, and 2 feet high?

5. If, in purchasing a load of wood, the seller should say that it contains 3 cord feet, how many solid feet must there be in the load? How many solid feet to contain 4 cord feet? 5 cord feet? 6 cord feet? 7 cord feet? 8 cord feet? 9 cord feet?

6. How many cord feet in a pile of wood 8 feet long, 1 foot wide, and 4 feet high?

In performing this last example, we multiply 4 feet (the height) by 1 foot (the width), making 4; then, this 4 by 8 feet (the length), making $32 \div 16$ (cord feet), $= 2$ cord feet, *Ans.* But, instead of multiplying the 4 by the 8 feet in length, and dividing by 16, we may simply divide by 2, without multiplying; for the divisor, 16, is 2 times as large as the multiplier, 8; consequently, it will produce the same result as before, thus: $4 \times 1 = 4 \div 2 = 2$ cord feet, *Ans., as before.*

Q. *When, then, a load of wood is 8 feet long, or contains two lengths, each 4 feet (which is the usual length of wood prepared for market,) what easy method is there of finding how many cord feet such a load contains?* A. Multiply the height and breadth together, and divide the product by 2.

7. How much wood in a load 8 feet long, 3 feet high, and 2 feet wide? $3 \times 2 = 6 \div 2 = 3$ cord feet, *Ans.*

8. How many cord feet in a load of wood 2 feet high, 2 feet wide, and of the usual length? 3 feet high and 2 feet wide? 3 feet wide and 3 feet high? 4 feet wide and 4 feet high? 4 feet wide and 6 feet high? How many cords in a load 4 feet high, 4 feet wide?

9. How wide must a load of wood be, which is 8 feet long and 1 foot high, to make 1 cord foot? How wide to make 2 cord feet? 3 cord feet? 6 cord feet? 10 cord feet?

10. What will a load of wood 8 feet long, 3½ feet wide, and 4 feet high, cost, at $1 per foot?

The foregoing remarks and illustrations may now be embraced in the following

RULES.

I. *How do you find the contents of any solid or cube?* *A.* Multiply the length, breadth and depth together.

II. *When the length of wood is* 8 *feet, how can you find the number of cord feet it contains, without multiplying by* 8 *and dividing by* 16? *A.* Multiply the breadth and height together, and divide the product by 2; the quotient will be cord feet.

III. *How do you bring cord feet into cords?* *A.* Divide by 8.

Note. If the wood is only 4 feet in length, proceed as last directed; then, as 8 feet in length is 2 times as much wood as only 4 feet in length, hence ½ the result found, as above, will be the answer in cord feet; that is, divide by 2 twice, or once by 4.

Exercises for the Slate.

1. How many solid feet in a load of wood 8 feet long, 4 feet wide, and 3½ feet high? $4 \times 3\frac{1}{2} = 14 \div 2 = 7$ cord feet, *Ans.*

2. How many feet in a load of wood 5 ft. 6 in. high, 3 ft. 9 in. wide, and of the usual length?

(*Reduce the inches to the decimal of a foot.*) *A.* $10\frac{3125}{10000} = 10\frac{5}{16}$ ft.

Perform this last example by reducing the inches of a foot to a common fraction. This method, in most cases, will be found preferable: thus, taking the last example:—

5 *ft.* 6 *in.* $= 5\frac{1}{2}$ *ft.* $= \frac{11}{2}$; then, 3 *ft.* 9 *in.* $= 3\frac{3}{4}$ *ft.* $= \frac{15}{4} \times \frac{11}{2} = \frac{165}{8} \div 2 = \frac{165}{16} = 10\frac{5}{16}$, *Ans., as before.*

3. In a block 8 ft. 6 in. in length, 3 ft. 3 in. wide, and 2 ft. 9 in. thick, how many solid feet? *A. Decimally* 75,96875 *feet* $= 75\frac{31}{32}$ feet. *By common fractions;* $\frac{17}{2} \times \frac{13}{4} \times \frac{11}{4} = \frac{2431}{32} = 75\frac{31}{32}$ feet, *Ans., as before.*

4. If a load of wood is 8 feet long and 3 feet wide, how high must it be to make 1 cord?

In this example, we know that the height multiplied by the width, and this product divided by 2, *must make* 8 *cord feet, that is,* 1 *cord or load; hence,* $8 \times 2 = 16 \div 3 = 5\frac{1}{3}$ *feet, height, Ans.*

5. If a load of wood is 5⅓ feet high, and 8 feet long, how wide must it be to make 2 cords?

2 *cords* = 16 *cord feet; then,* $16 \times 2 = 32 \div 5\frac{1}{3} = 6$ *feet wide, Ans.*

6. If a load of wood is 5⅓ feet high and 8 feet long, how wide must it be to make 3 cords? (9) 4 cords? (12) 8 cords? (24) *A.* 45 feet.

7. How many solid feet of timber in a stick 8 feet long, 10 inches thick, and 6 inches wide? ($3\frac{1}{3}$) 10 feet long, 12 inches thick, and 1 ft. 3 in. wide? ($12\frac{1}{2}$) 20 ft. 6 in. long, 24 inches wide, and 1 ft. 9 in. thick? ($71\frac{3}{4}$) *A.* $87\frac{7}{12}$ ft.

8. In a pile of wood 10 feet wide, 3 ft. 3 in. high, and 1 mile long, how many cord feet, and how many cords?

A. 10725 cord feet = $1340\frac{5}{8}$ cords.

9. How many tons of timber in 2 sticks, each 30 feet long, 20 inches wide, and 12 inches thick? *A.* 100 feet ÷ 50 = 2 tons.

10. How many bricks 8 inches long, 4 inches wide, and $2\frac{1}{4}$ inches thick, will build a wall in front of a garden, which is to be 240 feet long, 6 feet high, and 1 foot 6 inches wide?

A. 51840 bricks.

DUODECIMALS.

¶ **LXXXI.** *Q. From what is the word* duodecimals *derived?* *A.* From the Latin word *duodecim*, signifying *twelve.*

Q. In common decimals, we are accustomed to suppose any whole thing, as a foot, for instance, to be divided into ten equal parts; but how is a foot divided in duodecimals? and what are the parts called? *A.* Into twelve equal parts, called *inches* or *primes*, and each of these parts into twelve other equal parts, called *seconds;* also each second into twelve equal parts, called *thirds*, and each third into twelve equal parts, called *fourths*, and so on to any extent whatever.

Q. What, then, are duodecimals? *A.* Fractions of a foot.

Q. What fraction of a foot is 1 inch? *A.* $\frac{1}{12}$ ft.

Q. What fraction of a foot is 1 second? *A.* $\frac{1}{12}$ of $\frac{1}{12} = \frac{1}{144}$ ft.

Q. What fraction of a foot is 1 third? *A.* $\frac{1}{12}$ of $\frac{1}{12}$ of $\frac{1}{12} = \frac{1}{1728}$ ft.

Q. What fraction of a foot is 1 fourth?

A. $\frac{1}{12}$ of $\frac{1}{12}$ of $\frac{1}{12}$ of $\frac{1}{12} = \frac{1}{20736}$ ft.

Q. Now, since 12ths multiplied by 12ths make 144ths, and $\frac{12}{144}$ make $\frac{1}{12}$, also, 144ths multiplied by 12ths make 1728ths, and $\frac{12}{1728}$ make $\frac{1}{144}$, it is plain that we may write the fractions without their denominators, by making some mark to distinguish them. What marks are generally used for this purpose? *A.* 12ths, inches, or primes, are distinguished by an accent, thus; $8'$ signifies $\frac{8}{12}$, 8 inches, or 8 primes; $7'' = \frac{7}{144}$, or 7 seconds; $6''' = \frac{6}{1728}$, or 6 thirds, &c.

Q. We have seen that 12ths multiplied by 12ths produce 144ths; what, then, is the product of $5'$ (inches or primes) multiplied by $7'$ (inches)? *A.* $35''$, that is, 35 seconds, or $\frac{35}{144}$.

Q. What is the product of $5''$ (seconds) multiplied by $7'$ (inches)? *A.* $35'''$, that is, 35 thirds.

Q. What is the product of $5''$ (seconds) multiplied by $7''$ (seconds)?

A. $35''''$, that is, 35 fourths.

Q. *How may the value of the product always be determined?* *A.* By placing as many marks or accents at the right of the product as there are marks at the right of both multiplier and multiplicand counted together.

Q. *What, then, would* 7′′′′′ *(fifths) multiplied by* 8′′′′′′ *(sixths) produce?* *A.* 56′′′′′′′′′′′, that is, 56 elevenths.

Q. *What would* 7′′ *(seconds) multiplied by* 5′′′ *(thirds) produce?* *A.* 35′′′′′, that is, 35 fifths.

Q. *What would* 8′′′ *multiplied by* 3′ *produce?* *A.* 24′′′′, (fourths.)

Q. *From the preceding, what appears to be the value of feet multiplied by primes or inches, or what do feet multiplied by primes give?* *A.* Primes.

Q. *What do primes multiplied by primes give?* *A.* Seconds.

Q. *What do primes multiplied by seconds give?* *A.* Thirds.

Q. *What do seconds multiplied by seconds give?* *A.* Fourths.

Q. *What do seconds multiplied by thirds give?* *A.* Fifths.

Q. *What do thirds multiplied by thirds give?* *A.* Sixths.

Note. This might be extended in the same manner to any indefinite length. The following table contains a few of these denominations.

Repeat the

TABLE.

12′′′′ *(fourths)* make 1′′′ *(third.)*
12′′′ *(thirds)* . . . . 1′′ *(second.)*
12′′ *(seconds)* . . . 1′ *(inch or prime.)*
12′ *(inches or primes)* 1 *foot.*

Q. *How may duodecimals be added and subtracted?* *A.* In the same manner as compound numbers; 12 of a less denomination always making 1 of a greater, as in the foregoing table.

MULTIPLICATION OF DUODECIMALS.

Q. *What are duodecimals used for?* *A.* For measuring any thing respecting which length and breadth, also depth, are considered.

1. How many square feet in a board 10 ft. 8 in. long, and 1 ft. 5 in. broad?

We have seen how such an example may be performed by common decimals; we will now perform it by duodecimals.

OPERATION.

Length,	10 ft.	8′	
Breadth,	1	5′	
	4	5′	4′′
	10	8′	
Ans.,	15	1′	4′′

8 inches or primes $= \frac{8}{12}$ of a foot, and 5′ *(primes)* $= \frac{5}{12}$ of a foot; then, $\frac{8}{12} \times \frac{5}{12} = \frac{40}{144}$ of a foot, that is, 40′′ *(seconds)* = 3′ *(inches)* and 4′′ *(seconds)*: we now write down 4′′ at the right of the inches, reserving the 3′ to be carried to the inches. In multiplying 10 feet by the 5′, we say, $10 \times \frac{5}{12} = \frac{50}{12}$ or 50′ *(inches)*, and the 3′ we reserved makes 53′, = 4 feet and 5′, which we place under feet and

inches in their proper places. Then, multiplying 10 ft. 8′ by 1 ft. makes 10 ft 8′, which we write under the 4 ft. 5′. We now proceed to add these two products together, which, by carrying 12, after the manner of compound rules, make 15 ft. 1′ (*inch*) 4″ (*seconds*), the *Answer*.

It will be found most convenient in practice to begin by multiplying the multiplicand [illegible] by the feet, or highest denomination of the multiplier, then by the inches, &c., thus:

OPERATION.

10 ft.	8′	
1	5′	
10	8′	
4	5′	4″
15	1′	4″

1 × 8′ = 8′, and 1 ft. × 10 ft. = 10 ft. Then, 5′ × 8′ = 40″ = 3′, (to carry,) and 4″, (to write down); 10 × 5′ = 50′ + 3′ (to carry) = 53′ = 4 ft. and 5′, which we write down underneath the 10 and 8′. Then, the sum of these two products, added together as before, is 15 ft. 1′ 4″ *Ans.*, *the same result as the other.*

Note. Had we been required to multiply 15 ft. 1′ 4″ by feet and inches again, we should have proceeded in the same manner, carrying ‴ (thirds) one place further towards the right, and ⁗ (fourths) another place still, and so on.

From these examples we derive the following

RULE.

How do you multiply in Duodecimals?

A. Begin with the highest denomination of the multiplier and the lowest denomination of the multiplicand, placing the first figure in each product one place further towards the right than the former, recollecting to carry by 12, as in compound rules.

More Exercises for the Slate.

2. How many feet in a board 2 ft. 6′ wide, and 12 ft. 3′ long? *Ans.* 30 ft. 7′ 6″.

3. In a load of wood 8 ft. 4′ long, 2 ft. 6′ high, and 3 ft. 3′ wide, how many solid feet? *A.* 67 ft. 8′ 6″.

Note. Artificers compute their work by different measures. Glazing and mason's flat work are computed by the square foot; painting, paving, plastering, &c. by the square yard; flooring, roofing, tiling, &c. by the square of 100 feet; brick work by the rod of 16½ feet, whose square is 272¼; the contents of bales, cases, &c. by the ton of 40 cubic feet; and the tonnage of ships by the ton of 95 feet.

4. What will be the expense of plastering the walls of a room 8 ft. 6′ high, and each side 16 ft. 3′ long, at $,50 per square yard? *A.* $30,694+.

5. How many cubic feet in a block 4 ft. 3′ wide, 4 ft. 6′ long, and 3 ft. thick? *A.* 57 ft. 4′ 6″.

6. How much will a marble slab cost, that is 7 ft. 4′ long, and 1 ft. 3′ wide, at $1 per foot? *A.* $9,16⅔.

7. How many square feet in a board 17 ft. 7′ long, 1 ft. 5′ wide? *A.* 24 ft. 10′ 11″.

8. How many cubic feet of wood in a load 6 ft. 7′ long, 3 ft. 5′ high, and 3 ft. 8′ wide? *A.* 82 ft. 5′ 8″ 4‴.

9. A man built a house consisting of 3 stories; in the upper story there were 10 windows, each containing 12 panes of glass, each pane 14′ long, 12′ wide; the first and second stories contained 14 windows, each 15 panes, and each pane 16′ long, 12′ wide: how many square feet of glass were there in the whole house? *A.* 700 sq. ft.

10. What will the paving of a court yard, which is 70 ft. long, and 56 ft. 4′ wide, come to, at $,20 per square?

A. $788,66⅔.

11. How many solid feet are there in a stick of timber 70 ft. long, 15′ thick, and 18′ wide? *A.* 131 ft. 3′.

Questions on the foregoing.

1. How many pence are there in 1 s. 6 d.? How many cents?

2. What will 4 yards of cloth cost, in cents, at 1 s. 6 d. per yard? At 3 s. per yard? At 4 s. 6 d.? At 6 s.? At 9 s.? At 10 s. 6 d.?

3. If a man consume 1 lb. 9 oz. of bread in a week, how much would he consume in 1 month?

4. At 4 cents for 1 oz., what would 1 lb. cost?

5. At 4 cents for 2 oz., what would 1 lb. cost?

6. At 4 cents for 8 oz., what would 2 lbs. cost?

7. If a man spend $2¼ per day, how many days would he be in spending $4½? $6¾? $12¼? $20?

8. How many marbles, at 4 cents apiece, must be given for 24 apples, at 2 cents apiece?

9. How many yards of cloth, at $4 per yard, must be given for 6 bbls. of cider, at $2 per bbl.? For 8 bbls.? For 12 bbls.? For 18 bbls.?

10. What part of 1 month is 1 day? 2 days? 4 days? 5 days? 6 days? 7 days? 10 days? 20 days? 29 days?

11. What is the interest of $1 for 12 mo.? 10 mo.? 9 mo.? 6 mo.? 3 mo.? 1 mo.? 15 days?

12. What is the interest of $6 for 1 yr. 2 mo.? 2 yrs.? 1 yr. 1 mo.? 9 mo.? 2 mo.? 1 mo.? 15 days? 10 days? 6 days? 5 days? 1 day?

13. What is the amount of $1 for 6 mo.? 3 mo.? 2 mo.? 1 mo.? 15 days?

14. Suppose I owe a man $115, payable in 1 yr. 6 mo., without interest, and I wish to pay him now, how much ought I to pay him?

15. What is the discount of $115, for 2 yrs. 6 mo.?

16. William has ⅓ of an orange, and Thomas ½; what part of an orange do both own?

17 Harry had $\frac{3}{4}$ of an orange, which he wished to divide equally between his two little sisters; can you tell me what part of an orange each one would receive?

18. Which is the most, ,5 of 20, or ,25 of 40?

19. How many times can you draw $\frac{1}{2}$ of a gallon of cider from a barrel containing 30 gallons? How many times $\frac{1}{4}$ of a gallon? How many times $\frac{1}{3}$ of a gallon? How many times $\frac{1}{8}$ of a gallon?

20. A man, failing in trade, is able to pay his creditors only \$,33$\frac{1}{3}$ on the dollar; how much will he pay on \$3? On \$4? On \$12? On \$13? On \$300?

21. A man, failing in trade, was able to pay only \$,16$\frac{2}{3}$ on the dollar; how much would he pay on a debt of \$4? \$6? \$10? \$9? \$20? \$100? \$600?

22. Two men bought a barrel of flour for \$10; one gave \$3, and the other \$7: what part of the whole did each pay? What part of the flour must each have?

23. If 30 bushels of oats cost \$10, what is that a bushel? What will 5 bushels cost? What will 20 bushels?

24. If 3 men mow a field in 8 days, how many men will mow the same in 2 days? In 1 day? In 4 days?

25. Two men, A and B, hired a pasture; A paid \$3, and B \$5; what fractional part of the whole did each pay? The profits from the pasture were \$16; what was each man's share of the gain?

26. Three men, A, B, and C, are engaged in trade; A puts in \$4, B \$5, and C \$6; they gained \$60: what is each one's share of the gain?

27. Two men, A and B, hired a pasture for \$12; A put in 1 cow 4 months, and B 2 cows 3 months: what ought each to pay?

28. A merchant, having purchased a piece of broadcloth for \$2 per yard, wishes to make 20 per cent. on it; what price must he ask for it?

29. William has $\frac{1}{5}$ of a dollar, Thomas $\frac{1}{10}$, and Harry $\frac{1}{5}$; how many cents have they in all?

30. A merchant sold calico at \$,22 per yard, and thereby gained 10 per cent.; what did it cost him per yard?

31. Harry, having $\frac{3}{4}$ of an orange, gave $\frac{1}{5}$ to Thomas, who gave $\frac{1}{2}$ of his part to his little brother, and kept the remainder himself; what part did he keep? How much is $\frac{1}{5}$ of $\frac{3}{4}$? How much does $\frac{1}{2}$ of $\frac{1}{5}$ of $\frac{3}{4}$ from $\frac{3}{20}$ leave?

32. How much is $1 \times \frac{2}{3}$ of $\frac{3}{2}$? (1) How much is $1 \times \frac{6}{7}$ of $\frac{7}{6}$?

33. What is the quotient of $\frac{2}{3}$ divided by $\frac{3}{2}$?

34. How much does $\frac{3}{4}$ exceed ,75?

35. How much does $\frac{4}{5}$ exceed $\frac{7}{11}$?

36. How many strokes does a regular clock strike in 2 hours? 3 h.? 4 h.? 5 h.? 6 h.? 7 h.? 8 h.? 9 h.? 10 h.? 11 h.? 12 h.? 24 h.?

37. How many square feet in a board 12 inches wide, and 48 inches long? 36 in. long? 72 in. long?

38. What part of an acre of land is there in your father's garden, allowing it to be 4 rods long and 2 rods wide? 4 rods wide?

39. How many cord feet of wood are contained in a load 3 feet wide, 2 feet high, and of the usual length? How many feet in a load 6 feet high and 3 feet wide? 2 feet high and 6 feet wide? 4 feet high and $2\frac{1}{2}$ feet wide?

40. How many solid feet in a block 12 inches long, 12 inches thick, and 12 inches wide? 12 inches long, 12 inches wide, and 6 inches thick?

41. How long will it take to count $1000, at the rate of 50 a minute?

42. What is the difference between 4 square feet and 4 feet square? 10 miles square and 10 square miles? 3 rods square and 3 square rods?

A parenthesis, enclosing several numbers, signifies that these numbers are to be taken together, or as one whole number; but, when performing subtraction and addition with these and other numbers, they may be taken either as before, or one by one, thus:—

$(16+4)\times 3=60$, *read* $16+4=20\times 3=60$.

Or, $(16\times 3)=48$. $4\times 3=12$. *Then*, $48+12=60$, *the same as before.*

43. $(9+3)+8=$ how many? *A.* 20.
44. $(9-3)-4=$ how many? *A.* 2.
45. $(15+5)\times 4=$ how many? *A.* 80, or $60+20$.
46. $(15-5)\times 4=$ how many? *A.* 40, or $60-20$.
47. $(9+3)\div 3=$ how many? *A.* 4, or $3+1$.
48. $(9-3)\div 3=$ how many? *A.* 2, or $3-1$.
49. $(12-8)\div(3+1)=$ how many? *A.* 1, or $3-2$.

A line, or vinculum, drawn over numbers, is sometimes used instead of a parenthesis.

50. $\overline{4+8}\times 12=$ how many? *A.* 144, or $48+96$.

51. $\overline{2£\ 4\ s.}\times 2=$ how much? *A.* 4£ 8 s.
52. (2 s. 6 d.) $\times 2=$ how much?
53. (7 s. 6 d.) $\times 4=$ how much?

54. $\overline{3\times 3}\times\overline{2\times 10}=$ how many?

55. How many minutes of motion make 2 degrees of motion? How many seconds of motion make 3 minutes of motion?

56. How many degrees is the circumference of the earth?

57. The earth, you know, turns round once in every 24 hours, or, in common language, the sun moves round the earth in that time; in what time, then, will the sun travel over 15° (degrees)? and why? *A.* 1 hour for $360°\div 24$ h. $=15°$.

58. In what time will he travel over 1° (degree) of motion?

A. $\frac{1}{15}$ of 1 hour, or $\frac{1}{15}$ of 60 min., = 4 min.

59. In what time will he travel over 1′ (minute) of motion?

A. $\frac{1}{15}$ of 1 min., or $\frac{1}{15}$ of 60 sec., = 4 seconds.

Q. *By the foregoing we see that every degree of motion makes a difference in time of 4 minutes, and every minute of motion a difference of 4 seconds. Now, since longitude is reckoned in degrees, round the earth, can you tell me how to find the difference in time between one place and another, after knowing their difference in longitude?* *A.* Multiply the difference of longitude in degrees and minutes by 4, the product will be their difference in time, in minutes and seconds.

60. What is the difference in time between two places, whose difference in longitude is 2° 4′? *A.* 2° 4′ × 4 = 8 minutes and 16 seconds, the difference in time.

61. What is the difference in time between two places, whose difference in longitude is 5° 10′? *A.* 20 m. 40 sec. What, when the difference of longitude is 8°? *A.* 32 m. Is 10°? *A.* 40 m. Is 15°? *A.* 60 m. = 1 hour. Is 15° 15′? *A.* 1 h. 1 m.

Q. *The sun travels from east to west: which place, then, will have the earliest time?* *A.* The one most easterly.

62. There are two places, the one situated in 10° E. longitude, and the other in 4° E. longitude; what is the difference in time between these two places? When it is 24 minutes past 6 o'clock in the former, what hour is it in the latter? *A. 24 minutes, difference in time; then, 10° being the most easterly place, it is there 24 minutes earlier than at 4°; that is, when it is 24 minutes past 6 at 10°, it is only 6 at 4°.*

63. Boston is situated about 6° 40′ E. longitude from the city of Washington; when it is 2 o'clock at Washington, what o'clock is it at Boston? *A.* 26 m. 40 sec. past 2 o'clock.

64. "I recollect of reading a story once of a gentleman going to a foreign country, who had a fancy to look at a bright star every evening, at the same moment, with a certain lady whom he left behind, and they agreed to look at it at 9 o'clock;" but, it seems that, when the gentleman was in a different longitude, the time would, of course, be different; as, for instance, when he was in longitude differing 30° W. from where the lady was, she most probably had retired to rest, and was, perhaps, asleep, while he was gazing at the star. Can you tell me what o'clock it was, then, where she was? When he was 60° of W. longitude from her, what hour of the night was it at the place where the lady resided?

Exercises for the Slate.

1. Write down three millions, three hundred and three thousand, three hundred and three.

2. What is the difference between 50 eagles and 4599 dimes? *A.* $40,10.

3. What number is that, which, being divided by 65, the quotient will be 42? *A.* 2730.

4. A captain, 2 lieutenants, and 30 seamen, take a prize worth $7002, which they divide into 100 shares, of which the captain takes 12, the two lieutenants each 5, and the remainder is to be divided equally among the sailors; how much will each man receive? *A.* Captain's share, $840,24, each lieutenant's, $350,10, and each seaman's, $182,052.

5. Bring $400 into crowns, at 6 s. 8 d. each. *A.* 360 crowns.

6. Washington was born A. D. 1732; how many years old would he have been, had he lived until the end of the year 1827? and how many seconds old, allowing the year to contain $365\frac{1}{4}$ days? *A.* 95 years, 2997972000 seconds.

7. The wheels of a cart are 5 feet in circumference, and that of a wheelbarrow 27 inches; how many more times will the latter turn round than the former, in going round the earth? How many more times in going through the earth, allowing the diameter to be $\frac{1}{3}$ of the circumference? *A.* 32292480 times round the earth, 10764160 times through it.

8. How many minutes is it from the commencement of the Christian era to the end of the year 1827? *A.* 960928920.

9. Jacob, by contract, was to serve Laban for his two daughters 14 years; when he had accomplished 10 years, 10 mo., 10 weeks, 10 days, 10 hours, 10 minutes, how many minutes had he then to serve? *A.* 1416350.

10. Reduce $\frac{492}{744}$ to its lowest terms. *A.* $\frac{41}{62}$.

11. What is the value of $\frac{3}{7}$ of a cwt.? *A.* 1 qr. 20 lbs.

12. What is the amount of ,5 ,05 ,005 ,555 ,18765 and 8567? *A.* 8568,29765.

13. Divide $\frac{2}{81}$ by $\frac{3}{41}$. *A.* $\frac{82}{243}$.

14 From $17\frac{5}{7}$ take $\frac{1}{2}$ of $\frac{2}{3}$ of $14\frac{1}{2}$. *A.* $12\frac{74}{84}$.

15. Reduce $\frac{550}{12}$ to a mixed number. *A.* $45\frac{10}{12}$.

16. What is the value of ,425 of a pound? *A.* 8 s. 6 d.

17. Reduce 14 s. 6 d. 3 qrs. to the decimal of a pound. *A.* ,7281+.

18. From ,1 of a pound, take ,0678 of a pound. *A.* ,0322.

19. If you give $60 for 25 yards of cloth, what will 1 yard cost? *A.* $2,40.

20. A merchant sold 8 bales of linen, 6 of which contained 15 pieces each, and in each piece were 40 yards; the other 2 bales contained 12 pieces each, and in each piece were 27 yards; what did the whole amount to, at $\$1\frac{1}{4}$ per yard? *A.* $5310.

21. A man, dying, left $10024 to his wife and 2 sons, to be divided as follows: to his wife $\frac{3}{8}$, to his eldest son $\frac{2}{5}$ of the remainder, and to his youngest son the rest: what is the share

of each? *A.* $3759 *to his wife*, $2506 *to his eldest son*, $3759 *to his youngest.*

22. A farmer sold a grocer 20 bushels of rye, at $,75 per bushel; 200 lbs. of cheese, at 10 cents per lb.; in exchange for which he received 20 gallons of molasses, at 22 cents per gallon, and the balance in money; how much money did he receive? *A.* $30,60.

23. A has 150 yards of linen, at 25 cents per yard, which he wishes to exchange with B for muslin at 50 cents per yard; how much muslin must A receive? *A.* 75 yds.

24. A gave B 500 yards of broadcloth, at $2,50 per yard, for 600 umbrellas; what were the umbrellas apiece? *A.* $2,083+.

25. A farmer sold a grocer 20 bbls. of apples, each barrel containing 3 bushels, at 40 cents per bushel; (24) 30 bushels of corn, at 90 cents per bushel; (27) 500 lbs. of cheese, at 8 cents per pound; (40) 200 lbs. of butter, at 17 cents per pound; (34) 75 bushels of turnips, at 19 cents per bushel; (1425) 40 bushels of barley, at $1¼ per bushel; (50) 25 bushels of rye at 95 cents per bushel; (2375) and, in exchange, the farmer has received 2 bbls. of cider brandy, at 42 cents per gallon; (2646) 4 bbls. of flour, at $8¾ per barrel; (35) 60 gallons of molasses, at 34 cents per gallon; (2040) 40 gallons of wine, at $1,50 per gallon; (60) 10 lbs. of tea, at 73 cents per pound; (730) $40 in cash; and he agrees to take up what is still due him in rice at 7 cents per pound; how many pounds of rice must the grocer give the farmer to balance the account?

A. 340 lbs. 9 oz. 2$\frac{2}{7}$ dr.

26. What quantity of cider, at $1,20 per barrel, will buy 2 barrels of rum, at $2 per gallon? *A.* 105 bbls.

27. A man exchanged 40 bushels of salt, at $1,50 per bushel, for 200 bushels of oats, at 25 cents per bushel; how much was the balance in his favour? *A.* $10.

28. A sold B 16 cwt. of sugar, at 8¼ cents per pound; (14784) 20 bbls. of flour, at $11¼ per barrel; (225) 17 chests of tea, each containing 8 cwt., at 53 cents per pound; (807296) 30 tierces of rice, at $36 per tierce; (1080) for which B gave up A his note of $400, that had been on interest 6 yrs. 7 mo. 15 days; (552) in addition to which B gave A 700 dozen of wax candles, at $1,14 per dozen; (798) and for the balance A consents to take B's note, payable in 2½ years, without interest; but B, unexpectedly receiving some money, wishes to advance the cash, instead of giving his note; what sum of ready money ought B to pay A, discounting at 5½ per cent.? *A.* $7181,362+.

29. A had 200 bbls. of flour, at $10,50 per bbl., for which B gave him $1090 in money, and the rest in molasses, at 20 cents per gallon; how many hogsheads of molasses did he receive?

A. 80 hhds. 10 gals

30. A has linen cloth, worth $,20 per yard, but, in bartering, he will receive $,30; B has broadcloth, worth $4,60 per yard, ready money: at what price ought B to rate his broadcloth, to be in proportion with A's bartering price?

$,20 : $4,60 : : $,30 : $6,90, *Ans.*

Or, multiply $,30 by the ratio of 20 to $4,60, that is, 23; thus, 23 × 30 = $6,90; for, $4,60 being 23 times as much as 20, it is plain that 23 times $,30, A's bartering price, will give B's bartering price, A. $6,90.

31. A merchant, in bartering with a farmer for wood at $5 per cord, rated his molasses at 25 dollars per hhd., which was worth no more than $20; what price ought the farmer to have asked for his wood to be equal to the merchant's bartering price? *A.* $6,25.

The last ten examples are proper questions in a rule usually called *Barter.*

32. What number is that, which, being multiplied by 15, will make $\frac{3}{4}$? *A.* $\frac{1}{20}$.

33. What number is that, which, being divided by 15, will make $\frac{1}{20}$? *A.* $\frac{3}{4}$.

34. What decimal is that, which, being multiplied by ,625, will make ,25? *A.* ,4.

35. At $,75 per bushel, how much rye can be bought for $150? (200) For $600? (800) For $75? (100) *A.* 1100 bushels.

36. (800 + 12 + 88) ÷ (50-5) = how many? *A.* 20.

37. ($\frac{80}{16}$ + $\frac{144}{72}$) ÷ (3,55,-55) = how many? *A.* $2\frac{1}{3}$.

38. Two persons depart from one place at the same time; the one travels 35, and the other 40 miles a day; how far are they distant at the end of 10 days, if they both travel the same road? and how far, if they travel contrary directions? *A.* 50, and 750 miles.

39. Two men, A and B, traded in company; A put in $700 for 8 months, and B $1280 for 10 months; they gained $500: what was the share of each? *A.* A's share, $152,173; B's share, $347,826 +.

40. How many cord-feet of wood are contained in a load 8 feet long, 4 ft. 6′ wide, and 5 ft. 3′ high? *A.* $11\frac{13}{16}$ ft. *or* 11 ft. 9′ 9″.

41. What is the difference in time between two places, whose difference of longitude is 40°? (2,40) 50°? (3,20) 60°? (4) *A.* 10 h.

42. What time is it in 15° W. longitude, when it is 6 o'clock in 15° E. longitude? *A.* 4 o'clock.

43. If a cow yield 16 qts. of milk in a day for 240 days, and 20 qts. make 1 pound of butter, or 10 pounds of cheese, how much more profitable is the making of cheese than butter, the price of butter being 25 cts. per pound, and that of cheese $8\frac{1}{4}$ cts. per pound? *A.* $110,40.

44. If a field will feed 10 cows four weeks, how long will it feed 40 cows? *A.* 1 week.

45. A man bought a cask of wine, containing 126 gallons, for $315, and sold it at the rate of $2,75 per gallon; how much was his whole gain? how much per gallon? and how much per cent.? *A.* His whole gain, $31,50; per gallon, $,25; then, $\frac{,25}{2,50}=01$ per cent.

46. The rent of a certain farm is $500; the tenant employs 2 men; to each he pays $$11\frac{1}{2}$ a month for 8 months; (184) also a boy by the year, who is to have 2 suits of clothes, each worth $8,75 (1750), besides his board, while attending school, 3 months, or 12 weeks of the year, which is worth $,93 per week (1116); in the course of the year, the tenant loses 40 good merino sheep, valued at $5 per head (200); the skins brought him $$1,12\frac{1}{2}$ apiece (45); the other expenses of the year are calculated to average about $,39 (14235) per day; the sales of the farm are as follows, viz:—

1000 bushels of	Potatoes, ... at	$0,42	per bushel	(420)
150	Barley, ...	$0,99	. . . . .	(14850)
16	White Beans, . . .	$2,20	. . . . .	(3520)
400	Corn, ...	$0,53	. . . . . .	(212)
300	Rye, ...	$0,92	. . . . . .	(276)
$418\frac{1}{2}$	Buckwheat, ...	$0,38	. . . . .	(15903)
200	Oats, ...	$0,28	. . . . . .	(56)
10000 pounds . .	Cheese, ...	$$0,5\frac{3}{4}$	. . pound	(575)
500	Butter, ...	$0,18	. . . .	(90)
1 bushel	Clover, ...	$0,25	. . quart .	(8)
20	Turnips, ...	$0,43	. . bushel	(860)
70	Winter Apples, . .	$0,48	. . . . .	(3360)
10	Flax Seed, ...	$1,75	. . . . .	(1750)
400 pounds . . .	Merino Wool, . . .	$0,45	. . pound	(180)

and 20 calves, at $$4\frac{1}{2}$ per head (90); he carries 70 barrels of cider to the distillery, for one half of which he is to receive 1 qt. of cider-brandy for 4 qts. of cider, the rest brings him $1,13 per barrel (3955); how much cider-brandy will he receive? and how much will he clear, after deducting all the expenses of managing the farm, including the rent and loss, from the total amount of sales? *A.* $1338,97; $1102\frac{1}{2}$ qts. cider-brandy.

47. What difference is there between the compound interest of $10000 for 8 years, and the simple interest of the same sum for the same time? *A.* $1138,48.

48. What is the difference between the compound interest of $500 for 4 years, and the discount of the same sum for the same time? *A.* $34,463.

49. What is the difference between the amount of $1800, at compound interest, for 3 years, and the present worth of the same sum, for the same time? *A.* $618,405.

50. If 120 gallons of water, in one hour, fall into a cistern *containing* 600 gallons, and, by one pipe in the cistern, 35 gal-

lons run out, and, by another pipe, 65 gallons run out, in what time will the cistern be filled? *A.* 30 hours.

51. A certain clerk, in a country store, purchased the whole stock in trade, the quantity and price being as follows:—

INVENTORY

Quantity	Article	Price	Per	Amount
3	bbls. of Sugar, each 118 lbs. at	$0,06¼	per lb.	(22125)
4	canisters of Tea,	$5,00	canister,	(20)
10	bags of Coffee, each 20 lbs.	$0,35	lb.	(70)
10	bbls. of Pork, each 200 lbs.	$0,08		(160)
13	bbls. of Beef, each 200 lbs.	$0,07		(182)
40	Hams, 36 lbs. each,	$0,13		(18720)
200	lbs. of Tallow,	$0,10		(20)
2	hhds. of Rum,	$0,26	per gallon,	(3276)
1	hhd. of Molasses,	$0,24		(1512)
1	bbl. of Brandy, lacking 5 gals.,	$1,12		(2968)
2	bbls. of Brandy,	$1,16		(7308)
1	pipe of Wine, lacking 15 gals.,	$0,85		(9435)
2	do. do.	$1,00		(252)
4	bbls. of Gin,	$0,60		(7560)
1	bbl. of Vinegar,	$0,25		(5906)
40	empty Barrels,	$0,75	apiece,	(30)
63	empty Hogsheads,	$1,12		(7056)
2	pieces of Calico, 14 yds. each,	$0,22	per yard,	(616)
1	piece of Silk, 28	$0,89		(2492)
1	..... Cotton, 11½	$0,14		(161)
1	..... Cotton Plaid, 12	$0,19		(228)
1	..... Linen, 10	$0,46		(460)
1	..... Broadcloth, 27	$1,19		(3213)
1	.. blue . do. 15	$3,75		(5625)
1	.. mixed do. 10	$1,10		(11)
1	..... Satin, 53	$0,87½		(46375)
1	..... Vesting, 4 patterns	$0,80	per pattern,	(320)
4	Hats,	$2,17	apiece,	(868)
6	pair of Shoes,	$1,80	a pair,	(1080)
1	dozen pair of Children's Shoes,	$0,22		(264)
14	Whips,	$1,14	apiece,	(1596)
14	Hoes,	$0,83		(1162)
12	Axes,	$1,17		(1404)
15	Axe-helves,	$0,07		(105)
27	wooden Pails,	$0,23		(621)
45	..... Tubs,	$0,46		(2070)
10	Kettles,	$2,95		(2950)
2	dozen of Knives,	$0,17		(408)
1	bladder of Snuff, 4 lbs.,	$0,36	per lb.	(144)
2	Self-sharpening Ploughs,	$3,50	each,	(7)
4	Rakes,	$0,22		(88)
2	Hymn-Books,	$0,38	apiece,	(76)
4	Perry's Spelling-Books,	$0,18		(72)
2	Dwight's Geographies,	$0,16		(32)
1	Morse's Geography,	$1,20		(120)
2	Great Coats,	$2,00		(4)
1	Vest,	$0,50		(50)

Amount of stock, **$1671 +.**

By agreement, 15 per cent. was to be deducted from the amount of stock. For the above goods, the clerk has paid an follows: his services for the two last years, at $26 per month; turned in a note on James Spencer, of 500 dollars, with 5 years' compound interest due on it; (669112) and, for the balance, he was to give his note, payable in 6 years from the date of the transaction, without interest: now the question is, what sum of ready money will discharge said note? *A.* $93,557.

After the clerk had purchased the above stock, and settled for the same, he commenced business for himself. The rent of his store costs him $29 a year; his clerk-hire $27,814 precisely; in addition to which is the interest of his capital, $1420,35, (that is, the interest of the amount of stock, after the 15 per cent. is deducted, $85221). He next considered what price he must put on each article, to make a certain per cent He recollected that the goods were already rated in the inventory at 15 per cent. more than their actual cost. Now, said he, if I can make 5 per cent. in advance on their present valuation, clear of all expenses, I shall be satisfied. The question, then, is, at what price he must mark each article, commencing with the first on the inventory, so as to clear the 5 per cent.

The pupil will find, by calculation, that the expenses amount to 10 per cent. on the actual cost of all the articles; this, added to the 5 per cent., makes 15 per cent. advance; that is, each article must be marked 15 per cent. higher than its present valuation in the inventory. The answer to each follows in the same order as the articles stand in the above inventory, commencing with the 3 bbls. of sugar, and finding the selling price of each per lb. &c.

Note. In marking goods, it is customary to neglect the mills, if under 5; if exactly 5, add ½ of a cent, and if over 5, add 1 cent to the cents.

Sugar, at	. (7)	*per lb.*	*Hhds. . at*	(129)	*each.*	*Axes, at*	(134½)	*each.*
Tea, . . .	(575)	*can.*	*Calico,* . .	(25)	*pr. yd.*	*Axe-helves,*	(8)	. . . .
Coffee, . .	(40)	*per lb.*	*Silk,*	(102)	. . . .	*Pails,* . . .	(26)	. . . .
Pork, ...	(9)	. . . .	*Cotton,* . .	(16)	. . . .	*Tubs,* . . .	(53)	. . .
Beef, ...	(8)	. . . .	*Cotton Plaid,*	(22)	. . . .	*Kettles,* . .	(339)	. . . .
Hams, . . .	(15)	. . . .	*Linen,* . . .	(53)	. . . .	*Knives,* . .	(19½)	. . . .
Tallow, . .	(11½)	. . . .	*Broadcloth,*	(137)	. . . .	*Snuff,* . . .	(41)	*per lb.*
Rum, . . .	(30)	. *gal.*	*Blue do.* .	(431)	. . . .	*Ploughs,*	(402½)	*each.*
Molasses, .	(28)	. . . .	*Mixed do.*	(126½)	. . . .	*Rakes,* . .	(25)	. . . .
Brandy, . .	(129)	. . . .	*Satin,* . . .	(101)	. . . .	*Hymn-Books,*	(44)	. . .
Brandy, . .	(133)	. . . .	*Vestings,* .	(92)	*patt.*	*Spelling-B.* .	(21)	. . . .
Wine, . . .	(98)	. . . .	*Hats,* . .	(249½)	*each.*	*Geographies,*	(18)	. . .
Wine, . . .	(115)	. . . .	*Shoes,* . . .	(207)	*a pair.*	*Geography,*	(138)	. . .
Gin, ...	(69)	. . . .	*Shoes,* . . .	(25)	. . .	*Great Coat,*	(230)	. . . .
Vinegar, .	(29)	. . . .	*Whips,* . .	(131)	*each.*	*Vest,* . .	(57½)	. . . .
Barrels, . .	(86)	*each.*	*Hoes,* . . .	(95)	. . . .	*Ans.* .	$48,81½.	

52. Bought 42 gallons of rum for $37,80; how much water must be mixed with it, that it may be afforded for $,80 per gallon?

$,80 : $37,80 : : 1 *gal.* : 47¼ *galls.*; *then,* 47¼ = 42 + 5¼ *galls., Ans.*

53. A thief, having 30 miles the start of an officer, makes off at the rate of 8 miles an hour; the officer presses on after him at the rate of 10 miles an hour: how much does he gain on the

thief in one hour? how long before he will overtake the thief? *A.* 2 miles; 15 hours.

54. A person, looking at his watch, was asked what o'clock it was; he replied it was between 4 and 5; but, a more particular answer being requested, he said the hour and minute hands were then exactly together. What was the time? *In 1 hour the minute hand passes over 12 spaces, while the hour hand passes over only 1 space; that is, the minute hand gains upon the hour hand 11 spaces in 1 hour; consequently, it must gain 12 spaces before both will be together.* 60 *m.* $\div$ 11 $=$ $5\frac{5}{11}$ *m., gain in 1 hour; then,* $5\frac{5}{11}$ *m.* $\times$ 4 *hours* $=$ $21\frac{9}{11}$ *m. past 4 o'clock, Ans. Or, the ratio being* $\frac{1}{11}$ *of an hour, hence, 4 hours* $\times \frac{1}{11} = 21\frac{9}{11}$ *minutes past 4, Ans., as before.*

55. At 12 o'clock the hour and minute hands of a clock are exactly together; when will they be together again?

Ans. 1 h. 5 m. $27\frac{3}{11}$ sec.

56. If 10 men can perform a piece of work in 25 days, how many men will accomplish another piece of work, four times as large, in a fifth part of the time? *A.* 200.

57. A can do a piece of work in 8 days, and B in 12; in what time would both finish it by working together?

days, day, work, work,

$$\left.\begin{array}{l} 8 : 1 :: 1 : \frac{1}{8} \\ 12 : 1 :: 1 : \frac{1}{12} \end{array}\right\} \frac{1}{8} + \frac{1}{12} = \frac{5}{24}. \quad \textit{Then, } \frac{5}{24} : 1 :: 1 : 4\frac{4}{5}, \textit{ Ans.}$$

58. What number is that, from which, if you take $\frac{2}{3}$, the remainder will be $\frac{1}{4}$? $\frac{2}{3} + \frac{1}{4} = \frac{11}{12}$, *Ans.*

59. What number is that, from which, if you take $\frac{2}{3}$, the remainder will be $\frac{1}{6}$? *Ans.* $\frac{5}{6}$.

60. What number is that, from which, if you take $\frac{1}{2}$ of $\frac{3}{4}$ of $2\frac{1}{2}$, the remainder will be $\frac{5}{8}$? *Ans.* $\frac{25}{16} = 1\frac{9}{16}$.

61. What number is that, which, being divided by $\frac{2}{3}$, the quotient will be $\frac{8}{9}$? *A.* $\frac{16}{27}$.

62. What number is that, which, being multiplied by $\frac{4}{5}$, the product will be $3\frac{1}{8}$? *A.* $3\frac{29}{32}$.

63. What number is that, from which, if you take $\frac{3}{4}$ of itself, the remainder will be 12?

1, *or* $\frac{4}{4}$, $-\frac{3}{4}$, $=\frac{1}{4}$, *remainder. Then, the remainder 12, being* $4 \times 12 = 48$ *times greater than the remainder* $\frac{1}{4}$, *the number itself will be 48 times greater than* 1. *A.* 48.

64. What number is that, to which, if you add $\frac{1}{2}$ of $\frac{3}{5}$ of itself, the sum will be 39?

$\frac{1}{2}$ *of* $\frac{3}{5} = \frac{3}{10}$, *and the number itself* $\frac{10}{10}$; *then,* $\frac{3}{10} + \frac{10}{10} = \frac{13}{10}$; *if to the whole number* $\frac{1}{2}$ *of* $\frac{3}{5}$ *of it be added, the sum will be* $\frac{13}{10}$; *consequently 39 is* $\frac{13}{10}$ *of the number. Ans.* 30.

65. What number is that, to which if you add $\frac{1}{2}$ of itself, the sum will be 18? *A.* 12.

66. A owns $\frac{1}{8}$ of a vessel, B $\frac{2}{8}$, C $\frac{4}{8}$, and D the remainder; D's part is $100: can you tell me how many dollars is each man's part, and what part of the vessel D owns?

Ans. A's part, $100; B's, $200; C's, $400; and D's part is $\frac{1}{8}$.

67. There is a beam, $\frac{5}{16}$ of which is in the ground, $\frac{9}{16}$ in the water, and the rest, being 2 feet, out of water; how long is the beam? *A.* 16 feet.

68. The third part of an army was killed, the fourth part taken prisoners, and 1000 fled; how many were in this army? how many killed? how many taken captives?

$\frac{1}{3}+\frac{1}{4}=\frac{7}{12}$, of the whole army; then, as $\frac{5}{12}$ more makes $\frac{12}{12}$, or the whole army, $\frac{5}{12}=1000$; and if $\frac{5}{12}$ be 1000, how much is $\frac{12}{12}$, or the whole? Ans. 2400, the whole army; 800 killed, 600 captives.

69. Suppose that there is a mast erected, so that $\frac{1}{9}$ of its length stands in the ground, 12 feet in the water, and $\frac{5}{6}$ of its length in the air, or above water; I demand the whole length.

Reducing the fractions to the least common denominator, gives $\frac{2}{18}+\frac{15}{18}=\frac{17}{18}$; therefore 12 feet $=\frac{1}{18}$. *A.* 216 feet.

70. In an orchard of fruit-trees, $\frac{1}{2}$ of them bear apples, $\frac{1}{4}$ pears, $\frac{1}{6}$ plums, 40 of them peaches, and 10 cherries; how many trees does the orchard contain? $\frac{1}{12}=50+10$. *Ans.* 600.

71. A man spent one third of his life in England, one fourth in Scotland, and the remainder, which was 20 years, in the United States; to what age did he live? *A.* 48 years.

72. The number of scholars in a certain school is as follows: $\frac{1}{8}$ of the pupils study geography, $\frac{1}{3}$ grammar, $\frac{1}{2}$ arithmetic, and 10 learn to read: what number is pursuing each branch of study? *A.* 30 in geography, 80 in grammar, 120 in arithmetic, and 10 learn to read.

73. The double and the half of a certain number, increased by $7\frac{1}{2}$ more, make 100; what is that number? *A.* 37.

74. A man, having purchased a drove of cattle, was driving them to market, when he was met by a gentleman, who inquired of him where he was going with his 100 head of cattle? Sir, said he, I have not near 100, but if I had as many more as I now have, $\frac{1}{2}$ as many more, and 7 cattle and $\frac{1}{2}$, I should have a hundred. How many had he? *A.* 37.

75. Five eighths of a certain number exceed $\frac{2}{5}$ of the same number by 36; what is that number?

$\frac{5}{8}-\frac{2}{5}=\frac{9}{40}$; hence 36 is $\frac{9}{40}$ of the number sought. *A.* 160.

76. What number is that, which, being increased by $\frac{1}{5}$, $\frac{1}{3}$, and $\frac{1}{4}$ of itself, the sum will be 131? *Ans.* $73\frac{49}{107}$.

The eleven foregoing questions are usually performed by a rule called *Position*, but this method of solving them by fractions is preferable.

77. A hare starts up 12 rods before a hunter, and scuds away at the rate of 10 miles an hour; now, if the hunter does not change his place, how far will the hare get from the hunter in 45 seconds? *A.* 52 rods.

78. If a dog, by running 16 miles in one hour, gain on a hare, 6 miles every hour, how long will it take him to overtake her, provided she has 52 rods the start? *A.* $97\frac{1}{2}$ seconds.

79. A hare starts 12 rods before a greyhound, but is not perceived by him till she has been up 45 seconds; she scuds away at the rate of 10 miles an hour, and the dog after her at the rate of 16 miles an hour; what space will the dog run before he overtakes the hare? *A.* 138 rods, 3 yards, 2 feet.

80. A gentleman has an annuity of $2000 per annum; I wish to know how much he may spend daily, that, at the year's end, he may lay up 90 guineas, and give 20 cents per day to the poor of his own neighbourhood? *A.* $4,128.

81. What is the interest of $600 for 120 days? (12) For 2 days? (20) For 10 years, 10 mo. and 10 days? (391) For 5 years, 5 mo. and 5 days? (19550) For 6 years, 6 mo., and 6 days? (23460) For 4 years, 4 mo. and 4 days? (15640)

A. Total, $989,70.

82. What is the present worth of $3000, due $2\frac{1}{2}$ years hence, discounting at 6 per cent. per annum? *A.* $2608,695+.

83. Suppose A owes B $1000, payable as follows; $200 in 4 mo., $400 in 8 mo., and the rest in 12 mo.; what is the equated time for paying the whole? *A.* $8\frac{4}{5}$ months.

84. How many bricks, 8 inches long, 4 inches wide, and $2\frac{1}{2}$ inches thick, will it take to build a house 84 feet long, 40 feet wide, 20 feet high, and the walls to be 1 foot thick?

The pupil will perceive that he must deduct the width of the wall, that is, 1 foot, from the length of each side, because the inner sides are 1 foot less in length than the outer sides.

A. 105408 bricks.

APPENDIX.

ALLIGATION.

¶ LXXXII. Alligation is the method of mixing several simples of different qualities, so that the compound, or composition, may be of a mean or middle quality.

When the quantities and prices of the several things or simples are given, to find the mean price or mixture compounded of them, the process is called

ALLIGATION MEDIAL.

1. A farmer mixed together 2 bushels of rye, worth 50 cents a bushel, 4 bushels of corn, worth 60 cents a bushel, and 4 bushels of oats, worth 30 cents a bushel: what is a bushel of this mixture worth?

In this example, it is plain, that, if the cost of the whole be divided by the whole number of bushels, the quotient will be the price of one bushel of the mixture.

2 *bushels at* $,50 *cost* $1,00
4 . . . . . . $,60 . . . $2,40
4 . . . . . . $,30 . . . $1,20
—— ——
10 $4,60 $4,60 ÷ 10 = 46 *cts., Ans*

RULE. *Divide the whole cost by the whole number of bushels, &c., the quotient will be the mean price or cost of the mixture.*

2. A grocer mixed 10 cwt. of sugar at $10 per cwt., 4 cwt. at $4 per cwt., and 8 cwt. at $7½ per cwt.: what is 1 cwt. of this mixture worth? and what is 5 cwt. worth? *A.* 1 cwt. is worth $8, and 5 cwt. is worth $40.

3. A composition was made of 5 lbs. of tea, at $1⅓ per lb., 9 lbs. at $1,80 per lb., and 17 lbs. at $1⅛ per lb.: what is a pound of it worth?

A. $1,546$\frac{7}{10}$+.

4. If 20 bushels of wheat, at $1,35 per bushel, be mixed with 15 bushels of rye, at 85 cents per bushel, what will a bushel of this mixture be worth?

A. $1,135$\frac{7}{10}$+.

5. If 4 lbs. of gold of 23 carats fine be melted with 2 lbs. 17 carats fine, what will be the fineness of this mixture? *A.* 21 carats.

ALLIGATION ALTERNATE.

¶ LXXXIII. The process of finding the proportional quantity of each simple from having the mean price or rate, and the mean prices or rates of the several simples given, is called *Alligation Alternate*; consequently, it is the reverse of *Alligation Medial*, and may be proved by it.

1. A farmer has oats, worth 25 cents a bushel, which he wishes to mix with corn, worth 50 cents per bushel, so that the mixture may be worth 30 cents per bushel; what proportions or quantities of each must he take?

In this example, it is plain, that, if the price of the corn had been 35 cents, *that is, had it* exceeded the price of the mixture, (30 cents,) just as much as it falls short, he must have taken equal quantities of each sort; but, since the difference between the price of the corn and the mixture price is 4 times as much as the difference between the price of the oats and the mixture price, consequently, 4 times as much oats as corn must be taken, that is, 4 to 1, or 4 bushels of oats to 1 of corn. But since we determine this proportion by the differences, hence these differences will represent the same proportion.

These are 20 and 5, that is, 20 bushels of oats to 5 of corn, which are the quantities or proportions required. In determining these differences, it will be found convenient to write them down in the following manner:

OPERATION.

	cts.	bushels.	
30	$,25	20	Ans.
	$,50	5	

It will be recollected, that the difference between 50 and 30 is 20, that is, 20 bushels of oats, which must, of course, stand at the right of the 25, the price of the oats, or, in other words, opposite the price that is connected or linked with the 50; likewise the difference between 25 and 30 = 5, that is, 5 bushels of corn, opposite the 50, (the price of the corn.)

The answer, then, is 20 bushels of oats to 5 bushels of corn, or in that proportion.

By this mode of operation, it will be perceived, that there is precisely as much gained by one quantity as there is lost by another, and, therefore, the gain or loss on the whole is equal.

The same will be true of any two ingredients mixed together in the same way. In like manner the proportional quantities of any number of simples may be determined; for, if a less be linked with a greater than the mean price, there will be an equal balance of loss and gain between every two, consequently an equal balance on the whole.

It is obvious, that this principle of operation will allow a great variety of answers, for, having found one answer, we may find as many more as we please, by only multiplying or dividing each of the quantities found by 2, or 3, or 4, &c.; for, if 2 quantities of 2 simples make a balance of loss and gain, as it respects the mean price, so will also the double or treble, the ½, or ⅓ part, or any other ratio of these quantities, and so on to any extent whatever.

PROOF. We will now ascertain the correctness of the foregoing operation by the last rule, thus:

20 bushels of oats, at 25 cents per bushel, = $5,00
5 . . . . . *corn, at 50* . . . . , = $2,50

25   25)7,50(30
       75
       0

Ans. 30 cts., the price of the mixture.

Hence we derive the following

RULE.

I. *Reduce the several prices to the same denomination.*

II. *Connect, by a line, each price that is less than the mean rate, with one or more that is greater, and each price greater than the mean rate with one or more that is less.*

III. *Place the difference between the mean rate and that of each of the simples opposite the price with which they are connected.*

IV. *Then, if only one difference stands against any price, it expresses the quantity of that price; but if there be several, their sum will exvress the quantity.*

2. A merchant has several sorts of tea, some at 10 s., some at 11 s., some at 13 s., and some at 24 s. per lb.; what proportions of each must be taken to make a composition worth 12 s. per lb.?

OPERATIONS

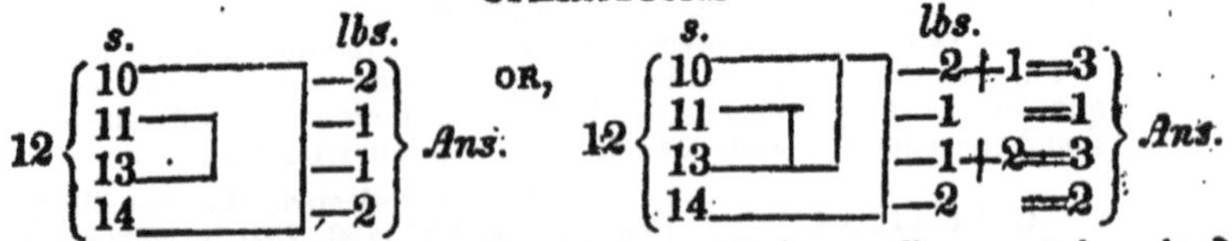

```
     s.          lbs.                       s.          lbs.
   ⎧10 ──────┐  —2 ⎫                      ⎧10 ──┐ ┐  —2+1=3 ⎫
12 ⎨11 ──┐   │  —1 ⎬ Ans.   OR,       12  ⎨11 ──┼─┤  —1   =1 ⎬ Ans.
   ⎪13 ──┘   │  —1 ⎪                      ⎪13 ──┘ │  —1+2=3 ⎪
   ⎩14 ──────┘  —2 ⎭                      ⎩14 ────┘  —2   =2 ⎭
```

3. How much wine, at 5 s. per gallon, and 3 s. per gallon, must be mixed together, that the compound may be worth 4 s. per gallon?

A. An equal quantity of each sort.

4. How much corn, at 42 cents, 60 cents, 67 cents, and 78 cents, per bushel, must be mixed together, that the compound may be worth 64 cents per bushel? *A.* 14 bushels at 42 cents, 3 bushels at 60 cents, 4 bushels at 67 cents, and 22 bushels at 78 cents.

5. A grocer would mix different quantities of sugar; viz. one at 20, one at 23, and one at 26 cents per lb.; what quantity of each sort must be taken to make a mixture worth 22 cents per lb.?

A. 5 at 20 cents, 2 at 23 cents, and 2 at 26 cents.

6. A jeweller wishes to procure gold of 20 carats fine, from gold of 16, 19, 21, and 24 carats fine; what quantity of each must he take?

A. 4 at 16, 1 at 19, 1 at 21, and 4 at 24.

We have seen that we can take 3 times, 4 times, ½, ¼, or any proportion of each quantity, to form a mixture. Hence, when the quantity of one simple is given, to find the proportional quantities of any compound whatever, after having found the proportional quantities by the last rule, we have the following

RULE.

As the PROPORTIONAL QUANTITY *of that price whose quantity is given* : *is to* EACH PROPORTIONAL QUANTITY : : *so is the* GIVEN QUANTITY : *to the* QUANTITIES *or* PROPORTIONS *of the compound required.*

1. A grocer wishes to mix 1 gallon of brandy, worth 15 s. per gallon, with rum, worth 8 s., so that the mixture may be worth 10 s. per gallon; how much rum must be taken?

By the last rule, the differences are 5 to 2; that is, the proportions are 2 of brandy to 5 of rum; hence he must take 2½ gallons of rum for every gallon of brandy. *A.* 2½ gallons.

2. A person wishes to mix 10 bushels of wheat, at 70 cents per bushel, with rye at 48 cents, corn at 36 cents, and barley at 30 cents per bushel, so that a bushel of this mixture may be worth 38 cents; what quantity of each must be taken?

We find by the last rule, that the proportions are 8, 2, 10, and 32.

Then, *as* 8 : 2 : : 10 : 2½ *bushels of rye.*
8 : 10 : : 10 : 12½ *bushels of corn.*
8 : 32 : : 10 : 40 *bushels of barley.* } *Ans.*

3. How much water must be mixed with 100 gallons of rum, worth 90 cents per gallon, to reduce it to 75 cents per gallon? *A.* 20 gallons.

4. A grocer mixes teas at $1,20, $1, and 60 cents, with 20 lbs. at 40 cents per lb.; how much of each sort must he take to make the composition worth 80 cents per lb.? *A.* 20 at $1,20, 10 at $1, and 10 at 60 cents.

5. A grocer has currants at 4 cents, 6 cents, 9 cents, and 11 cents per lb.; and he wishes to make a mixture of 240 lbs., worth 8 cents per lb.; how many currants of each kind must he take? In this example we can find the *proportional quantities* by linking, as before; then it is plain that their sum will be

in the same proportion to any part of their sum, as the whole compound is to any part of the compound, which exactly accords with the principle of Fellowship.

Hence we have the following

RULE.

As the sum of the PROPORTIONAL QUANTITIES *found by linking, as before : is to* EACH PROPORTIONAL QUANTITY : : *so is the* WHOLE QUANTITY *or compound required : to the* REQUIRED QUANTITY *of each.*

We will now apply this rule in performing the last question.

8	4	—3	*Then,*	10 : 3 : : 240 : 72 *lbs., at* 4 *cents.*
	6	—1		10 : 1 : : 240 : 24 *lbs., at* 6 *cents.*
	9	—2		10 : 2 : : 240 : 48 *lbs., at* 9 *cents.*
	11	—4		10 : 4 : : 240 : 96 *lbs., at* 11 *cents.* *Ans*
		10		

1. A grocer, having sugars at 8 cents, 12 cents, and 16 cents per pound, wishes to make a composition of 120 lbs., worth 13 cents per pound, without gain or loss; what quantity of each must be taken?

A. 30 lbs. at 8, 30 lbs at 12, and 60 lbs at 16.

2. How much water, at 0 per gallon, must be mixed with wine, at 80 cents per gallon, so as to fill a vessel of 90 gallons, which may be offered at 50 cents per gallon? *A.* $56\frac{2}{8}$ gallons of wine, and $33\frac{6}{8}$ gallons of water.

3. How much gold, of 15, 17, 18, and 22 carats fine, must be mixed together, to form a composition of 40 ounces of 20 carats fine?

A. 5 oz. of 15, of 17, of 18, and 25 oz. of 22.

INVOLUTION.

¶ **LXXXIV.** Q. How much does 2, multiplied into itself, or by 2, make?

Q. How much does 2, multiplied into itself, or by 2, and that product by 2, make?

Q. When a number is multiplied into itself once or more, in this manner, what is the process called? *A. Involution, or the Raising of Powers.*

Q. What is the number, before it is multiplied into itself, called? *A.* The first power, or root.

Q. What are the several products called? *A.* Powers.

Q. In multiplying 6 by 6, that is, 6 into itself, making 36, we use 6 twice: what, then, is 36 called? *A.* The second power, or square of 6.

Q. What is the 2d power or square of 8? 10? 12? *A.* 64, 100, 144.

Q. In multiplying 3 by 3, making 9, and the 9 also by 3, making 27, we use the three 3 times; what, then, is the 27 called? *A.* The 3d power, or cube of 3.

Q. What is the 3d power of 2? 3? 4? *A.* 8, 27, 64.

Q. What is the figure, or number, called which denotes the power, as, 3d power, 2d power, &c.? *A.* The index or exponent.

Q. When it is required, for instance, to find the 3d power of 3, what is the index, and what is the power? *A.* 3 is the index, 27 the power.

Q. This index is sometimes written over the number to be multiplied, thus, 3^2; what, then, is the power denoted by 2^4? *A.* $2 \times 2 \times 2 \times 2 = 16$.

Q. When a figure has a small one at the right of it, thus, 6^5, what does it mean? *A.* The 5th power of 6, or that 6 must be raised to the 5th power.

1. How much is 12^2, or the square of 12? *A.* 144

2. How much is 4^2, or the square of 4? *A.* 16

3. How much is 10^2, or the square of 10? *A.* 100.

4. How much is 4^3, or the cube of 4? *A.* 64.

5. How much is 1^4, or the 4th power of 1? *A.* 1.
6. What is the biquadrate or 4th power of 3? *A.* 81.
7. What is the square of $\frac{1}{2}$? $\frac{2}{3}$? *A.* $\frac{1}{4}$, $\frac{4}{9}$.
8. What is the cube of $\frac{1}{2}$? $\frac{2}{3}$? $\frac{1}{4}$? *A.* $\frac{1}{8}$, $\frac{8}{27}$, $\frac{1}{64}$
9. What is the square of ,5? 1,2? *A.* ,25; 1,44.
10. Involve 2 to the 2d power; 2 to the 3d power. *A.* 4, 8
11. Involve $\frac{1}{8}$ to the 2d power; $\frac{2}{10}$ to the 2d power. *A.* $\frac{1}{64}$,
12. Involve $\frac{4}{8}$, to the 2d power. *A.* $\frac{16}{64}=\frac{1}{4}$.
13. Involve $\frac{2}{3}$ to the 2d power. *A.* $\frac{4}{9}$.

14. What is $\frac{1}{2}^2$, or the square of $\frac{1}{2}$? *A.* $\frac{1}{4}$.

15. What is the value of $\frac{2}{10}^2$? *A.* $\frac{1}{25}$.

16. What is the value of $\frac{1}{3}^4$? *A.* $\frac{1}{81}$.

Exercises for the Slate.

1. What is the square of 900? *A.* 810000.
2. What is the cube of 211? *A.* 9393931.
3. What is the biquadrate or 4th power of 80? *A.* 40960000.
4. What is the sursolid or the 5th power of 7? *A.* 16807.
5. Involve $\frac{11}{12}$, $\frac{8}{9}$, $\frac{4}{39}$, each to the 3d power. *A.* $\frac{13}{17}$ $\frac{64}{59319}$.
6. What is the square of $5\frac{1}{2}$? *A.* $30\frac{1}{4}$.
7. What is the square of $16\frac{1}{2}$? *A.* $272\frac{1}{4}$.

8. What is the value of 8^5? *A.* 32768.

9. What is the value of 10^4? *A.* 10000

10. What is the value of 6^4? *A.* 1296.
11. What is the cube of 25? *A.* 15625.

EVOLUTION.

¶ **LXXXV.** *Q.* What number, multiplied into itself, v

that is, what is the first power or root of 16^2? *A.* 4.

Q. Why? *A.* *Because* $4\times4=16$.

Q. What number, multiplied into itself three times, will mal

what is the 1st power or root of 27^3? *A.* 3.

Q. Why? *A.* *Because* $3\times3\times3=27$.

Q. What, then, is the method of finding the first powers or r

&c. powers called? *A.* Evolution, or the Extraction of Roots

Q. In Involution we were required, with the first power or root being given, to find higher powers, as 2d, 3d, &c. powers; but now it seems, that, with the 2d, 3d, &c. powers being given, we are required to find the 1st power or root again; how, then, does Involution differ from Evolution? *A. It is exactly the opposite of Involution.*

Q. How, then, may Evolution be defined? *A. It is the method of finding the root of any number.*

1. What is the square root of 144? *A.* 12.

Q. Why? *A.* Because $12 \times 12 = 144$.

2. What is the cube root of 27? *A.* 3.

Q. Why? *A.* Because $3 \times 3 \times 3 = 27$.

3. What is the biquadrate root of 81? *A.* 3.

Q. Why? *A.* Because $3 \times 3 \times 3 \times 3 = 81$.

We have seen, that any number may be raised to a perfect power by Involution; but there are many numbers of which precise roots cannot be obtained; as, for instance, the square root of 3 cannot be exactly determined, there being no number, which, by being multiplied into itself, will make 3. By the aid of decimals, however, we can come nearer and nearer; that is, approximate towards the root, to any assigned degree of exactness. Those numbers, whose roots cannot exactly be determined, are called SURD ROOTS, and those, whose roots can exactly be determined, are called RATIONAL ROOTS.

To show that the square root of a number is to be extracted, we prefix this character, $\sqrt{\ }$. Other roots are denoted by the same character with the index of the required root placed before it. Thus $\sqrt{9}$ signifies that the square root of 9 is to be extracted; $\sqrt[3]{27}$ signifies that the cube root of 27 is to be extracted; $\sqrt[4]{64}$ = the 4th root of 64.

When we wish to express the power of several numbers that are connected together by these signs, +, —, &c., a vinculum or parenthesis is used, drawn from the top of the sign of the root, and extending to all the parts of it; thus, the cube root of 30 — 3 is expressed thus, $\sqrt[3]{30-3}$, &c.

EXTRACTION OF THE SQUARE ROOT.

¶ **LXXXVI.** Q. We have seen (¶ LXXXV.) that the root of any number is its 1st power; also that a square is the 2d power: what, then, is to be done, in order to find the 1st power; that is, to extract the square root of any number?

A. It is only to find that number, which, being multiplied into itself, will produce the given number.

Q. We have seen (¶ LXXIX.) that the process of finding the contents of a square consists in multiplying the length of one side into itself; when, then, the contents of a square are given, how can we find the length of each side; or, to illustrate it by an example, If the contents of a square figure be 9 feet, what must be the length of each side? *A.* 3 feet.

Q. Why? *A.* Because 3 ft. × 3 ft. = 9 square feet.

Q. What, then, is the difference in contents between a square figure whose sides are each 9 feet in length, and one which contains only 9 square feet?

$9 \times 9 = 81 - 9 = 72$, *Ans.*

Q. What is the difference in contents between a square figure containing 3 square feet, and one whose sides are each 3 feet in length? *A.* 6 square feet.

Q. What is the square root of 144? or what is the length of each side of a figure, which contains 144 square feet? *A.* 12 square feet.

Q. Why? *A.* Because $12 \times 12 = 144$.

Q. How, then, may we know if the root or answer be right? *A. By multiplying the root into itself; if it produces the given number, it is right.*

Q. If a square garden contains 16 square rods, how many rods does it measure on each side? and why?

A. 4 rods. Because 4 rods $\times$ 4 rods $=$ 16 square rods

1. What is the square root of 64? and why?
2. What is the square root of 100? and why?
3. What is the square root of 49? and why?
4. Extract the square root of 144.
5. Extract the square root of 36.
6. What is the square root of 3600?
7. What is the square root of ,25? *A.* ,5.
8. What is the square root of 1,44? *A.* 1,2.
9. What is the value of $\sqrt{25}$? or, what is the square root of 25?
10. What is the value of $\sqrt{,4}$? *A.* ,2.
11. What is the square root of $\frac{1}{4}$? *A.* $\frac{1}{2}$.
12. What is the value of $\sqrt{\frac{4}{9}}$? *A.* $\frac{2}{3}$.
13. What is the square root of $\frac{1}{4}$ of $\frac{1}{9}$? *A.* $\frac{1}{6}$.
14. What is the square root of $6\frac{1}{4}$? $\sqrt{6\frac{1}{4}} = \sqrt{\frac{25}{4}} = \frac{5}{2} = 2\frac{1}{2}$, *Ans*
15. What is the value of $\sqrt{\frac{2}{3}}$ of $\frac{3}{8}$? *A.* $\frac{1}{2}$.
16. What is the square root of $30\frac{1}{4}$?
17. What is the difference between the square root of 4 and the square of 4? or, which is the same thing, what is the difference between $\sqrt{4}$ and 4^2?

$\sqrt{4} = 2$, *and* $4^2 = 16$; *then*, $16 - 2 = 14$, *Ans*

18. What is the difference between $\sqrt{9}$ and 9^2?
19. What is the difference between $\sqrt{16}$ and $\sqrt{9}$?
20. What is the difference between $\sqrt{\frac{1}{81}}$ and $\frac{1}{3}^2$? *A.* 0.
21. There is a square room, which is calculated to accommodate 100 scholars; how many can sit on one side?
22. If 400 boys, having collected together to perform some military evolutions, should wish to march through the town in a solid phalanx, or square body, of how many must the first rank consist?
23. A general has 400 men; how many must be placed in rank and file to form them into a square?
24. A certain square pavement contains 1600 square stones, all of the same size; I demand how many are contained in one of its sides? *A.* 40.
25. A man is desirous of making his kitchen garden, containing $2\frac{1}{2}$ acres, or 400 rods, a complete square; what will be the length of one side?
26. A square lot of land is to contain $22\frac{1}{2}$ acres, or 3600 rods of ground; but, for the sake of fruit, there is to be a smaller square within the larger, which is to contain 225 rods: what is the length of each side of both squares?

A. 60 rods the outer, 15 rods the inner.

Exercises for the Slate.

1. If a square field contains 6400 square rods, how many rods in length does it measure on each side? *A.* 80 rods.
2. How many trees in each row of a square orchard, which contains 2500 trees? *A.* 50 trees
3. A general has a brigade consisting of 10 regiments, each regiment of 10 companies, and each company of 100 men: how many must be placed in rank and file, to form them in a complete square? *A.* 100 men.
4. What is the square root of 2500? *A.* 50.
5. What is the 1st power of 1000000^2? *A.* 1000.
6. What is the value of $\sqrt{360000}$? *A.* 600.

7. What is the difference between the square root of 36 and the square of 36? *A.* 1290.

8. What is the difference between $\sqrt{4900}$ and 4900^2? *A.* 24009930.

9. What is the difference between $\sqrt{81}$ and 81^2? *A.* 6552.

10. What is the difference between $\sqrt{\frac{4}{36}}$ and $\frac{4}{36}^2$?

$\sqrt{\frac{4}{36}} = \frac{2}{6}$, *and* $\frac{4}{36}^2 = \frac{16}{1296}$; *then,* $\frac{2}{6} - \frac{16}{1296} = \frac{416}{1296} = \frac{26}{81}$, *Ans.*

11. What is the difference between $\sqrt{\frac{9}{16}}$ and $\frac{9}{16}^2$? *A.* $\frac{111}{256}$.

12. What is the amount of $\sqrt{4}$ and $\sqrt{9}$? *A.* 5.

13. What is the sum of $\sqrt{4}$ and 9^2? *A.* 83.

14. What is the amount of $\sqrt{30\frac{1}{4}}$ and $\sqrt{272\frac{1}{4}}$? *A.* 22.

15. What is the length of one side of a square garden, which contains 1296 square rods? in other words, what is the square root of 1296?

In this example we have a little difficulty in ascertaining the root. This, perhaps, may be obviated by examining the following figure, (which is in the form of the garden, and supposed to contain 1296 square rods,) and carefully noting down the operation as we proceed.

OPERATIONS.

1st.	2d.
square rods.	*square rods.*
30)1296(30	3)1296(36
900	9
$60+6=66$)396(6	66)396
396	396
0	0

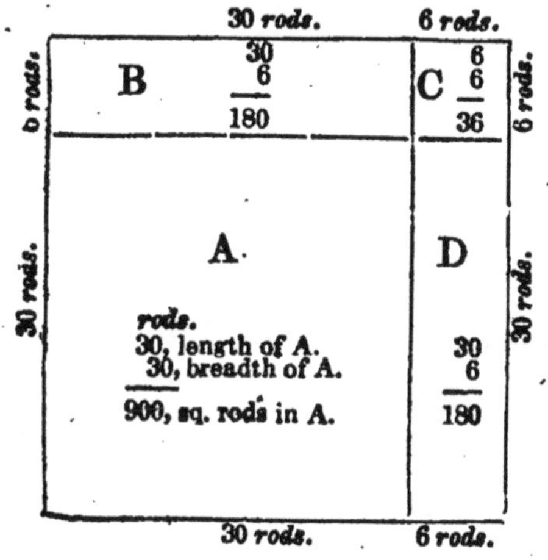

In this example we know that the root, or the length of one side of the garden, must be greater than 30, for $30^2 = 900$, and less than 40, for $40^2 = 1600$, which is greater than 1296; therefore, we take 30, the less, and, for convenience' sake, write it at the left of 1296, as a kind of divisor, likewise at the right of 1296, in the form of a quotient in division; (*See Operation 1st.*); then, subtracting the square of 30, $= 900$ sq. rods, from 1296 sq. rods, leaves 396 sq. rods.

FIG.

The pupil will bear in mind, that the FIG. on the left is in the form of the garden, and contains the same number of square rods, viz. 1296. This figure is divided into parts, called A, B, C, and D. It will be perceived, that the 900 square rods, which we deducted, are found by multiplying the length of A, being 30 rods, by the breadth, being also 30 rods, that is, $30^2 = 900$.

To obtain the square rods in B, C, and D, the remaining parts of the figure, we may multiply the length of each by the breadth of each, thus; $30 \times 6 = 180$, $6 \times 6 = 36$, and $30 \times 6 = 180$; then, $180 + 36 + 180 = 396$ square rods: or, add the length of B, that is, 30, to the length of D, which is also 30, making 60; or, which is the same thing, we may double 30, making 60; to this add the length of C, 6 rods, and the sum is 66.—Now, to obtain the square rods in the whole length of B, C, and D, we multiply their length, 6 rods, by the breadth of each side, thus; $66 \times 6 = 396$ square rods, the same as before.

We do the same in the operation; that is, we first double 30 in the quotient, and add the 6 rods to the sum, making 66, for a divisor; next, multiply 66, the divisor, by 6 rods, the width, making 396; then, taking 396 from 396 leaves 0.

The pupil will perceive, the only difference between the 1st and 2d operation (which see) is, that in the 2d we neglect writing the ciphers at the right of the numbers, and use only the significant figures. Thus, for 30 + 6, we write 3 (tens) and 6 (units), which, joined together, make 36; for 900, we write 9, (hundreds). This is obvious from the fact, that the 9 retains its place under the 2, (hundreds). Instead of 60 + 6, we write 66; omitting the ciphers in this manner cannot reasonably make any difference, and, in fact, it does not, for the result is the same in both.

By neglecting the ciphers, we may, perhaps, be at a loss, sometimes, to determine where we must place the square number. In the last example, we knew where the square of the root 3 (tens) = 9 (hundreds) should be placed, for the ciphers, at the right, indicate it; but had these ciphers been dropped, we should, doubtless, have hesitated in assigning the 9 its proper place. This difficulty will be obviated by observing what follows.

The square of any number never contains but twice as many, or at least but one figure less than twice as many, figures as are in the root. Thus, the square of the root 30 is 900; now, in 900 there are but 3 figures, and in 30, 2 figures; that is, the square of 30 contains but 1 figure more than 30. We will take 99, whose square is 9801, in which there are 4 figures, and in its root, 99, but 2; that is, there are exactly twice as many figures in the square 9801 as are in its root, 99. This will be equally true of any numbers whatever.

Hence, to know where to place the several square numbers, we may point off the figures in the given number into periods of two figures each, commencing with the units, and proceeding towards the left. And, since the value of both whole numbers and decimals is determined by their distance from the units' place, consequently, when there are decimals in the given number, we may begin at the units' place, and point off the figures towards the right, in the same manner as we point off whole numbers towards the left.

By each of the preceding operations, then, we find that the root of 1296 is 36, or, in other words, the length of each side of the garden is 36 rods.

PROOF. This work may now be proved by adding together all the square rods contained in the several parts of the figure, thus:—

A *contains* 30 × 30 = 900 *square rods*
B ...... 30 × 6 = 180 ....
C ...... 6 × 6 = 36 ....
D ...... 30 × 6 = 180 ....
1296 *square rods.*

OR, *by Involution,*
36 × 36 = 1296, *Ans., as before.*

From these illustrations we derive the following

RULE.

I. *Point off the given number into periods of two figures each, by putting a dot over the units, another over the hundreds, and so on; and, if there are decimals, point them in the same manner, from units towards the right hand. These dots show the number of figures of which the root will consist.*

II. *Find the greatest square number in the left hand period, and write its root as a quotient in division; subtract the square number from the left hand period, and to the remainder bring down the next right hand period for a dividend.*

III. *Double the root, (quotient figure,) already found, and place it at the left of the dividend for a divisor.*

IV. *Write such a figure at the right hand of the divisor; also the same figure in the root, as, when multiplied into the divisor thus increased, the product shall be equal to, or next less than the dividend. This quotient figure will be the second figure in the root.*

Note. The figure last described, at the right of the divisor, in the second operation, is the 6 rods, the width, which we add to 60, making 66; or, omitting the 0 in 60, and annexing 6, then multiplying 66 by 6, we wrote the 6 in the quotient, at the right of 3, making 36.

V. *Multiply the whole increased divisor by the last quotient figure, and write the product under the dividend.*

VI. *Subtract this product from the dividend, and to the remainder bring down the next period, for a new dividend.*

VII. *Double the quotient figures, that is, the root already found, and continue the operation as before, till all the periods are brought down.*

More Exercises for the Slate.

16. What is the square root of 65536?

OPERATION.

```
2)65536(256, Ans.
  4
  ---
45)255
   225
   ----
506)3036
    3036
    ----
```

PROOF.

```
  256
  256
 ----
 1536
1280
512
-----
65536
```

17. What is the square root of 6480,25?

OPERATION.

```
8)6480,25(80,5, Ans.
  64
  ---
1605)80,25
     80,25
     -----
     0000
```

PROOF.

```
 80,5
 80,5
 ----
 4025
6440
-------
6480,25
```

18. What is the square root of 470596? *A.* 686.
19. What is the square root of 1048576? *A.* 1024.
20. What is the square root of 2125764? *A.* 1458
21. What is the square root of 6718464? *A.* 2592.
22. What is the square root of 23059204? *A.* 4802.
23 What is the square root of 4294967296? *A.* 65536.
24. What is the square root of 40?

In this example we have a remainder, after obtaining one figure in the root. In such cases we may continue the operation to decimals, by annexing two ciphers for a new period, and thus continue the operation to any assignable degree of exactness. But since the last figure, in every dividend thus formed, will always be a cipher, and as there is no figure under 10 whose square number ends in a cipher, there will, of course, be a remainder; consequently, the pupil need not expect, should he continue the operation to any extent, ever to obtain an exact root. This, however, is by no means necessary; for annexing only one or two periods of ciphers will obtain a root sufficiently exact for almost any purpose. *A.* 6,3245+.

25. What is the square root of 30? *A.* 5,4772.

26. What is the square root of $\frac{64}{144}$? *A.* $\frac{8}{12}=\frac{2}{3}$.

Or, we may reduce the given fraction to its lowest terms before the root is extracted.

Thus, $\sqrt{\frac{64}{144}}=\sqrt{\frac{4}{9}}=\frac{2}{3}$ *Ans., as before*

27. What is the square root of $\frac{450}{2048}$? *A.* $\frac{15}{32}$.

28. What is the square root of $\frac{224}{350}$? *A.* $\frac{4}{5}$.

29. What is the square root of $\frac{121}{12345321}$? *A.* $\frac{1}{101}$.

If the fraction be a surd, the easiest method of proceeding will be to reduce it to a decimal first, and extract its root afterwards.

30. What is the square root of $\frac{70}{84}$? *A.* ,9128+.

31. What is the square root of $\frac{11}{12}$? *A.* ,9574+

32. What is the square root of $\frac{9}{13}$? *A.* ,83205.

33. What is the square root of $420\frac{1}{4}$?

In this example, it will be best to reduce the mixed number to an improper fraction, before extracting its root, after which it may be converted into a mixed number again. *A.* $20\frac{1}{2}$.

34. What is the square root of $912\frac{1}{25}$? *A.* $30\frac{1}{5}$.

35. A general has an army of 5625 men; how many must he place in rank and file, to form them into a square? $\sqrt{5625}=75$, *Ans.*

36. A square pavement contains 24336 square stones of equal size; how many are contained in one of its sides? *A.* 156.

37. In a circle, whose area, or superficial contents, is 4096 feet, I demand what will be the length of one side of a square containing the same number of feet? *A.* 64 feet.

38. A gentleman has two valuable building spots, one containing 40 square rods, and the other 60, for which his neighbour offers him a square field, containing 4 times as many square rods as the building spots; how many rods in length must each side of this field measure? $\sqrt{40+60\times4}=20$, *Ans.*

39. How many trees in each row of a square orchard, containing 14400 trees? *A.* 120 trees.

40. A certain square garden spot measures 4 rods on each side; what will be the length of one side of a garden containing 4 times as many square rods? *A.* 8 rods.

41. If one side of a square piece of land measure 5 rods, what will the side of one measure, which is 4 times as large? 16 times as large? 36 times as large? *A.* 10, 20, 30.

42. A man is desirous of forming a tract of land, containing 140 acres, 2 roods and 20 rods into a square, what will be the length of each side? *A.* 150 rods.

43. The distance from Providence to Norwich, (Conn.) is computed to be 45 miles; now, allowing the road to be 4 rods wide, what will be the length of one side of a square lot of land, the square rods of which shall be equal to the square rods contained in said road? *A.* 240 rods.

EXTRACTION OF THE CUBE ROOT.

¶ **LXXXVII.** Q. Involution, (¶ LXXXIV.,) you doubtless recollect, is the raising of powers; can you tell me what is the 3d power of 3, and what the power is called? *A.* 27, called a cube.

Q. Evolution, (¶ LXXXVII.,) was defined to be the extracting the 1st power, or roots of higher powers; can you tell me, then, what is the cube root of 27? *A.* 3.

Q. Why? *A.* Because $3 \times 3 \times 3$, or, expressed thus, $3^3 = 27$.

Q. What, then, is it to extract the cube root of any number? *A.* It is only to find that number, which, being multiplied into itself three times, will produce the given number.

Q. We have seen, (¶ LXXX.,) that, to find the contents of solid bodies, such as wood, for instance, we multiply the length, breadth and depth together. These dimensions are called cubic, because, by being thus multiplied, they do in fact contain so many solid feet, inches, &c. as are expressed by their product; but what do you suppose the shape of a solid body is, which is an exact cube? *A.* It must have six equal sides, and each side must be an exact square. *See block A, which accompanies this work.*

Q. Now, since the length, breadth and thickness of any regular cube are exactly alike, as, for instance, a cubical block, which contains 27 cubic feet, can you inform me what is the length of one side of this block, and what the length may be called? *A.* Each side is 3 feet, and may be called the cube root of 27.

Q. Why? *A.* Because $3^3 = 27$.

Q. What is the length of each side of a cubical block containing 64 cubic inches? *A.* 4 inches.

Q. Why? *A.* Because $4 \times 4 \times 4$, or $4^3 = 64$ cubic inches.

Q. What is the cube root of 64, then? *A.* 4.

Q. Why? *A.* Because $4^3 = 64$.

Q. What is the length of each side of a cubical block containing 1000 cubic feet? *A.* 10.

Q. Why? *A.* Because $10^3 = 1000$.

1. In a square box which will contain 1000 marbles, how many will it take to reach across the bottom of the box, in a straight row? *A.* 10.

2. What is the difference between the cube root of 27 and the cube of 3? *A.* 24.

3. What is the difference between $\sqrt[3]{8}$ and 2^3? *A.* 6.

4. What is the difference between $\sqrt[3]{1}$ and 1^3? *A.* 0.

5. What is the difference between the cube root of 27 and the square root of 9? *A.* 0.

6. What is the difference between $\sqrt[3]{8}$ and $\sqrt{4}$? *A.* 0.

Operation by Slate illustrated.

7. A man, having a cubical block containing 13824 cubic feet, wishes to know the length of each side, without measuring it; what is the length of each side of said block?

Should we attempt to illustrate the reason of the rule for extracting the cube root by exhibiting the picture of the cube and its various parts on paper, it would tend rather to confuse than illustrate the subject. The best method of doing it is, by making several small blocks, which may be supposed to contain a certain proportional number of feet, inches, &c. corresponding with the operation of the rule. They may be made in a few minutes, from a small strip of a pine board, with a common penknife, at the longest, in less time than the teacher can make the pupil comprehend the reason, from merely seeing the picture on paper. In demonstrating the rule in this way, it will be an amusing and instructive exercise, to both teacher and pupil, and may be comprehended by any pupil, however young, who is so fortunate as to have progressed as far as this rule. It will give him distinct ideas respecting the different dimensions of square and cubic measures, and indelibly fix on his mind the reason of the rule, consequently the rule itself. But for the convenience of teachers, blocks, illustrative of the operation of the foregoing example, will accompany this work.

The following are the supposed proportional dimensions of the several blocks used in the demonstration of the above example, which, when put together, *ought* to make an exact cube, containing 13824 cubic feet.

One block, 20 feet long, 20 feet wide, and 20 feet thick; this we will call A.

Three small blocks, each 20 feet long, 20 feet wide, and 4 feet thick; each of these we will call B.

Three smaller blocks, each 20 feet long, 4 feet wide, and 4 feet thick; each of these we will call C.

One block, and the smallest, 4 feet long, 4 feet wide, and 4 feet thick; this we will call D.

We are now prepared to solve the preceding example.

In this example, you recollect, we were to find the length of one side of the cube, containing 13824 cubic feet.

OPERATION 1st.

```
                        ft.     13824(20, root.
                        20³ =  8000
2×2×300 = divisor, 1200)5824(4
                     quot. 4|
                        4800|
        2×30×4×4 =  960|
           4×4×4 =   64|
                        5824 deducted.
                        0000
```

Or,

The same operation, by neglecting the ciphers, may be performed thus:—

OPERATION 2d.

```
              13824(20+4, or 24, { root,
               8                  { Ans.
2×2×300 = 1200)5824, dividend.
               |4800
2×30×4×4 =     | 960
   4×4×4 =     |  64
               5824, subtrahend.
               0000
```

In this example, we know that one side cannot be 30 feet, for $30^3 = 27000$ solid feet, being more than 13824, the given sum; therefore, we will take 20 for the length of one side of the cube.

Then, $20 \times 20 \times 20 = 8000$ solid feet, which we must, of course, deduct from 13824, leaving 5824. (See *operation 1st.*) These 8000 solid feet, the pupil will perceive, are the solid contents of the cubical block, marked A. This corresponds with the operation; for we write 20 feet, the length of the cube A, at the right of 13824, in the form of a quotient; and its square, 8000, under 13824; from which subtracting 8000, leaves 5824, as before.

As we have 5824 cubic feet remaining, we find the sides of the cube A are not so long as they ought to be; consequently we must enlarge A; but in doing this, we must enlarge the three sides of A, in order that we may preserve the cubical form of the block. We will now place the three blocks, each of which is marked B, on the three sides of A Each of these blocks, in order to fit, must be as long and as wide as A; and, by examining them, you will see that this is the case; that is, 20 feet long and 20 feet wide; then $20 \times 20 = 400$, the square contents in one B, and $3 \times 400 = 1200$, square contents in 3 Bs; then it is plain, that 5824 solid contents, divided by 1200, the square contents, will give the thickness of each block. But an easier method is, to square the 2, (tens,) in the root 20, making

4, and multiply the product, 4, by 300, making 1200, a divisor, the same as before.

We do the same in the operation, (which see;) that is, we multiply the square of the quotient figure, 2, by 300, thus, $2 \times 2 = 4 \times 300 = 1200$; then the divisor, 1200, (the square contents,) is contained in 5824 (solid contents) 4 times; that is, 4 feet is the thickness of each block marked B. This quotient figure, 4, we place at the right of 5824, and then 1200 square feet $\times$ 4 feet, the thickness, $= 4800$ solid feet.

If we now examine the block, thus increased by the addition of the 3 Bs, we shall see that there are yet 3 corners not filled up: these are represented by the 3 blocks, each marked C, and each of which, you will perceive, is as long as either of the Bs, that is, 20 feet, being the length of A, which is the 20 in the quotient. Their thickness and breadth are the same as the thickness of the Bs, which we found, by dividing, to be 4 feet, the last quotient figure. Now, to get the solid contents of each of these Cs, we multiply their thickness (4 feet) by their breadth (4 feet), $= 16$ square feet; that is, the square of the last quotient figure, 4, $= 16$; these 16 square contents must be multiplied by the length of each, (20 feet,) or, as there are 3, by $3 \times 20 = 60$; or, which is easier in practice, we may multiply the 2, (tens,) in the root, 20, by 30, making 60, and this product by $4^2 = 16$, the square contents $= 960$ solid feet.

We do the same in the operation, by multiplying the 2 in 20 by $30 = 60 \times 4 \times 4 = 960$ solid feet, as before; this 960 we write under the 4800, for we must add the several products together by and by, to know if our cube will contain all the required feet.

By turning over the block, with all the additions of the blocks marked B and C, which are now made to A, we shall spy a little square space, which prevents the figure from becoming a complete cube. The little block for this corner is marked D, which the pupil will find, by fitting it in, to exactly fill up this space. This block, D, is exactly square, and its length, breadth, and thickness are alike, and, of course, equal to the thickness and width of the Cs, that is, 4 feet, the last quotient figure; hence, 4 ft. $\times$ 4 ft. $\times$ 4 ft. $= 64$ solid feet in the block D; or, in other words, the cube of 4, (the quotient figure,) which is the same as $4^3 = 64$, as in the operation. We now write the 64 under the 960, that this may be reckoned in with the other additions.

We next proceed to add the solid contents of the Bs, Cs and D together, thus, $4800 + 960 + 64 = 5824$, precisely the number of solid feet which we had remaining after we deducted 8000 feet, the solid contents of the cube A.

If, in the operation, we subtract the amount, 5824, from the remainder or dividend, 5824, we shall see that our additions have taken all that remained after the first cube was deducted, there being no remainder.

The last little block, when fitted in, as you saw, rendered the cube complete, each side of which we have now found to be $20 + 4 = 24$ feet long, which is the cube root of 13824 (solid feet); but let us see if our cube contains the required number of solid feet.

PROOF.

1 A, $= 8000$ *solid feet.*
3 Bs, $= 4800$ *solid feet.*
3 Cs, $= 960$ *solid feet.*
1 D, $= 64$ *solid feet.*

13824 { *solid feet in the given cube.*

OR,
by Involution,

$24 \times 24 \times 24 = 13824$ { *solid feet, as before.*

In operation 2d, we see, by neglecting the ciphers at the right of 8, the 8 is still 8000, by its standing under 3 (thousands); hence, we may point off three figures by placing a dot over the units, and another over the thousands, and so on.

From the preceding example and illustrations we derive the following

RULE.

I. *Divide the given number into periods of three figures each, by placing a point over the unit figure, and every third figure from the place of units to the left, in whole numbers, and to the right in decimals.*

II. *Find the greatest cube in the left hand period, and place its root in the quotient.*

III. *Subtract the cube thus found from the said period, and to the remainder bring down the next period, and call this the dividend.*

IV. *Multiply the square of the quotient by 300, calling it the divisor.*

V. *Find how many times the divisor is contained in the dividend, and place the result in the root (quotient) ; then multiply the divisor by this quotient figure, placing the product under the dividend.*

VI. *Multiply the former quotient figure, or figures, by 30, and this product by the square of the last quotient figure, and place the product under the last ; under these two products place the cube of the last quotient figure, and call their amount the subtrahend.*

VII. *Subtract the subtrahend from the dividend, and to the remainder bring down the next period for a new dividend, with which proceed as before, and so on, until the whole is finished.*

Note 1. When the subtrahend happens to be larger than the dividend, the quotient figure must be made one less, and we must find a new subtrahend. The reason why the quotient figure will be sometimes too large is, because this quotient figure merely shows the width of the three first additions to the original cube ; consequently, when the subsequent additions are made, the width (quotient figure) may make the solid contents of all the additions more than the cubic feet in the dividend, which remain after the solid contents of the original cube are deducted.

2. When we have a remainder, after all the periods are brought down, we may continue the operation by annexing periods of ciphers, as in the square root.

3. When it happens that the divisor is not contained in the dividend, a cipher must be written in the quotient, (root,) and a new dividend formed by bringing down the next period in the given sum

More Exercises for the Slate.

8. What is the cube root of 9663597?

OPERATION.

9663597(213, *Ans.*

$2^3 = 8$

$2^2 \times 300 = 1200$)1663 *first dividend.*

1200

$2 \times 30 \times 1^2 = 60$

$1^3 = 1$

1261 *first subtrahend.*

$21^2 \times 300 = 132300$)402597 *second dividend.*

396900

$21 \times 30 \times 3^2 = 5670$

$3^3 = 27$

402597 *second subtrahend.*

000000

9. What is the cube root of 17576? *A.* 26.
10. What is the cube root of 571787? *A.* 83.
11. What is the cube root of 970299? *A.* 99.
12. What is the cube root of 2000376? *A.* 126.
13. What is the cube root of 3796416? *A.* 156.
14. What is the cube root of 94818816? *A.* 456.
15. What is the cube root of 175616000? *A.* 560.
16. What is the cube root of 748613312? *A.* 908.
17. What is the cube root of 731189187729? *A.* 9009.
18. What is the cube root of $\frac{27}{64}$? *A.* $\frac{3}{4}$.
19. What is the cube root of $\frac{216}{343}$? *A.* $\frac{6}{7}$.
20. What is the cube root of $\frac{4913}{9261}$? *A.* $\frac{17}{21}$.

If the root be a surd, reduce it to a decimal before its root is extracted, as in the square root.

21. What is the cube root of $\frac{1}{400}$? *A.* .13+.
22. What is the cube root of $\frac{1}{25}$? *A.* .34+.
23. What is the length of one side of a cubical block, which contains 1728 solid or cubic inches? *A.* 12.
24. What will be the length of one side of a cubical block, whose contents shall be equal to another block 32 feet long, 16 feet wide, and 8 feet thick?

$\sqrt[3]{32 \times 16 \times 8} = 16$ feet, *Ans.*

25. There is a cellar dug, which is 16 feet long, 12 feet wide, and 12 feet deep; and another, 63 feet long, 8 feet wide, and 7 feet deep; how many solid or cubic feet of earth were thrown out, and what will be the length of one side of a cubical mound, which may be formed from said earth? *A.* 5832; 18.

26. How many solid inches in a cubical block, which measures 1 inch on each side? How many in one measuring 2 inches on each side? 3 inches on each side? 4 inches on each side? 6 inches on each side? 10 inches on each side? 20 inches on each side? *A.* 1, 8, 27, 64, 216, 1000, 8000.

27. What is the length of one side of a cubical block, which contains 1 solid or cubic inch? 8 solid inches? 27 solid inches? 64 solid inches? 125 solid inches? 216 solid inches? 1000 solid inches? 8000 solid inches?

A. 1, 2, 3, 4, 5, 6, 10, 20.

By the two preceding examples we see that the sides of the cube are as the cube roots of their solid contents, and their solid contents as the cubes of their sides. It is likewise true, that the solid contents of all similar figures are in proportion to each other as the cubes of their several sides or diameters.

Note. The relative length of the sides of cubes, when compared with their solid contents, will be best illustrated by reference to the cubical blocks, accompanying this work.

28. If a ball, 3 inches in diameter, weigh 4 pounds, what will a ball of the same metal weigh, whose diameter is 6 inches?

$3^3 : 6^3 :: 4 : 32$: *Ratio*, $2^3 \times 4 = 32$ lbs., *Ans.*

29. If a globe of silver, 3 inches in diameter, be worth $160, what is the value of one 6 inches in diameter? $3^3 : 6^3 ::$ $160 : $1280, *Ans.*

30. There are two little globes; one of them is 1 inch in diameter, and the other 2 inches; how many of the smaller globes will make one of the larger?

A. 8

31. If the diameter of the planet Jupiter is 12 times as much as the diameter of the earth, how many globes of the earth would it take to make one as large as Jupiter? *A.* 1728.

32. If the sun is 1000000 times as large as the earth, and the earth is 8000 miles in diameter, what is the diameter of the sun? *A.* 800000 miles.

Note. The roots of most powers may be found by the square and cube roots only; thus the square root of the square root is the biquadrate, or 4th root, and the sixth root is the cube of this square root

ARITHMETICAL PROGRESSION.

¶ **LXXXVIII.** Any rank or series of numbers more than 2, increasing by a constant addition, or decreasing by a constant subtraction of some given number, is called an *Arithmetical Series*, or *Progression*.

The number which is added or subtracted continually is called the *common difference.*

When the series is formed by a continual addition of the common difference, it is called an *ascending series*; thus,

2, 4, 6, 8, 10, &c., is an ascending arithmetical series; but

10, 8, 6, 4, 2, &c., is called a *descending arithmetical series*, because it is formed by a continual subtraction of the common difference, 2.

The numbers which form the series are called the *terms* of the series or progression. The first and last terms are called the *extremes*, and the other terms the *means*.

In Arithmetical Progression there are reckoned 5 terms, any three of which being given, the remaining two may be found, viz.

1. *The first term.*
2. *The last term.*
3. *The number of terms.*
4. *The common difference.*
5. *The sum of all the terms.*

The first term, the last term, and the number of terms, being given, to find the common difference;—

1. A man had 6 sons, whose several ages differed alike; the youngest was 3 years old, and the oldest 28; what was the common difference of their ages?

The difference between the youngest son and the eldest evidently shows the increase of the 3 years by all the subsequent additions, till we come to 28 years; and, as the number of these additions are, of course, 1 less than the number of sons, (5), it follows, that, if we divide the whole difference (28—3 =), 25, by the number of additions, (5), we shall have the difference between each one separately, that is, the common difference.

Thus, 28—3 = 25; *then,* 25 ÷ 5 = 5 *years, the common difference.* *A.* 5 yrs.

Hence, to find the common difference;—

Divide the difference of the extremes by the number of terms, less* 1, *and the quotient will be the common difference.

2. If the extremes be 3 and 23, and the number of terms 11, what is the common difference? *A.* 2.

3. A man is to travel from Boston to a certain place in 6 days, and to go only 5 miles the first day, increasing the distance travelled each day by an equal excess, so that the last day's journey may be 45 miles; what is the daily increase, that is, the common difference? *A.* 8 miles.

4. If the amount of $1 for 20 years, at simple interest, be $2,20, what is the rate per cent.?

In this example we see the amount of the first year is $1,06, and the last year $2,20; consequently, the extremes are 106 and 220, and the number of terms 20. *A.* $,06 = 6 per cent.

5. A man bought 60 yards of cloth, giving 5 cents for the first yard, 7 for the second, 9 for the third, and so on to the last; what did the last cost?

Since, in this example, we have the common difference given, it will be easy to find the price of the last yard; for, as there are as many additions as there are yards, less 1, that is, 59 additions of 2 cents, to be made to the first yard, it follows, that the last yard will cost 2 × 59 = 118 cents more than the first, and the whole cost of the last, reckoning the cost of the first yard, will be 118 + 5 = $1,23. *A.* $1,23.

Hence, when the common difference, the first term, and the number of terms, are given, to find the last term;—

Multiply the common difference by the number of terms, less* 1, *and add the first term to the product.

6. If the first term be 3, the common difference 2, and the number of terms 11, what is the last term? *A.* 23.

7. A man, in travelling from Boston to a certain place in 6 days, travelled the first day 5 miles, the second 8 miles, travelling each successive day 3 miles farther than the former; what was the distance travelled the last day? *A.* 20.

8. What will $1, at 6 per cent., amount to, in 20 years, at simple interest?

The common difference is the 6 per cent.; for the amount of $1, for 1 year, is $1,06, and $1,06 + $,06 = $1,12 the second year, and so on. *A.* $2,20.

9. A man bought 10 yards of cloth, in arithmetical progression; for the 1st yard he gave 6 cents, and for the last yard he gave 24 cents; what was the amount of the whole?

In this example it is plain that ½ the cost of the first and last yards will be the average price of the whole number of yards; thus, 6 cts. + 24 cts. = 30 ÷ 2 = 15 cts., average price; then, 10 yds. × 15 = 150 cts. = $1,50, whole cost. *A.* $1,50.

Hence, when the extremes, and the number of terms are given, to find the sum of all the terms;—

Multiply ½ the sum of the extremes by the number of terms, and the product will be the answer.

10. If the extremes be 3 and 273, and the number of terms 40, what is the sum of all the terms? *A.* 5520.

11. How many times does a regular clock strike in 12 hours? *A.* 78.

12. A butcher bought 100 oxen, and gave for the first ox $1, for the second $2, for the third $3, and so on to the last; how much did they come to at that rate? *A.* $5050.

13. What is the sum of the first 1000 numbers, beginning with their natural order, 1, 2, 3, &c.? *A.* 500500.

14. If a board, 18 feet long, be 2 feet wide at one end, and come to a point at the other, what are the square contents of the board? *A.* 18 feet.

15. If a piece of land, 60 rods in length, be 20 rods wide at one end, and at the other terminate in an angle or point, what number of square rods does it contain? *A.* 600.

16. A person, travelling into the country, went 3 miles the first day, and increased every day's travel 5 miles, till at last he went 58 miles in one day; how many days did he travel?

We found, in example 1, the difference of the extremes, divided by the number of terms, less 1, gave the common difference; consequently, if, in this example, we divide (58 — 3 =) 55, the difference of the extremes, by the common difference, 5, the quotient, 11, will be the number of terms, less 1; then, 1 + 11 = 12, the number of terms. *A.* 12.

Hence, when the extremes and common difference are given, to find the number of terms;—

Divide the difference of the extremes by the common difference, and the quotient, increased by 1, will be the answer.

17. If the extremes be 3 and 45, and the common difference 6, what is the number of terms? *A.* 8.

18. A man, being asked how many children he had, replied, that the youngest was 4 years old, and the eldest 32, the increase of the family having been 1 in every 4 years; how many had he? *A.* 8.

GEOMETRICAL PROGRESSION.

¶ **LXXXIX.** Any rank or series of numbers, increasing by a constant multiplier, or decreasing by a constant divisor, is called *Geometrical Progression.*

Thus, 3, 9, 27, 81, &c., is an *increasing geometrical series;*

And 81, 27, 9, 3, &c., is a *decreasing geometrical series.*

There are five terms in Geometrical Progression; and, like Arithmetical Progression, any three of them being given, the other two may be found, viz.

1. *The first term.*
2. *The last term.*
3. *The number of terms.*
4. *The sum of all the terms.*
5. *The ratio*; that is, the multiplier or divisor, by which we form the series.

1. A man purchased a flock of sheep, consisting of 9; and, by agreement, was to pay what the last sheep came to, at the rate of \$4 for the first sheep, \$12 for the second, \$36 for the third, and so on, trebling the price to the last; what did the flock cost him?

We may perform this example by multiplication; thus, $4 \times 3 \times 3 \times 3 \times 3 \times 3 \times 3 \times 3 \times 3 = \$26244$, *Ans.* But this process, you must be sensible, would be, in many cases, a very tedious one; let us see if we cannot abridge it, thereby making it easier.

In the above process we discover that 4 is multiplied by 3 eight times, one time less than the number of terms; consequently, the 8th power of the ratio 3, expressed thus, 3^8, multiplied by the first term, 4, will produce the last term. But, instead of raising 3 to the 8th power in this manner, we need only raise it to the 4th power, then multiply this 4th power into itself; for, in this way, we do, in fact, use the 3 eight times, raising the 3 to the same power as before; thus, $3^4 = 81$; then, $81 \times 81 = 6561$; this, multiplied by 4, the first term, gives \$26244, the same result as before. *A.* \$26244.

Hence, when the first term, ratio, and number of terms, are given, to find the last term;—

I. *Write down some of the leading powers of the ratio, with the numbers* 1, 2, 3, *&c. over them, being their several indices.*

II. *Add together the most convenient indices to make an index less by* 1 *than the number of terms sought.*

III. *Multiply together the powers, or numbers standing under those indices; and their product, multiplied by the first term, will be the term sought.*

2. If the first term of a geometrical series be 4, and the ratio 3, what is the 11th term?

1, 2, 3, 4, 5, *indices.* } *Note.* The pupil will notice that the series
3, 9, 27, 81, 243, *powers.* } does not commence with the first term, but with the ratio.

The indices $5 + 3 + 2 = 10$, and the powers under each, $243 \times 27 \times 9 = 59049$; which, multiplied by the first term, 4, makes 236196, the 11th term required. *A.* 236196.

3. The first term of a series, having 10 terms, is 4, and the ratio 3; what is the last term? *A.* 78732.

4. A sum of money is to be divided among 10 persons; the first to have \$10, the second \$30, and so on, in threefold proportion; what will the last have? *A.* \$196830.

5. A boy purchased 18 oranges, on condition that he should pay only the price of the last, reckoning 1 cent for the first, 4 cents for the second, 16 cents for the third, and in that proportion for the whole; how much did he pay for them? *A.* \$171798691,84.

6. What is the last term of a series having 18 terms, the first of which is 3, and the ratio 3? *A.* 387420489.

7. A butcher meets a drover, who has 24 oxen. The butcher inquires the price of them, and is answered, \$60 per head; he immediately offers the drover \$50 per head, and would take all. The drover says he will not take that; but, if he will give him what the last ox would come to, at 2 cents for the first, 4 cents for the second, and so on, doubling the price to the last, he might have the whole. What will the oxen amount to at that rate?

A. \$167772,16.

8. A man was to travel to a certain place in 4 days, and to travel at whatever rate he pleased; the first day he went 2 miles, the second 6 miles, and so on to the last, in a threefold ratio; how far did he travel the last day, and how far in all?

In this example, we may find the last term as before, or find it by adding each day's travel together, commencing with the first, and proceeding to the last, thus: 2 + 6 + 18 + 54 = 80 miles, the whole distance travelled, and the last day's journey is 54 miles. But this mode of operation, in a long series, you must be sensible, would be very troublesome. Let us examine the nature of the series, and try to invent some shorter method of arriving at the same result.

By examining the series 2, 6, 18, 54, we perceive that the last term, (54,) less 2, (the first term,) = 52, is 2 times as large as the sum of the remaining terms; for 2 + 6 + 18 = 26; that is, 54 − 2 = 52 ÷ 2 = 26; hence, if we produce another term, that is, multiply 54, the last term, by the ratio 3, making 162, we shall find the same true of this also; for 162 − 2, (the first term,) = 160 ÷ 2 = 80, which we at first found to be the sum of the four remaining terms, thus: 2 + 6 + 18 + 54 = 80. In both of these operations it is curious to observe, that our divisor, (2,) each time, is 1 less than the ratio, (3.)

Hence, when the extremes and ratio are given, to find the sum of the series, we have the following easy

RULE.

Multiply the last term by the ratio, from the product subtract the first term, and divide the remainder by the ratio, less 1; the quotient will be the sum of the series required.

9. If the extremes be 5 and 6400, and the ratio 6, what is the whole amount of the series?

$$\frac{\overline{6400 \times 6 - 5}}{6 - 1} = 7679, \textit{Ans.}$$

10. A sum of money is to be divided among 10 persons in such a manner, that the first may have $10, the second $30, and so on, in threefold proportion; what will the last have, and what will the whole have?

The pupil will recollect how he found the *last* term of the series by a foregoing rule; and, in all cases in which he is required to find the *sum* of the series, when the last term is not given, he must first find it by that rule, and then work for the sum of the series, by the present rule.

A. The last, $196830; and the whole, $295240.

11. A hosier sold 14 pair of stockings, the first at 4 cents, the second at 12 cents, and so on in geometrical progression; what did the last pair bring him, and what did the whole bring him? *A.* Last, $63772,92; whole, $95659,36.

12. A man bought a horse, and, by agreement, was to give a cent for the first nail, three for the second, &c.; there were four shoes, and in each shoe eight nails; what did the horse come to at that rate?

A. $9265100944259,20

13. At the marriage of a lady, one of the guests made her a present of a half-eagle, saying, that he would double it on the first day of each succeeding month throughout the year, which he said would amount to something like $100; now, how much did his estimate differ from the true amount?

A. $20375.

14 If our pious ancestors, who landed at Plymouth, A. D. 1620, being 101 in number, had increased so as to double their number in every 20 years, how great would have been their population at the end of the year 1840?

A. 206747.

ANNUITIES AT SIMPLE INTEREST.

¶ XC. An annuity is a sum of money, payable every year, for a certain number of years, or forever.

When the annuity is not paid at the time it becomes due, it is said to be in *arrears*.

The sum of all the annuities, such as rents, pensions, &c., remaining unpaid, with the interest on each, for the time it has been due, is called the *amount* of the annuity.

Hence, to find the amount of an annuity;—

Calculate the interest on each annuity, for the time it has remained unpaid, and find its amount; then the sum of all these several amounts will be the amount required.

1. If the annual rent of a house, which is $200, remain unpaid, (that is, in arrears,) 8 years, what is the amount?

In this example, the rent of the last (8th) year being paid when due, of course, there is no interest to be calculated on that year's rent.

The amount of $200 *for* 7 *years* = $284
The amount of $200 .... 6 .... = $272
The amount of $200 .... 5 .... = $260
The amount of $200 .... 4 .... = $248
The amount of $200 .... 3 .... = $236
The amount of $200 .... 2 .... = $224
The amount of $200 .... 1 .... = $212
The eighth year, paid when due, = $200

$1936, *Ans.*

2. If a man, having an annual pension of $60, receive no part of it till the expiration of 8 years, what is the amount then due? *A.* $580,80.

3. What would an annual salary of $600 amount to, which remains unpaid (or in arrears) for 2 years? (1236) For 3 years? (1908) For 4 years? (2616) For 7 years? (4956) For 8 years? (5808) For 10 years? (7620)

Ans. $24144.

4. What is the present worth of an annuity of $600, to continue 4 years?

The present worth, (¶ LXVII.,) is such a sum as, if put at interest, would amount to the given annuity; hence,

$600 ÷ $1,06 = $566,037, *present worth, 1st year.*
$600 ÷ $1,12 = $535,714, ... 2d .....
$600 ÷ $1,18 = $508,474, ... 3d .....
$600 ÷ $1,24 = $483,870, ... 4th .....

Ans., $2094,095, *present worth required.*

Hence, to find the present worth of an annuity;—

Find the present worth of each year by itself, discounting from the time it becomes due, and the sum of all these present worths will be the answer.

5. What sum of ready money is equivalent to an annuity of $200, to continue 3 years, at 4 per cent.? *A.* $556,063.

6. What is the present worth of an annual salary of $800 to continue 2 years? (1469001) 3 years? (2146067) 5 years? (3407512) *A.* $7023,48.

ANNUITIES AT COMPOUND INTEREST.

¶ **XCI.** The amount of an annuity, at simple and compound interest, is the same, excepting the difference in interest.

Hence, to find the amount of an annuity at compound interest;—

Proceed as in ¶ XC., reckoning compound, instead of simple interest.

1. What will a salary of $200 amount to, which has remained unpaid for 3 years?

The amount of $200 *for* 2 *years* = $224,72
The amount of $200 *for* 1 *year* = $212,00
The 3*d year,* . . . . = $200,00

A. $636,72

2. If the annual rent of a house, which is $150, remain in arrears for 3 years, what will be the amount due for that time? *A.* $477,54.

Calculating the amount of the annuities in this manner, for a long period of years, would be tedious. This trouble will be prevented, by finding the amount of $1, or 1£, annuity, at compound interest, for a number of years, as in the following

TABLE I.

Showing the amount of $1 or 1£ annuity, at 6 per cent. compound interest, for any number of years, from 1 to 50.

Yrs.	6 per cent.	Yrs.	6 per cent.	Yrs.	6 per cent.	Yrs.	6 per cent.	Yrs.	6 per cent.
1	1,0000	11	14,9716	21	39,9927	31	84,8016	41	165,0467
2	2,0600	12	16,8699	22	43,3922	32	90,8897	42	175,9495
3	3,1836	13	18,8821	23	46,9958	33	97,3431	43	187,5064
4	4,3746	14	21,0150	24	50,8155	34	104,1837	44	199,7568
5	5,6371	15	23,2759	25	54,8645	35	111,4347	45	212,7423
6	6,9753	16	25,6725	26	59,1563	36	119,1208	46	226,5068
7	8,3938	17	28,2123	27	63,7057	37	127,2681	47	231,0972
8	9,8974	18	30,9056	28	68,5281	38	135,9042	48	245,9630
9	11,4913	19	33,7599	29	73,6397	39	145,0584	49	261,7208
10	13,1807	20	36,7855	30	79,0581	40	154,7619	50	278,4241

It is evident, that the amount of $2 annuity is 2 times as much as one of $1, and one of $3, 3 times as much; hence,

To find the amount of an annuity, at 6 per cent.;—

Find by the Table the amount of $1, *at the given rate and time, and multiply it by the given annuity, and the product will be the amount required.*

3 What is the amount of an annuity of $120, which has remained unpaid 15 years?

The amount of $1, *by the Table, we find to be* $23,2759; *therefore,* $23,2759 × 120 = $2793,108, *Ans.*

4 What will be the amount of an annual salary of $400, which has been in

arrears 2 years? (824) 3 years? (127344) 4 years? (174984) 6 years. (279012) 12 years? (674796) 20 years? (147142) *Ans.* $28099,56.

5. If you lay up $100 a year from the time you are 21 years of age till you are 70, what will be the amount at compound interest? *A.* $26172,08.

6. What is the present worth of an annual pension of $120, which is to continue 3 years?

In this example, the present worth is evidently that sum, which, at compound interest, would amount to as much as the amount of the given annuity for the 3 years? *Finding the amount of $120 by the Table, as before, we have $382,032; then, if we divide $382,032 by the amount of $1, compound interest, for 3 years, the quotient will be the present worth.* This is evident from the fact, that the quotient, multiplied by the amount of $1, will give the amount of $120, or, in other words, $382,032. *The amount of $1 for 3 years, at compound interest, is $1,19101;*

then, $382,032 ÷ $1,19101 = $320,763, *Ans.*

Hence, to find the present worth of an annuity;—

Find its amount in arrears for the whole time; this amount, divided by the amount of $1 for said time, will be the present worth required.

Note. The amount of $1 may be found ready calculated in the Table of compound interest, ¶ LXXI.

7. What is the present worth of an annual rent of $200, to continue 5 years? *A.* $842,472.

The operations in this rule may be much shortened by calculating the present worth of $1 for a number of years, as in the following

TABLE II.

Showing the present worth of $1 or 1£ annuity, at 6 per cent. compound interest for any number of years, from 1 to 32.

Years.	6 per cent.	Years.	6 per cent.	Years.	6 per cent.	Years.	6 per cent.
1	0,94339	9	6,80169	17	10,47726	25	12,78335
2	1,83339	10	7,36008	18	10,82760	26	13,00316
3	2,67301	11	7,88687	19	11,15811	27	13,21053
4	3,46510	12	8,38384	20	11,46992	28	13,40616
5	4,21236	13	8,85268	21	11,76407	29	13,59072
6	4,91732	14	9,29498	22	12,04158	30	13,76483
7	5,58238	15	9,71225	23	12,30338	31	13,92908
8	6,20979	16	10,10589	24	12,55035	32	14,08398

To find the present worth of any annuity, by this Table, we have only to multiply the present worth of $1, found in the Table, by the given annuity, and the product will be the present worth required.

8. What sum of ready money will purchase an annuity of $300, to continue 10 years?

The present worth of $1 annuity, by the Table, for 10 years, is $7,36008; then 7,36008 × 300 = $2208,024, *Ans.*

9. What is the present worth of a yearly pension of $60, to continue 2 years? (1100034) 3 years? (1603806) 4 years? (207906) 8 years? (3725874) 20 years? (6881952) 30 years? (8258898) *A.* $2364,9624.

10. What salary, to continue 10 years, will $2208,024 purchase?

This example is the 8th example reversed; consequently, $2208,024 ÷ 7,36008 = 300, the annuity required. *A.* $300.

Hence, to find that annuity which any given sum will purchase;—

Divide the given sum by the present worth of $1 annuity for the given time, found by Table II.; the quotient will be the annuity required.

11. What salary, to continue 20 years, will $688,95 purchase? *A.* $60 +.

To divide any sum of money into annual payments, which, when due, shall form an equal amount, at compound interest;—

12. A certain manufacturing establishment, in Massachusetts, was actually sold for $27000, which was divided into 4 notes, payable annually, so that the principal and interest of each, when due, should form an equal amount, at compound interest, and the several principals, when added together, should make $27000; now, what were the principals of said notes?

It is plain, that, in this example, if we find an annuity to continue 4 years, which $27000 will purchase, the present worth of this annuity for 1 year will be the first payment, or principal of the note; the present worth for 2 years, the second, and so on to the last year.

The annuity which $27000 will purchase, found as before, is 7791,97032 +.

Note. To obtain an exact result, we must reckon the decimals, which were rejected in forming the tables. This makes the last divisor 3,4651056.

Ans.	*The* 1st *is* $7350,915,	*amount for*	1 *yr.*	$7791,97032
	...... 2*d* .. $6934,825,	....	2	$7791,97032
	...... 3*d* .. $6542,288,	....	3	$7791,97032
	...... 4*th* .. $6171,970,	....	4	$7791,97032
	Proof, $26999,998 +			

PERMUTATION.

¶ XCII. Permutation is the method of finding how many different ways any number of things may be changed.

1. How many changes may be made of the three first letters of the alphabet?

In this example, had there been but two letters, they could only be changed twice; that is, *a, b,* and *b, a;* that is, $1 \times 2 = 2$; but, as there are three letters, they may be changed $1 \times 2 \times 3 = 6$ times, as follows:—

1 a, b, c.
2 a, c, b.
3 b, a, c.
4 b, c, a.
5 c, b, a.
6 c, a, b.

Hence, to find the number of different changes or permutations, which may be made with any given number of different things;—

Multiply together all the terms of the natural series, from 1 up to the given number, and the last product will be the number of changes required.

2. How many different ways may the first 5 letters of the alphabet be arranged? *A.* 120.

3. How many changes may be rung on 15 bells, and in what time may they be rung, allowing 3 seconds to every round? *A.* 1307674368000 changes; 3923023104000 seconds.

4. What time will it require for 10 boarders to seat themselves differently every day at dinner, allowing 365 days to the year? *A.* 9941$\frac{335}{365}$ years.

5. Of how many variations will the 26 letters of the alphabet admit?
A. 403291461126605635584000000

POSITION

Is a rule which teaches, by the use of supposed numbers, to find true ones. It is divided into two parts, called Single and Double.

SINGLE POSITION.

¶ **XCIII.** This rule teaches to resolve those questions whose results are proportional to their suppositions.

1. A schoolmaster, being asked how many scholars he had, replied, "If I had as many more as I now have, one half as many more, one third, and one fourth as many more, I should have 296." How many had he?

Let us suppose he had	24
Then as many more =	24
$\frac{1}{2}$ *as many* =	12
$\frac{1}{3}$ *as many* =	8
$\frac{1}{4}$ *as many* =	6
	74

We have now found that we did not suppose the right number. If we had, the amount would have been 296. But 24 has been increased in the same manner to amount to 74, that some unknown number, the true number of scholars, must be, to amount to 296. Consequently, it is obvious, that 74 has the same ratio to 296 that 24 has to the true number. The question may, therefore, be solved by the following statement:

As 74 : 296 : : 24 : 96, *Ans.*

This answer we prove to be right by increasing it by itself, one half itself, one third itself, and one fourth itself;

Thus, 96 + 96 + 48 + 32 + 24 = 296

From these illustrations we derive the following

RULE.

I. *Suppose any number you choose, and proceed with it in the same manner you would with the answer, to see if it were right.*

II. *Then say, As this result : the result in the question : : the supposed number : number sought.*

More Exercises for the Slate.

2. James lent William a sum of money on interest, and in 10 years it amounted to $1600; what was the sum lent? *A.* $1000.

3. Three merchants gained, by trading, $1920, of which A took a certain sum, B took three times as much as A, and C four times as much as B; what share of the gain had each? *A.* A's share was $120; B's, $360, and C's, $1440.

4. A person, having about him a certain number of crowns, said, if a third, a fourth, and a sixth of them were added together, the sum would be 45; how many crowns had he? *A.* 60.

5. What is the age of a person, who says, that if $\frac{6}{10}$ of the years he has lived be multiplied by 7, and $\frac{2}{3}$ of them be added to the product, the sum would be 292? *A.* 60 years.

6. What number is that, which, being multiplied by 7, and the product divided by 6, the quotient will be 14? *A.* 12.

DOUBLE POSITION.

¶ **XCIV.** This rule teaches to solve questions by means of two supposed numbers.

In *Single Position*, the number sought is always multiplied or divided by some proposed number, or increased or diminished by itself, or some known part of itself, a certain number of times. Consequently, the result will be proportional to its supposition, and but one supposition will be necessary; but, in *Double Position* we employ two, for the results are not proportional to the suppositions.

1. A gentleman gave his three sons $10000, in the following manner: to the second $1000 more than to the first, and to the third as many as to the first and second. What was each son's part?

Let us suppose the share of the first, 1000
Then the second = 2000
Third = 3000
Total, 6000
This, subtracted from 10000, *leaves* 4000

The shares of all the sons will, if our supposition be correct, amount to $10000; but, as they amount to $6000 only, we call the error 4000.

Suppose, again, that the share of the first was 1500
Then the second = 2500
Third = 4000
8000
2000

We perceive the error in this case to be $2000.

The first error, then, is $4000, and the second $2000. Now, the difference between these errors would seem to have the same relation to the difference of the suppositions, as either of the errors would have to the difference between the supposition which produced it and the true number. We can easily make this statement, and ascertain whether it will produce such a result:

As the difference of errors, 2000 : 500, difference of suppositions : : either of the errors, (say the first) 4000 : 1000, the difference between its supposition and the true number. Adding this difference to 1000, the supposition, the amount is 2000 for the share of the first son; then $3000 that of the second, $5000 that of the third, *Ans.* For 2000 + 3000 + 5000 = 10000, the whole estate.

Had the supposition proved too great, instead of too small, it is manifest that we must have subtracted this difference.

The differences between the results and the result in the question are called *errors: these* are said to be *alike*, when both are either too great or too small; *unlike*, when one is too great, and the other too small.

From these illustrations we derive the following

RULE.

I. *Suppose any two numbers, and proceed with each according to the manner described in the question, and see how much the result of each differs from that in the question.*

II. *Then say, As the difference* of the errors : the difference of the suppositions : : either error : difference between its supposition and the number sought.*

More Exercises for the Slate.

2. Three persons disputing about their ages, says B, "I am 10 years older than A;" says C, "I am as old as you both:" now what were their several ages, the sum of all of them being 100? *Ans.* A's, 20; B's, 30; C's, 50.

3. Two persons, A and B, have the same income; A saves ⅕ of his yearly; but B, by spending $150 per annum more than A, at the end of 8 years finds himself $400 in debt; what is their income, and what does each spend per annum?

First, suppose each had $200; secondly, $300; then the errors will be 400 and 200. *A.* Their income is $400; A spends $300, B $450.

4. There is a fish whose head is 8 feet long, his tail is as long as his head and half his body, and his body is as long as his head and tail; what is the whole length of the fish?

First, suppose his body 30; secondly, 28; the errors will then be 1 and 2.

A. 32 feet

5. A labourer was hired 80 days upon this condition,—that for every day he was idle he should forfeit 50 cents, and for every day he wrought he should receive 75 cents; at the expiration of the time he received $25; now how many days did he work, and how many days was he idle?

A. He worked 52 days, and was idle 28.

MISCELLANEOUS EXAMPLES.

1. There is a room, one side of which is 20 feet long and 8 feet high; how many square feet are contained in that side?

This side is a regular parallelogram (¶ LXXIX.); and, to find the square contents, we have seen that we must multiply the length by the breadth; thus, 20 ft. × 8 ft. = 160 sq. ft., *Ans.* But, had we been required to find the square contents of half of this parallelogram, as divided in the figure on the left, it is plain that, if we should multiply (20) the whole length by ½ of (8) the width, or, in this case, the height, the product would be the square contents in this half, that is, in the figure B C D; thus, ½ of 8 = 4; then, 4 × 20 = 80 sq. ft., which is precisely ½ of 160, the square contents in the whole figure.

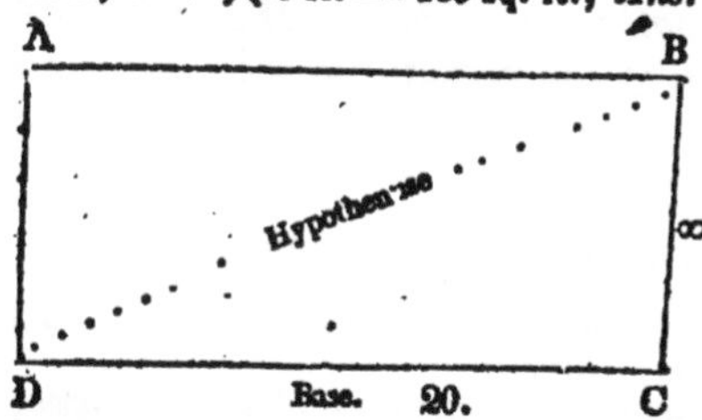

The half B C D is called a triangle, because it has, as you see, 3 sides and 3 angles, and because the line B C falls perpendicularly on C D; the angle at C is called a right angle; the whole angle, then, B C D may properly be called a right-angled triangle.

---

* The difference of the errors, when alike, will be one *subtracted* from the other; when unlike, one *added* to the other.

The line B C is called a *perpendicular*, C D the *base*, and D B the *hypothenuse*.

Note. Both the base and perpendicular are sometimes called the *legs* of the triangle.

Hence, to find the area of a right-angled triangle;—

Multiply the length of the base by ½ the length of the perpendicular; the product will be the area required.

2. What is the area of a triangular piece of land, one side of which is 40 rods, and the distance from the corner opposite that side to that side 20 rods?

Ans. $\frac{20}{2} \times 40 = 400$ rods.

Note. To find the area of any irregular figure, divide it into triangles.

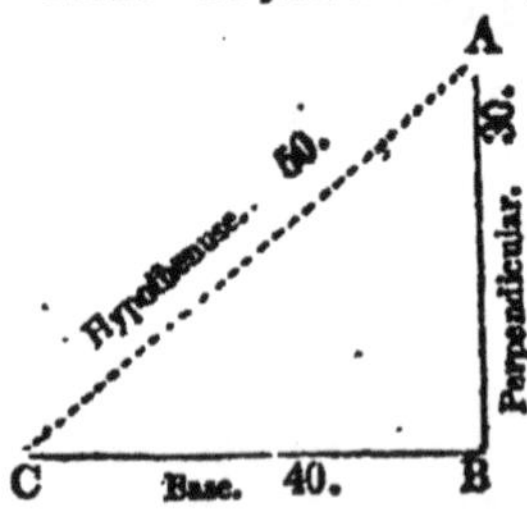

In any right-angled triangle, it has been ascertained, that the square of the hypothenuse is equal to the sum of the squares of the other two sides. Thus, in the adjacent figure, $40^2 = 1600$, and $30^2 = 900$; then, $\sqrt{900+1600} = 50$, the hypothenuse

I. Hence, to find the hypothenuse, when the legs are given;—

Add the squares of the two legs together, and extract the square root of their sum.

II. When the hypothenuse and one leg are given, to find the other leg;—

From the square of the hypothenuse subtract the square of the given leg, and the square root of the remainder will be the other.

3. A river 80 yards wide passes by a fort, the walls of which are 60 yards high; now, what is the distance from the top of the wall to the opposite bank of the river?

In this example we are to find the hypothenuse. *Ans.* 100 yards.

4. There is a certain street, in the middle of which, if a ladder 40 feet long be placed, it will reach a window 24 feet from the ground, on either side of said street; what is the width of the street?

In this example, we are to find the length of the base of two triangles, and then the sum of these will be the distance required. *Ans.* 64 feet.

5. There is a certain elm, 20 feet in diameter, growing in the centre of a circular island; the distance from the top of the tree to the water, in a straight line, is 120 feet; and the distance from the foot 90 feet; what is the height of the tree?

As the tree is 20 feet in diameter, the distance from its centre to the water is the length of the base, that is, $10 + 90 = 100$ feet. *A.* 66,332 ft. +.

6. Two ships sail from the same port; one goes due north 40 leagues, the other due east 30 leagues; how far are they apart?

We are here to find the hypothenuse. *A.* 50 leagues.

7. A man, in a hunting excursion, shot a squirrel from the top of a stately oak, 80 feet high, its diameter being 6 feet; the person stood 19 paces from the tree (3 feet being equal to one pace); now, how far was it from the squirrel to the place where the hunter stood, when he discharged his piece? *A.* 100 ft.

8. What is the circumference of a wheel, the diameter of which is 8 feet?

The circumference of a circle is greater, you are sensible, than the diameter, being a little more than 3 *times, or, more accurately,* 3,141592 *times the diameter.* *A.* 25,13 + ft.

9. What is the diameter of a wheel, or circle, whose circumference is 12 feet? *A.* 4 ft., nearly.

10. If the distance through the earth be 8000 miles, how many miles around it? *A.* 25132,7 miles, nearly.

11. What is the area or contents of a circle, whose diameter is 6 feet, and its circumference 19 feet?

NOTE. *The area of a circle may be found by multiplying half the diameter by half the circumference, or by multiplying the square of half the diameter by* 3,141592. *A.* 28½ ft.

12 What is the area of a circle, whose diameter is 20 feet?

$10^2 = 100 \times 3,141592 = 314,1592$, *Ans*

13. What is the diameter of a circle, whose area is 314,1592? *A.* 20 ft.

14. What is the area, or square contents, of the earth, allowing it to be 8000 miles in diameter, and 25000 in circumference?

NOTE. *The area of a globe or ball is* 4 *times as much as the area of a circle of the same diameter; therefore, if we multiply the whole circumference into the whole diameter, the product will be the area.* *A.* 200000000.

15. What are the solid contents of a globe or ball 12 inches in diameter?

The solid contents of a globe are found by multiplying its area by $\frac{1}{6}$ *of its diameter.* *A.* $904\frac{7}{10}$ + solid inches.

16. What are the solid contents of a round stick of timber, 10 inches in diameter, and 20 feet long?

In this example, we may first find the area of one end, as before directed for a circle; then multiply by 20 feet, the length. *A.* 11 feet, nearly.

Note. Solids of this form may be called *cylinders*.

17. What are the solid contents of a cylinder 4 feet in diameter, and 10 feet long? *A.* 125 + feet.

When solids, being either round or square, taper regularly till they come to a point, they contain just ⅓ as much as if they were all the way as large as they are at the largest end.

When solids decrease regularly, as last described, they are called *pyramids.* When the base is square, they are called *square pyramids;* when triangular, *triangular pyramids;* and when round, *circular pyramids*, or *cones.*

Hence, to find the solid contents of such figures;—

Multiply the area of the largest end by ⅓ of the perpendicular height.

What are the solid contents of a cone, the height of which is 30 feet, and its base 8 feet in diameter? *A.* 502,6 + ft.

18. There is a pyramid, whose base is 3 feet square, and its perpendicular height 9 feet; what are its solid contents? *A.* $3^2 \times \frac{9}{3} = 27$ ft.

19. What is the length of one side of a cubical block, which contains 9261 solid feet? *A.* 21 ft.

20. In a square lot of land, which contains 2648 acres, 3 roods, and 1 rod, what is the length of one side? *A.* 651 rods.

21. A grocer put 5 gallons of water into a cask containing 30 gallons of wine worth 75 cents per gallon; what is a gallon of this mixture worth?

A. 64$\frac{2}{7}$ cts.

22. The first term of a geometrical series is 4, the last 56984, and the ratio 6, what is the sum of all the terms? *A.* 68380

23. "*The great bet, and when it will be paid.*—The public mind has been considerably amused for a few days past with a singular bet, said to have been made between a friend of Mr. Adams and a friend of Gen. Jackson, on the eastern shore of Maryland. The bet was, that the Jackson man was to receive from the Adams man 1 cent for the first electoral vote that Jackson should receive over 130, 2 cents for the second, 4 for the third, and so on, doubling for every successive vote; and the Adams man was to have one hundred dollars if Jackson did not receive over 130 votes. According to the present appearances, Jackson will receive 173, 43 over 130, and the sum the Adams man will have to pay, in that event, will be $87960930222,07.

"But the joke does not appear to be all on the Jackson man's side. The money is to be counted, and it will take a pretty long lifetime of any common man to count out the 'shiners.' Let's see:—allowing that a man can count sixty dollars a minute, and that he continues to count without ceasing, either to sleep, to take refreshment, or to keep the Sabbath, it will take him *twenty-seven hundred and eighty-nine years, nearly;* but allow him to work eight hours a day, and rest on the Sabbath, he will be occupied 9789+ years; so that the Adams man, when he is called upon for the cash, may tell his Jackson friend, 'Sit down, sir; as soon as I can count the money you shall have it; even the banks take time to count the money, you know.'"

A PRACTICAL SYSTEM
OF
BOOK-KEEPING,
FOR
FARMERS AND MECHANICS.

Almost all persons, in the ordinary avocations of life, unless they adopt some method of keeping their accounts in a regular manner, will be subjected to continual losses and inconveniences; to prevent which the following plan or outline is composed, embracing the principles of Book-Keeping in the most simple form. Before the pupil commences this study, it will not be necessary for him to have attended to all the rules in the Arithmetic; but he should make himself acquainted with the subject of Book-Keeping before he is suffered to leave school. A few examples only are given, barely sufficient to give the learner a view of the manner of keeping books; it being intended that the pupil should be required to compose similar ones, and insert them in a book adapted to this purpose.

Book-Keeping is the method of recording business transactions. It is of two kinds—single and double entry; but we shall only notice the former.

Single entry is the simplest form of Book-Keeping, and is employed by retailers, mechanics, farmers, &c. It requires a Day-Book, Leger, and, when money is frequently received and paid out, a Cash-Book.

DAY-BOOK.

This should be a minute history of business transactions in the order of time in which they occur; it should be ruled with head lines, with one column on the left hand for post-marks and references, and two columns on the right for dollars and cents. The owner's name, the town or city, and the date of the first transaction, should stand at the head of the first page. It is the

FORM OF A DAY-BOOK.

EDWARD L. PECKHAM. *Boston, Jan.* 1, 1829.

				$	c.
1	×	***James Murray, Jr.*** . . . . *Dr.*			
		To 1 gall. Lisbon Wine, ...	$1,92		
		" 6 yds. Calico, *a* 37½ cts.	2,25		
		" 2 yds. Broadcloth, *a* $4,50	9,00	13	17
1	×	***Robert Hawkins,*** Blacksmith, ... *Dr.*			
		To 217 lbs. Iron, *a* 8 cts.		17	36
2	×	***Thomas Yeomans,*** . . . . *Cr.*			
		By Cash, ...		75	75
		3			
2	×	***Archibald Tracy,*** Salem, ... *Dr.*			
		To 1 piece Broadcloth, containing 29 yds., *a* $3 per yd., 90 days' credit, ...		87	00
2	×	***James Warren,*** Wartland, ... *Dr.*			
		To 1 cask Nails, 225 lbs., *a* 8 cts.		18	00
		Cr.			
		By 37 lbs. Cheese, *a* 10 cts.	$ 3,70		
		" 41 lbs. Feathers, *a* 70 cts.	28,70		
		Balance to be paid in Corn, at market price.		32	40
2	×	***Isaac Thomas,*** Brattle Square, . . . *Dr.*			
		To 32 galls. Molasses, *a* 50 cts.		16	00
		4			
2	×	***William Angell,*** Roxbury, ... *Dr.*			
		To 300 lbs. Pork, at 7 cts.	$21,00		
		" 30 bu. Corn, *a* 45 cts.	13,50	34	50
2	×	***Samuel Stone,*** . . . . *Dr.*			
		To 50 lbs. Harness Leather, *a* 30 cts.	$15,00		
		" 7 tons Hay, *a* $10	70,00	85	00
		5			
2	×	***George Carpenter,*** . . . . *Dr.*			
		To 17 Brooms, *a* 12 cts.	$2,04		
		" 7 lbs. Butter, *a* 20 cts.	1,40		
		" 4 lbs. Cheese, *a* 10 cts.	,40	3	84
3	×	***Jesse B. Sweet,*** Mendon, ... *Dr.*			
		To 1 hhd Molasses, 98 — 6 = 92 galls., *a* 30 cts.		27	60
		Cr.			
		By Cash, ...		15	00

Jan. 5, 1829.

				$	c.
3	✕	*Jesse Metcalf*, Tanner, ...	*Dr.*		
		To 20 Calf-Skins, *a* $5	$100,00		
		" 50 Dried Hides, *a* $4	200,00		
		60 days' credit.		300	00
1	✕	*James Murray, Jr.*	*Cr.*		
		By 20 bu. Corn, *a* 60 cts.	$12,00		
		" 4 bu. Oats, *a* 40 cts.	1,60	13	60
		6			
2	✕	*James Warren*, ...	*Dr.*		
		To 24 bu. Corn, *a* 60 cts.		14	40
2	✕	*Archibald Tracy*, ...	*Dr.*		
		To 1 cord Wood, ...	$6,00		
		" 30 lbs. Feathers, *a* 70 cts.	21,00	27	00
1	✕	*Robert Hawkins*, ...	*Cr.*		
		By shoeing my Horse, ...	$2,00		
		" " " Oxen, ...	3,00	5	00
		7			
2	✕	*Samuel Stone*, ...	*Dr.*		
		To 2 yds. Broadcloth, *a* $4	$8,00		
		" 4 pr. Shoes, *a* $1	4,00	12	00
2	✕	*Thomas Yeomans*, ...	*Dr.*		
		To 200 bu. Corn, *a* 70 cts.		140	00
		9			
3	✕	*Jesse B. Sweet*, ...	*Dr.*		
		To 30 quintals Fish, *a* $3,75		112	50
2	✕	*George Carpenter*, ...	*Dr.*		
		To 200 lbs. Cheese, *a* 8 cts.	$16,00		
		" 1 firkin Butter, 76 lbs., weight of tub, 10 lbs. = 66, *a* 20 cts.	13,20	29	20
2	✕	*Archibald Tracy*, ...	*Dr.*		
		To 2 bbls. Flour, *a* $10	$20,00		
		" 25 lbs. Lard, *a* 10 cts.	2,50		
		" 3 bu. Salt, *a* 66 cts.	1,98	24	48
2	✕	*Isaac Thomas*, ...	*Dr.*		
		To 50 yds. Calico, *a* 22 cts.	$11,00		
		" 75 yds. brown Sheeting, *a* 14 cts.	10,50	[illegible]	[illegible]
			Cr.		
		By Order on Goodrich & Lord, for $12,80			

Jan. 9, 1829.

			$	c
3 ×	*Jesse Metcalf,* . . . . *Dr.*			
	To 500 pr. Men's Shoes, *a* 95 cts.		475	00
	—10—			
2 ×	*Thomas Yeomans,* . . . . *Dr.*			
	To 3 bbls. Flour, *a* $9,50		28	50
1 ×	*Robert Hawkins,* . . . . *Dr.*			
	To 120 lbs. Blistered Steel, *a* 8 cts.	$9,60		
	" 100 lbs. Russia Iron, *a* 5 cts.	5,00		
			14	60
1 ×	*James Murray, Jr.* . . . . *Dr.*			
	To 10 lbs. Sugar, *a* 11 cts.	$1,10		
	" 20 lbs. Coffee, *a* 15 cts.	3,00		
	" 6 galls. Molasses, *a* 37 cts.	2,22		
			6	32
2 ×	*William Angell,* . . . . *Cr.*			
	By 200 lbs. Lard, *a* 6 cts.	$12,00		
	" 350 lbs. Bacon, *a* 12 cts.	42,00		
			54	00
	—12—			
3 ×	*James Hammond,* . . . . *Dr.*			
	To 1 bbl. Flour, ...	$10,00		
	" 3 bu. Corn, *a* 65 cts.	1,95		
	" 6 galls. Wine, *a* $1,25	7,50		
	" 3 lbs. Coffee, *a* 16 cts.	,48		
	" 4 bu. Salt, *a* 70 cts.	2,80		
	" 1 lb. Y. H. Tea, ...	1,25		
	" 14 lbs. Sugar, *a* 12 cts.	1,68		
	" 3 yds. Broadcloth, *a* $2,50	7,50		
	" 12 yds. Shirting, *a* 19 cts.	2,28		
			35	44
	—13—			
1 ×	*James Murray, Jr.* . . . . *Dr.*			
	To 6 lbs. Raisins, *a* 20 cts.	$1,20		
	" 5 galls. Currant Wine, *a* 75 cts.	3,75		
			4	95

LEGER.

This book is used to collect the scattered accounts of the Day-Book, and to arrange all that relates to each individual into one separate statement. The business of collecting these accounts from the Day-Book, and writing them in the Leger, is called *posting*. This should be done once a month or oftener. Debts due from others, and entered upon the Day-Book, are placed on the side of *Dr*; whatever is on the Day-Book as due to another is placed on the side of *Cr*.

When an account is posted, the page of the Leger, in which this account is kept, is written in the left hand column of the Day-Book.

Every Leger should have an alphabetical Index, where the names of the several persons, whose accounts are kept in the Leger, should be written, and *the page noted down.*

When one Leger is full, and a new one is opened, the accounts in the former *ld be all balanced*, and the balances transferred to the new Leger

EXPLANATION OF THE LEGER, AND THE MANNER OF POSTING.

It will be seen, that the name of *James Murray, Jr.* stands first on the Day-Book; of course, we shall post his account first. We enter his name on the first page of the Leger, in a large, fair hand, writing *Dr.* on the left, and *Cr.* on the right. At the top of the left hand column we enter the year, under which we write the month and day when the first charge was made in the Day-Book, and in the next column the page of the Day-Book where the charge stands. Then, as there are several articles in the first charge, instead of specifying each article, as in the Day-Book, we merely say, *To Sundries*, and enter the amount in the proper columns. This charge being thus posted, we write the page of the Leger, viz. 1, in the left hand column of the Day-Book, and opposite to it a ╳, to show more distinctly that the charge is posted. We then pass a finger carefully over the names, till we again come to the name of *James Murray, Jr.*, which we find on the second page; but, as this is credit, we enter it on the credit side, with the date and page in their proper columns. We then enter the Leger-page and cross, as before, and then proceed again in search of the same name, until every charge and credit is transferred into the Leger. The next name is to be taken and proceeded with in the same way as the first; and so continue till all the accounts are posted.

As it is uncertain how extensive an account may be when once opened, it is better to take a new page for every name, until all the Leger-pages are occupied. By this time, it is probable, several accounts will have been settled: we may then enter a second name on the same pages, and so continue till all the pages are full.

Whenever any account is settled, the amount or the balance is ascertained, and the settlement entered in the Leger. The settlement may also be entered in the Day-Book; and many practise this, although it is not essentially necessary. But it is essentially necessary that one, if not both the books, should show how every account is settled, whether by cash, note, order, goods, or whatever way the amount or balance is liquidated.

N. B. In making out bills, the Leger is used as a reference to the charges in the Day-Book, which must be exactly copied.

FORM OF A LEGER.

Dr.		*James Murray, Jr.*							*Cr.*
1829.			$	c.	1829			$	c.
Jan. 1.	1	To Sundries,	13	17	Jan. 5.	2	By Corn and Oats,	13	60
" 10.	3	do.	6	32	" 15		By Cash, to bal.	10	84
" 13.	3	do.	4	95					
			$24	44				$24	44

Dr.		*Robert Hawkins,*							*Cr.*
1829.			$	c.	1829.			$	c.
Jan. 1.	1	To Iron, . . .	17	36	Jan. 6.	2	By Work, . .	5	00
" 10.	3	" Sundries,	14	60	" 12.		" Note, a 60 days,	26	96
			$31	96				$31	96

Dr.		*Thomas Yeomans,*							*Cr.*
1829.			$	c.	1829.			$	c.
Jan. 7.	2	To Corn, . .	140	00	Jan. 1.	1	By Cash, . .	75	75
" 10.	3	" Flour, . .	28	50	" 11.		" Check, for bal.	92	75
			$168	50				$168	50

Dr.		*Archibald Tracy,*							*Cr.*
1829.			$	c.	1829.			$	c.
Jan. 3.	1	To Broadcloth,	87	00	Apr. 2.		By Cash, . .	138	48
" 6.	2	" Sundries, .	27	00					
" 9.	2	do. . .	24	48					
			$138	48					

Dr.		*James Warren,*							*Cr.*
1829.			$	c.	1829.			$	c
Jan. 3.	1	To Nails, . .	18	00	Jan. 3.	1	By Sundries, .	32	40
" 6.	2	" Corn, . .	14	40					
			$32	40					

Dr.		*Isaac Thomas,*							*Cr.*
1829.			$	c.	1829.			$	c.
Jan. 3.	1	To Molasses, .	16	00	Jan. 9.	2	By Order, . .	12	80
" 9.	2	" Sundries, .	21	50	" 20.		" Note, a 90 days,	24	70
			$37	50				$37	50

Dr.		*William Angell,*							*Cr.*
1829.			$	c.	1829.			$	c.
Jan. 4.	1	To Sundries, .	34	50	Jan. 10.	3	By Sundries, .	54	00
" 16.		" Cash, . .	19	50					
			$54	00					

Dr.		*Samuel Stone,*							*Cr.*
1829.			$	c.	1829.			$	c.
Jan. 4.	1	To Sundries, .	85	00	Jan. 30.		By Cash, . .	97	00
" 7.	2	do. . .	12	00					
			$97	00					

Dr.		*George Carpenter,*							*Cr.*
[illegible]			$	c.	1829.			$	c
Jan. 5.	1	To Sundries, .	3	84	Jan. 15.		By Note, a 60 days,	33	04
" 9.	2	do. . .	29	20					
			$33	04					

Dr. *Jesse B. Sweet,* **Cr.**

1829.			$	c.	1829.			$	c.
Jan. 5.	1	To Molasses, .	27	60	Jan. 5.	1	By Cash, . .	15	00
" 9.	2	" Fish, .	112	50	" 20.		" Cash, to bal.	125	10
			$140	10				$140	10

Dr. *Jesse Metcalf,* **Cr.**

1829.			$	c	1829.		$	c
Jan. 5.	2	To Sundries, .	300	00	Apr. 7.	By his Check, .	775	00
" 9.	3	" Shoes, . .	475	00				
			$775	00				

Dr. *James Hammond,* **Cr.**

1829.			$	c.	1829.		$	c.
Jan. 12.	3	To Sundries, .	35	44	Jan. 30.	By Order on Brown & Ives, . . .	35	44

INDEX TO THE LEGER.

A	Page.
Angell, William	2
C	
Carpenter, George	2
H	
Hawkins, Robert	1
Hammond, James	3
M	
Murray, James, Jr.	1
Metcalf, Jesse	3
S	
Sweet, Jesse B.	3
Stone, Samuel	2
T	
Thomas, Isaac	2
Tracy, Archibald	2
W	
Warren, James	2
Y	
Yeomans, Thomas	2

CASH-BOOK.

This book records the payments and receipts of cash

It is kept by making cash *Dr.* to cash on hand and what is received, and *Cr.* by whatever is paid out.

At the end of every day or week, as may best suit the nature of the business, the cash on hand is counted, and entered on the *Cr.* side.

If there is no error, this will make the sum of the *Dr.* equal to that of the *Cr.* A balance is then struck, and the cash on hand carried again upon the *Dr.* side.

FORM OF A CASH-BOOK.

Dr. CASH. **Cr.**

1827.		$	c.	1827.		$	c.
Jan. 1	To Cash on hand	637	50	Jan. 2	By rent of store for one quarter, paid Thomas Taylor	62	50
2	" J. Thompson	37	94				
"	" J. Hart, paid acc't.	65	43				
3	" H. Palmer on note	127	28	4	" Paid note to R. Thacher	127	83
4	" S. Snowdon	84	73				
5	" J. Mervin on acc't.	17	90	5	" Family expenses	27	61
6	" S. Crane	100	90	6	" Merchandise bo't of T. Thamor	614	27
"	Sales of Merchandise	311	18		Cash on hand	550	65
		1382	86			1382	86
8	Cash on hand	550	65				

Form of a Bill from the preceding Work.

Mr. James Murray

to Edward L. Peckham, *Dr.*

1829.			$	c.
Jan. 1.	To 1 gall. Lisbon Wine	$1,92		
" "	" 6 yds. Calico, *a* 37½ cts.	2,25		
" "	" 2 yds. Broadcloth, *a* $4,50, ...	9,00		
			13	17
" 10.	" 10 lbs. Sugar, *a* 11 cts.	1,10		
" "	" 6 galls. Molasses, *a* 37½ cts.	2,22		
" "	" 20 lbs. Coffee, *a* 15 cts.	3,00		
			6	32
" 12.	" 6 lbs. Raisins, *a* 20 cts.	1,20		
" "	" 5 galls. Currant Wine, *a* 75 cts.	3,75		
			4	95
			24	44
	Cr.			
" 5.	By 20 bu. Corn, *a* 60 cts.	12,00		
" "	" 4 bu. Oats, *a* 40 cts.	1,60		
" 15.	" Cash to balance, ...	10,84		
	Errors excepted.		24	44

EDWARD L. PECKHAM.

Boston, January 15, 1829.

2d Form.

Mr. Jesse Metcalf

to E. L. Peckham, *Dr.*

1829.		$	c.
Jan. 5.	To 20 Calf-Skins, *a* $5, ...	100	00
" "	" 50 Dried Hides, *a* $4, ...	200	00
" 9.	" 500 pair Men's Shoes, *a* 95 cts.	475	00
		$775	00

Received payment, by his check on N. E. Bank.

EDWARD L. PECKHAM

Boston, April 7, 1829

MERCANTILE FORMS.

No. 1. *Negotiable Note.*

$78,50.

Boston, May 5, 1827.

On Demand, I promise to pay Claude Lorraine, or Order, Seventy-eight Dollars Fifty Cents, with Interest, for value received.

JAMES HONESTUS.

No. 2. *Note payable to Bearer.*

$40.

Boston, Sept. 17, 1827

Six months from date, I promise to pay A. B., or Bearer, Forty Dollars for value received.

SIMEON PAYWELL.

No. 3. *Note by two Persons.*

$500.

Berlin, Oct. 28, 1827.

For value received, we, jointly and severally, promise to pay C. D. or Order, on demand, Five Hundred Dollars, with Interest.

HORACE WALCOTT.
JAMES HART.

No. 4. *Note at Bank.*

$150.

Boston, Feb. 25, 1819.

Ninety-five days from date, I promise to pay Thomas Andrews, or Order, at the Phœnix Bank, One Hundred and Fifty Dollars, for value received

JOHN REYNOLDS.

Remarks relating to Notes of Hand.

1. A negotiable note is one which is made payable to A. B. *or order*—It is otherwise, when these words are omitted.

2. By *endorsing a note* is understood, that the person to whom it is payable writes his name on the back of it. For additional security, any other person may afterwards endorse it.

3. If the note be made payable to A. B., *or order*, (*see No.* 1,) then A. B. can sell said note to whom he pleases, provided he endorses it; and whoever buys said note may lawfully demand payment of the signer of the note, and if the signer, through inability or otherwise, refuses to pay said note, the purchaser may lawfully demand payment of the endorser.

4. If the note be made payable to A. B., *or bearer*, (*see No.* 2,) then the signer only is responsible to any one who may purchase it.

5. Unless a note be written payable on some specific future time, it should be written *on demand*; but should the words *on demand* be omitted, the note is supposed to be recoverable by law.

6. When a note, payable at a future day, becomes due, it is considered on interest from that time till paid, though no mention be made of interest.

7. No mention need be made in a note of the *rate* of interest: that particular is settled by law, and will be collected according to the laws of the state where the note is dated. In some states it is 6 per cent.; in others, 7.

8. If two persons, jointly and severally, (*see No.* 3,) sign a note, it may be collected by law of either.

9. A note is not valid, unless the words for *value received* be expressed.

10. When a note is given, payable in any article of merchandise, or property other than money, deliverable on a specified time, such articles should be tendered in payment at said time; otherwise the holder of the note may demand the value in money.

Account with Interest.

Mr. Thomas I. Spencer

to H. Tisdall, Dr.

1816—Nov. 1.	To 3 yards Cloth, a $7,50 per yard,	$22,50
Dec. 2.	" 6 galls. Wine, 4,25 per gallon,	25,50
1819—Jan. 1.	" Balance of Interest,	5,80
		$53,80

Supra. Cr.

1817—Nov. 1.	By Cash,	$22,50
1819—Jan. 1.	Ditto in full,	31,30
		$53,80

Boston, Jan. 1, 1819. H. TISDALL

A Receipt for Money on Account.

Received of James Wardell Three Dollars on account.
Boston, June 21, 1816. SIMEON BRANDT

A General Receipt.

Received of Jonathan Andrews Fourteen Dollars in full of all accounts. HORACE RITTER.
Boston, Dec. 31, 1827.

Receipt for Money paid on a Note.

Received of Leonard Temple Seventy-two Pounds and Eleven Shillings, on his note for the sum of One Hundred and Seventy-two Pounds, and dated at Enfield, Oct. 27, 1826. D. THOMAS.
Boston, August 27, 1828.

An Order for Money.

Messrs. R. Potter & Co.

Pay James Thomas, or Order, Eleven Dollars, and this shall be your receipt for the same. SHEELAH SPENCER.
Boston, Sept. 16, 1828.

An Order for Goods.

Mr. Albion N. Olney,

Pay the Bearer Seventy-one Dollars, in goods from your store, and charge Your obedient servant,
Oxford, Dec. 31, 1827. R. RAYNAL.

Note. A receipt given in full of *all accounts* cuts off accounts only; but a receipt given in full of *all demands* cuts off not only all accounts, but all demands whatsoever.

An order, when paid, should be receipted on the back, by the person to whom it is made payable, or by some one duly authorized to sign for him; but when it is made payable to *bearer*, or to *A. B. or bearer*, it may be receipted by any one who presents it for payment.

THE END.